Nutritional Management of Cystic Fibrosis

Nutritional Management of Cystic Fibrosis

Editors

Maria R. Mascarenhas
Jessica Alvarez

MDPI • Basel • Beijing • Wuhan • Barcelona • Belgrade • Manchester • Tokyo • Cluj • Tianjin

Editors
Maria R. Mascarenhas
Univ Penn
USA

Jessica Alvarez
Emory University
USA

Editorial Office
MDPI
St. Alban-Anlage 66
4052 Basel, Switzerland

This is a reprint of articles from the Special Issue published online in the open access journal *Nutrients* (ISSN 2072-6643) (available at: https://www.mdpi.com/journal/nutrients/special_issues/CF_nutrition).

For citation purposes, cite each article independently as indicated on the article page online and as indicated below:

LastName, A.A.; LastName, B.B.; LastName, C.C. Article Title. *Journal Name* **Year**, *Volume Number*, Page Range.

ISBN 978-3-0365-6267-4 (Hbk)
ISBN 978-3-0365-6268-1 (PDF)

© 2023 by the authors. Articles in this book are Open Access and distributed under the Creative Commons Attribution (CC BY) license, which allows users to download, copy and build upon published articles, as long as the author and publisher are properly credited, which ensures maximum dissemination and a wider impact of our publications.

The book as a whole is distributed by MDPI under the terms and conditions of the Creative Commons license CC BY-NC-ND.

Contents

About the Editors vii

Preface to "Nutritional Management of Cystic Fibrosis" ix

Montserrat A. Corbera-Hincapie, Kristen S. Kurland, Mark R. Hincapie, Anthony Fabio, Daniel J. Weiner, Sandra C. Kim and Traci M. Kazmerski
Geospatial Analysis of Food Deserts and Their Impact on Health Outcomes in Children with Cystic Fibrosis
Reprinted from: *Nutrients* **2021**, *13*, 3996, doi:10.3390/nu13113996 1

Yiqing Peng, Malinda Wu, Jessica A. Alvarez and Vin Tangpricha
Vitamin D Status and Risk of Cystic Fibrosis-Related Diabetes: A Retrospective Single Center Cohort Study
Reprinted from: *Nutrients* **2021**, *13*, 4048, doi:10.3390/nu13114048 13

Margaret P. Marks, Sonya L. Heltshe, Arthur Baines, Bonnie W. Ramsey, Lucas R. Hoffman and Michael S. Stalvey
Most Short Children with Cystic Fibrosis Do Not Catch Up by Adulthood
Reprinted from: *Nutrients* **2021**, *13*, 4414, doi:10.3390/nu13124414 23

Kevin J. Scully, Laura T. Jay, Steven Freedman, Gregory S. Sawicki, Ahmet Uluer, Joel S. Finkelstein and Melissa S. Putman
The Relationship between Body Composition, Dietary Intake, Physical Activity, and Pulmonary Status in Adolescents and Adults with Cystic Fibrosis
Reprinted from: *Nutrients* **2022**, *14*, 310, doi:10.3390/nu14020310 31

Josie M. van Dorst, Rachel Y. Tam and Chee Y. Ooi
What Do We Know about the Microbiome in Cystic Fibrosis? Is There a Role for Probiotics and Prebiotics?
Reprinted from: *Nutrients* **2022**, *14*, 480, doi:10.3390/nu14030480 43

Rosara Milstein Bass, Alyssa Tindall and Saba Sheikh
Utilization of the Healthy Eating Index in Cystic Fibrosis
Reprinted from: *Nutrients* **2022**, *14*, 834, doi:10.3390/nu14040834 71

William B. Nicolson, Julianna Bailey, Najlaa Z. Alotaibi, Stefanie Krick and John D. Lowman
Effects of Exercise on Nutritional Status in People with Cystic Fibrosis: A Systematic Review
Reprinted from: *Nutrients* **2022**, *14*, 933, doi:10.3390/nu14050933 81

Dhiren Patel, Albert Shan, Stacy Mathews and Meghana Sathe
Understanding Cystic Fibrosis Comorbidities and Their Impact on Nutritional Management
Reprinted from: *Nutrients* **2022**, *14*, 1028, doi:10.3390/nu14051028 95

Julianna Bailey, Stefanie Krick and Kevin R. Fontaine
The Changing Landscape of Nutrition in Cystic Fibrosis: The Emergence of Overweight and Obesity
Reprinted from: *Nutrients* **2022**, *14*, 1216, doi:10.3390/nu14061216 113

Peter N. Freswick, Elizabeth K. Reid and Maria R. Mascarenhas
Pancreatic Enzyme Replacement Therapy in Cystic Fibrosis
Reprinted from: *Nutrients* **2022**, *14*, 1341, doi:10.3390/nu14071341 131

Montserrat A. Corbera-Hincapie, Samar E. Atteih, Olivia M. Stransky, Daniel J. Weiner,
Iris M. Yann and Traci M. Kazmerski
Experiences and Perspectives of Individuals with Cystic Fibrosis and Their Families Related to Food Insecurity
Reprinted from: *Nutrients* **2022**, *14*, 2573, doi:10.3390/nu14132573 . . . . **143**

Francisco José Sánchez-Torralvo, Nuria Porras, Ignacio Ruiz-García,
Cristina Maldonado-Araque, María García-Olivares, María Victoria Girón,
Montserrat Gonzalo-Marín, Casilda Olveira and Gabriel Olveira
Usefulness of Muscle Ultrasonography in the Nutritional Assessment of Adult Patients with Cystic Fibrosis
Reprinted from: *Nutrients* **2022**, *14*, 3377, doi:10.3390/nu14163377 . . . . **153**

Maret G. Traber, Scott W. Leonard, Vihas T. Vasu, Brian M. Morrissey, Huangshu (John) Lei,
Jeffrey Atkinson and Carroll E. Cross
α-Tocopherol Pharmacokinetics in Adults with Cystic Fibrosis: Benefits of Supplemental Vitamin C Administration
Reprinted from: *Nutrients* **2022**, *14*, 3717, doi:10.3390/nu14183717 . . . . **167**

Paulina Wysocka-Wojakiewicz, Halina Woś, Tomasz Wielkoszyński,
Aleksandra Pyziak-Skupień and Urszula Grzybowska-Chlebowczyk
Vitamin Status in Children with Cystic Fibrosis Transmembrane Conductance Regulator Gene Mutation
Reprinted from: *Nutrients* **2022**, *14*, 4661, doi:10.3390/nu14214661 . . . . **181**

About the Editors

Maria R. Mascarenhas

Maria R. Mascarenhas, PhD, RD, is an Associate Professor in the Division of Endocrinology, Lipids, and Metabolism at the Emory University School of Medicine. She is a registered dietitian with a PhD in Nutrition Sciences from the University of Alabama at Birmingham. Her research focuses on the role of nutrition and body composition on metabolism in chronic diseases, including cystic fibrosis and cystic-fibrosis-related diabetes. Her research integrates state-of-the-art metabolomics with many aspects of nutrition research, including rigorous clinical trials, body composition analysis, biomarker assessment, analysis of dietary intake, and physical activity. Dr. Alvarez served on the 2020 Academy of Nutrition and Dietetics Evidence Analysis Workgroup Expert Committee for Development of Nutrition Guidelines in Cystic Fibrosis. She has regularly been funded by the National Institutes of Health, the Cystic Fibrosis Foundation, and other granting mechanisms, and has maintained a strong publication record in clinical nutrition and endocrine research.

Jessica Alvarez

Jessica Alvarez , PhD, RD, is an Associate Professor in the Division of Endocrinology, Lipids, and Metabolism at the Emory University School of Medicine. She is a registered dietitian with a PhD in Nutrition Sciences from the University of Alabama at Birmingham. Her research focuses on the role of nutrition and body composition on metabolism in chronic diseases, including cystic fibrosis and cystic-fibrosis-related diabetes. Her research integrates state-of-the-art metabolomics with many aspects of nutrition research, including rigorous clinical trials, body composition analysis, biomarker assessment, analysis of dietary intake, and physical activity. Dr. Alvarez served on the 2020 Academy of Nutrition and Dietetics Evidence Analysis Workgroup Expert Committee for Development of Nutrition Guidelines in Cystic Fibrosis. She has regularly been funded by the National Institutes of Health, the Cystic Fibrosis Foundation, and other granting mechanisms, and has maintained a strong publication record in clinical nutrition and endocrine research.

Preface to "Nutritional Management of Cystic Fibrosis"

We are delighted to share this Special Issue targeted at all health care providers and researchers working and taking care of people with cystic fibrosis (CF). With the advent of CFTR modulator drugs, the landscape of CF care is changing, and non-pulmonary manifestations are emerging as important issues. Nutrition has always been a critical issue in CF care, but with increasing lifespans and unprecedented increases in overweight and obesity in this population, it is important to shift research and clinical focus and care plans to exploring the role that nutrition has on long-term health and quality of life. Our purpose for this Special Issue was to share updates and new nutrition topics with our readers. With this is mind, we present updates on classic topics such as pancreatic enzyme replacement therapy and growth, as well as new topics such as the microbiome and probiotics, food insecurity, and diet quality specific to individuals with CF. We cover new literature on vitamin D and CFRD, the interaction between vitamin E pharmacokinetics and vitamin C, and vitamin status in children with CFTR gene mutations. Achieving an optimal nutrition status encompasses more than body weight; thus, we include studies of body composition, including the use of muscle ultrasound, as well as the impact of exercise on nutrition status. We have enjoyed putting this Special Issue together and hope that the readers will find it enjoyable and enlightening too. We thank all of the excellent authors for their contributions to this issue.

Maria R. Mascarenhas and Jessica Alvarez
Editors

Article

Geospatial Analysis of Food Deserts and Their Impact on Health Outcomes in Children with Cystic Fibrosis

Montserrat A. Corbera-Hincapie [1,*], Kristen S. Kurland [2], Mark R. Hincapie [1], Anthony Fabio [3], Daniel J. Weiner [1], Sandra C. Kim [1] and Traci M. Kazmerski [1]

1. Department of Pediatrics, University of Pittsburgh School of Medicine, Pittsburgh, PA 15213, USA; hincapiemr@upmc.edu (M.R.H.); daniel.weiner@chp.edu (D.J.W.); sandra.kim@chp.edu (S.C.K.); traci.kazmerski@chp.edu (T.M.K.)
2. School of Architecture, Heinz College of Information Systems and Public Policy, Carnegie Mellon University, Pittsburgh, PA 15213, USA; kurland@andrew.cmu.edu
3. Department of Epidemiology, University of Pittsburgh Epidemiology Data Center, Pittsburgh, PA 15260, USA; anthony.fabio@pitt.edu
* Correspondence: corberahincapiema@upmc.edu; Tel.: +1-786-412-1300

Abstract: Food insecurity (FI) is defined as "the limited or uncertain access to adequate food." One root cause of FI is living in a food desert. FI rates among people with cystic fibrosis (CF) are higher than the general United States (US) population. There is limited data on the association between food deserts and CF health outcomes. We conducted a retrospective review of people with CF under 18 years of age at a single pediatric CF center from January to December 2019 using demographic information and CF health parameters. Using a Geographic Information System, we conducted a spatial overlay analysis at the census tract level using the 2015 Food Access Research Atlas to assess the association between food deserts and CF health outcomes. We used multivariate logistic regression analysis and adjusted for clinical covariates and demographic covariates, using the Child Opportunity Index (COI) to calculate odds ratios (OR) with confidence intervals (CI) for each health outcome. People with CF living in food deserts and the surrounding regions had lower body mass index/weight-for-length (OR 3.18, 95% CI: 1.01, 9.40, $p \leq 0.05$ (food desert); OR 4.41, 95% CI: 1.60, 12.14, $p \leq 0.05$ (600 ft buffer zone); OR 2.83, 95% CI: 1.18, 6.76, $p \leq 0.05$ (1200 ft buffer zone)). Food deserts and their surrounding regions impact pediatric CF outcomes independent of COI. Providers should routinely screen for FI and proximity to food deserts. Interventions are essential to increase access to healthy and affordable food.

Keywords: cystic fibrosis; food insecurity; food deserts; body mass index

1. Introduction

Food insecurity (FI) is an important barrier that has the potential to significantly impact the health and well-being of children and families. The US Department of Agriculture (USDA) defines FI as a household in which there is "limited or uncertain access to adequate food" [1,2]. Some root causes of FI include poverty and food deserts, or low-income census tracts with a substantial number of residents with poor access to retail outlets selling healthy and affordable foods [3]. The USDA further defines low income and low access within a census tract as "a poverty rate of 20 percent or greater, or a median family income at or below 80 percent of the statewide or metropolitan area median family income," and "at least 500 persons and/or at least 33 percent of the population lives more than 1 mile from a supermarket or large grocery store (10 miles, in the case of rural census tracts)," respectively [4].

Evidence suggests that FI rates among people with cystic fibrosis (CF) may be higher than the general United States (US) population [5]. In CF, optimal growth and nutrition, identified as body mass index (BMI) > 50%, are correlated with better lung function and overall health; therefore, adequate caloric intake is a mainstay of CF care [5–8]. As people with CF have increased caloric demands (1.5 to 2 times the energy needs of the general population), FI and/or poor access to healthy food options may contribute to inability to achieve and maintain appropriate weight gain and lead to malnutrition in CF [9]. Addressing FI is key because it can independently have detrimental effects on child health, including more frequent hospitalizations, developmental problems, nutritional deficiencies, chronic stress leading to depression/anxiety/toxic stress, and increased long-term mortality resulting from metabolic syndrome, particularly cardiovascular disease [2,10,11].

Although FI screening has increased across many CF centers, there remains limited data on the association between FI, particularly as it relates to food deserts, and CF health outcomes [12]. This study investigated the effects of food deserts and their surrounding regions on health outcomes, including BMI/weight-for-length, percent predicted forced expiratory volume in 1 s (ppFEV1) and hospitalizations secondary to pulmonary exacerbations, in children and adolescents with CF. Additionally, we examined the impact of other neighborhood features that might affect child health by utilizing the Child Opportunity Index (COI).

2. Materials and Methods

2.1. Setting and Study Design

We conducted a retrospective review of people with CF under 18 years of age at a tertiary level Children's Hospital center from January 2019 to December 2019. We collected patient demographic information, including age, sex, race/ethnicity, and home address, along with CF health parameters, including year-best BMI percentile or weight-for-length percentile, year-best ppFEV1 for children 6 years and older, CF modulator use, and number of hospitalizations due to pulmonary exacerbation from the US Cystic Fibrosis Foundation Patient Registry (CFFPR).

We geocoded the most recent participant home addresses using a Geographic Information System (GIS), a computerized system that can capture, store, analyze, manage, and present data that are linked to a location. GIS differs from other information systems to address location questions, as it uses multiple layers of geospatial data and advanced spatial statistics, networking, and analysis tools [3].

The Institutional Review Board (IRB) of the University of Pittsburgh approved this study with a waiver of informed consent (protocol number: 20050263).

2.2. Area Resources

We used the 2015 Food Access Research Atlas, a USDA database that provides food access data for populations within census tracts and identifies which census tracts are low income and low access, also known as food deserts [13]. It maps food access indicators, such as accessibility to healthy food sources measured by distance or availability of a vehicle, for each individual census tract "using $\frac{1}{2}$-mile and 1-mile demarcations to the nearest supermarket for urban areas, 10-mile and 20-mile demarcations to the nearest supermarket for rural areas" [13]. For the purposes of this study, we used the 1- and 10-mile demarcation for urban and rural areas, which align with the USDA food desert definition. Data from the 2015 Atlas came from the 2010 Decennial Census and the 2014–2018 American Community Survey (ACS).

We used the Child Opportunity Index (COI) 2.0 to examine the impact of neighborhood factors on CF health and to adjust for any confounders in our analysis related to food deserts. COI 2.0 measures neighborhood resources and conditions that play a role in child development and is specific to children. It includes 29 indicators that are sorted into three domains: (a) education; (b) health and environment; and (c) social and economic. Examples of the indicators include, but are not limited to, early childhood education, green space, social and economic resources, employment rate, poverty rate, median household income, and public assistance rate. The COI 2.0 provides an overall opportunity score for each census tract using 2015 US Census Bureau and the American Community Survey data with the lower scores being suggestive of less opportunity. Each domain is also assigned an opportunity score for each census tract. The census tracts are then grouped into high, very high, moderate, low, very low opportunity scores for each domain and overall category. For the purposes of this study, we grouped the COI scores into high/very high versus moderate/low/very low [14].

Using GIS software (ArcGIS Pro, Esri, Inc., Redlands, CA, USA), we conducted a spatial overlay analysis using the 2015 Food Access Research Atlas (Figure 1a). We also included buffer zones with a 600-foot (600 ft) and 1200-foot (1200 ft) walkability distance from food deserts to assess if living near a food desert (in addition to living within a food desert) contributed to poor health outcomes (Figure 1b,c). By using a reverse spatial join in ArcGIS Pro, we applied the food desert layer and COI score layer onto the geocoded patient addresses and their health parameters.

2.3. Outcome Variables

Our primary health outcome was year-best BMI percentile (for patients above 2 years of age) or weight-for-length percentile (for patients under 2 years of age). Secondary outcomes were lung function (as measured by year-best ppFEV1 for patients that were 6 years and older, and who could adequately and reproducibly participate in the testing) and number of hospitalizations secondary to pulmonary exacerbations. We defined ideal BMI as >50th percentile, ideal ppFEV1 as >90th percentile and significant pulmonary exacerbations if requiring 2 or more hospitalizations in one year. The Cystic Fibrosis Foundation recommends that children and adolescents maintain a BMI or weight for length at or above the 50th percentile, as weight above this threshold has been associated with better ppFEV1 [15]. The CF pulmonary guidelines categorize severity of lung disease with normal FEV1 being greater than 90% predicted [16].

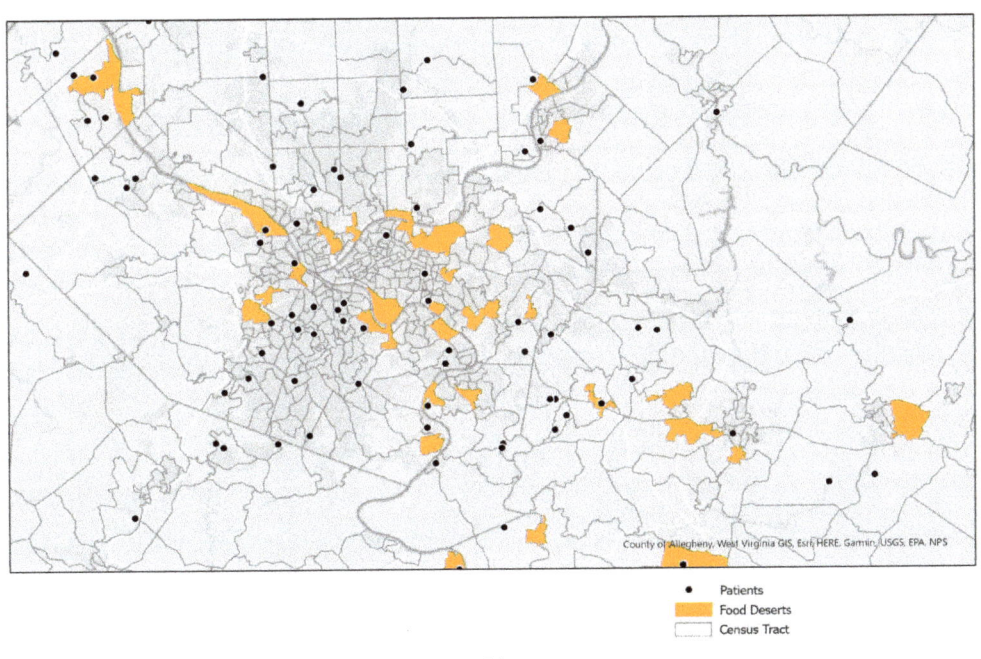

(a)

Figure 1. *Cont.*

Figure 1. (a) Zoomed in spatial overlay of food deserts (Allegheny County, PA, USA) and patient sample. (b) Zoomed in spatial overlay of food deserts (Allegheny County, PA, USA) plus a 600 ft buffer and patient sample. (c) Zoomed in spatial overlay of food deserts (Allegheny County, PA, USA) plus a 1200 ft buffer and patient sample.

2.4. Statistical Analysis

We used GIS software (ArcGIS Pro, Esri, Inc., Redlands, CA, USA) for the geographic data analysis and IBM SPSS Statistics software (Version 28.0, Armonk, NY, USA) for the statistical analysis. We estimated odds ratios and 95% confidence intervals (CI) for associations between food deserts, buffer zones, COI scores, and the three health outcomes (BMI/weight-for-length, ppFEV1, and pulmonary exacerbations requiring hospitalization) using logistic regression.

We created multivariate logistic regression models for each health outcome to assess the individual effects of food deserts (and their buffer zones) and COI scores and both combined. BMI/weight-for-length, ppFEV1, and hospitalizations secondary to pulmonary exacerbations were dependent variables, and food deserts and COI scores were independent variables in the models. Model 1 examined food deserts individually. Model 2 examined COI scores individually and their potential effects on health outcomes. Model 3 examined food deserts adjusting for COI scores. We adjusted each model for covariates, including age, sex (female/male), race/ethnicity (non-Hispanic White/other), and modulator use (yes/no). We repeated this methodology for food deserts plus 600 ft and 1200 ft buffers. We used cross tabulation to observe differences between the several variables that were applied. We defined statistical significance as $p \leq 0.05$.

3. Results

The study included 206 children and adolescents with CF residing in Pennsylvania, West Virginia, and Ohio. Table 1 presents the demographic and clinical characteristics of the patient sample. The average age of the patient sample was 9.5 +/- 5.6 years. About half of the patient population was female (47%). Most of the participants in the sample were White (92.7%), which is consistent with the national US CF population [17].

Table 1. Characteristics of the CF patient sample.

Characteristics	Overall	Food Desert	Food Desert + 600 ft Buffer	Food Desert + 1200 ft Buffer
Total	206	17	20	29
Demographics				
Age (range 0–18 years)	9.5 * (5.6)	8.7 * (5.5)	9.5 * (5.5)	9.8 * (4.9)
Female, %	47%	47%	45%	52%
White, %	92.7%	94%	95%	93%
Modulator Use, %	11.6%	18%	20%	14%
Clinical Outcomes				
BMI/weight-for-length, percentile	63 * (24.4)	56.6 * (29.7)	53.5 * (28.5)	56.8 * (28.0)
ppFEV1 (>6 years old)	95.9 * (17.1) n = 141	92.5 * (17.2) n = 11	87.2 * (18.8) n = 14	90 * (16.9) n = 22
Pulmonary Exacerbations	0.5 * (1.2) Range: 0–8	0.7 * (2.2) Range: 0–6	0.8 * (1.4) Range: 0–6	0.6 * (1.2) Range: 0–6

* Mean (standard deviation).

3.1. Model 1: The Individual Impact of Food Deserts and Surrounding Regions on CF Health Outcomes

When accounting for food deserts alone, children and adolescents with CF living in a food desert had 2.9 times the odds (95% CI: 1.1, 8.21, $p \leq 0.05$) of having a non-ideal BMI as those not living in a food desert. There were no significant increased odds of a non-ideal ppFEV1 or increased hospitalizations from pulmonary exacerbations in food deserts (Table 2).

Children and adolescents with CF living in a food desert or within a 600 ft walkability distance from a food desert had four times the odds (95% CI: 1.54, 10.69, $p \leq 0.05$) of having a non-ideal BMI as those not living in or 600 ft from a food desert. Additionally, those living in a food desert or within a 600 ft walkability distance from a food desert had three times the odds (95% CI: 0.99, 9.66, $p \leq 0.05$) to have a non-ideal FEV1 as those not living in or 600 ft from a food desert. There were no significant increased odds of increased hospitalizations from pulmonary exacerbations in food deserts or within a 600 ft walkability distance (Table 2).

Table 2. Multiple Regression Models of CF Outcomes: Food Deserts, Surrounding Regions, and COI.

Health Outcome	COI	Food Desert	Food Desert + 600 ft Buffer	Food Desert + 1200 ft Buffer
BMI				
Model 1 [a] OR (95% CI)		2.9 (1.1, 8.21) * p-value 0.039	4.06 (1.54, 10.69) * p-value 0.005	2.65 (1.16, 6.07) * p-value 0.021
Model 2 [b] OR (95% CI)	1.07 (0.59, 1.94) p-value 0.832			
Model 3 [c] OR (95% CI)		3.18 (1.01, 9.40) * p-value 0.036	4.41 (1.60, 12.14) * p-value 0.004	2.83 (1.18, 6.76) * p-value 0.020
ppFEV1 (6 years and older)				
Model 1 [a] OR (95% CI)		1.79 (0.51, 6.32) p-value 0.363	3.09 (0.99, 9.66) * p-value 0.050	1.90 (0.74, 4.92) p-value 0.184
Model 2 [b] OR (95% CI)	0.99 (0.49, 2.02) p-value 0.984			
Model 3 [c] OR (95% CI)		1.90 (0.51, 7.14) p-value 0.340	3.33 (1.03, 10.84) * p-value 0.045	2.02 (0.75, 5.46) p-value 0.166
Hospitalizations for Pulmonary Exacerbations				
Model 1 [a] OR (95% CI)		2.22 (0.53, 9.25) p-value 0.273	2.49 (0.69, 8.95) p-value 0.163	1.19 (0.35, 4.05) p-value 0.778
Model 2 [b] OR (95% CI)	1.01 (0.41, 2.53) p-value 0.976			
Model 3 [c] OR (95% CI)		2.42 (0.52, 11.15) p-value 0.258	2.63 (0.69, 10.01) p-value 0.156	1.20 (0.34, 4.31) p-value 0.777

[a] Model 1: Food Desert +/- Buffer Zone Alone; [b] Model 2: COI Alone; [c] Model 3: Food Desert +/- Buffer Zone adjusted for COI Overall Score; OR: odds ratio; CI: confidence interval; * Significant estimate at $p \leq 0.05$.

Children and adolescents with CF living in a food desert or within a 1200 ft walkability distance from a food desert had 2.7 times the odds (95% CI: 1.16, 6.07, $p \leq 0.05$) of having a non-ideal BMI as those not living in or 1200 ft from a food desert. There were no significant increased odds of a non-ideal FEV1 or increased hospitalizations from pulmonary exacerbations in food deserts or within a 1200 ft walkability distance (Table 2).

3.2. Model 2: The Individual Impact of Childhood Opportunity (COI Scores) on CF Health Outcomes

When accounting for COI scores individually, there were no significant increased odds of a non-ideal BMI, non-ideal FEV1, or increased hospitalization from pulmonary exacerbations in children and adolescents with CF that lived in census tracts that had a moderate/low/very low opportunity score versus a high/very high opportunity score (Table 2).

3.3. Model 3: The Impact of Food Deserts and Surrounding Regions on CF Health Outcomes, Adjusting for COI Scores

Children and adolescents with CF living in a food desert had 3.18 times the odds (95% CI: 1.01, 9.4, $p \leq 0.05$) of having a non-ideal BMI as those not living in a food desert when adjusting for the overall COI score. When adjusting for each individual COI domain, children and adolescents with CF living in a food desert continued to have increased odds of having a non-ideal BMI as those not living in a food desert. There were no significant increased odds of a non-ideal FEV1 or increased hospitalization from pulmonary exacerbations in food deserts when accounting for the overall COI score or any of the three domains (Table 2).

Children and adolescents with CF living in a food desert or within a 600 ft walkability distance from a food desert had 4.4 times the odds (95% CI: 1.6, 12.14, $p \leq 0.05$) of having a non-ideal BMI and 3.3 times the odds (95% CI: 1.03, 10.84, $p \leq 0.05$) of having a non-ideal FEV1 as those not living in or 600 ft from a food desert when adjusting for overall COI scores. When adjusting for each individual domain, children and adolescents with CF living in a food desert or within a 600 ft walkability distance continued to have increased odds of having a non-ideal BMI and a non-ideal FEV1 as those not living in or 600 ft from a food desert. There were no significant increased odds of increased hospitalizations from pulmonary exacerbations in food deserts or within a 600 ft walkability distance when accounting for the overall COI score or any of the three domains (Table 2).

Children and adolescents with CF living in a food desert or within a 1200 ft walkability distance from a food desert had 2.8 times the odds (95% CI: 1.18, 6.76, $p \leq 0.05$) of having a non-ideal BMI as those not living in or 1200 ft from a food desert when adjusting for overall COI scores. When adjusting for each individual domain, children and adolescents with CF living in a food desert or within a 1200 ft walkability distance continued to have increased odds of having a non-ideal BMI as those not living in or 1200 ft from a food desert. There were no significant increased odds of a non-ideal FEV1 or increased hospitalization from pulmonary exacerbations in food deserts or within a 1200 ft walkability distance when accounting for the overall COI score or any of the three domains (Table 2).

4. Discussion

Children and adolescents with CF living in a food desert or within a 600 ft or 1200 ft walkability distance from a food desert have increased likelihood of having a non-ideal BMI/weight-for-length. Patients living within a 600 ft walkability distance from a food desert also had increased odds of having suboptimal lung function. These findings are independent of other childhood opportunity factor or social determinants of health, including healthcare access, education quality, employment, poverty rate, and neighborhood environment, suggesting that the unfavorable health outcomes are driven by food access rather than by alternate social, environmental or health indicators.

Prior studies have demonstrated that the social, economic, and neighborhood environment play a role in the health [18]. Structural factors play a significant role in health inequities and furthermore suggest that "environment restricts freedom of choice, or that behavior is chosen to compensate for unfavorable circumstances" [19]. Poverty is a barrier to healthy foods, a safe environment, and better employment, all of which are predictors of better health [20,21]. Children are not immune from these effects. Children with limited access to healthy foods develop poorer eating habits, placing them at increased risk for developing chronic disease, such as mental health ailments, asthma, and diabetes. Additionally, children, who live in a food insecure household, have increased odds of having fair/poor health reported and of being hospitalized since birth [22].

Our findings are consistent with current literature suggesting that living in a food desert has a negative impact on health [23,24]. It has also been noted that FI contributes to greater healthcare costs, particularly when looking at patients with chronic disease [25]. A systematic review of food deserts found that in addition to limited access to supermarkets, small independent stores and convenience stores charge higher prices for food, compounding the effects of food deserts. Individuals, including children, living in food deserts are at increased risk of developing obesity, in part due to lower fruit and vegetable access [18,21,26–29]. Some characteristics of neighborhoods that are affected by poor access to grocery stores and healthy foods include lower-income and minority neighborhoods [28]. A cross-sectional study comparing lower income versus higher income neighborhoods found that certain neighborhood characteristics may predispose children to obesity regardless of other demographic or socioeconomic factors [30].

There have been a few studies examining the impact of social determinants of health on health outcomes in people with CF; however, no prior study has looked at the impact of food deserts and their surrounding regions on CF health outcomes. A recent study by Oates and colleagues linked respiratory health in people with CF to state- and area-level characteristics, particularly an association with area resource deprivation and overall state child health [31]. Another study looking at the associations of socioeconomic status with CF health outcomes found that medically indigent individuals with CF suffer more adverse outcomes than the general CF population, including higher mortality and worse pulmonary function and growth [32].

While similar GIS techniques have been used in other chronic disease population studies, this study is one of the first to apply GIS methodology in CF [21,31,33,34]. Importantly, a strength of this study was our use of full home addresses, leading to more precise estimates of geographic linkages to health outcomes. Most studies, including the ones referenced above, use participants' postal ZIP codes as opposed to full addresses. ZIP codes are helpful in assessing food desert status or level of geography for indices; however, they can span across multiple census tracts consisting of potentially demographically diverse areas and may be less precise due to misclassification of socioeconomic status [35].

Given these findings, it is important to address how we can increase food access for people with CF living in or near food deserts, in addition to identifying individuals at risk. Screening for FI is one potential option; however, some people with CF and families might not answer truthfully due to the stigma associated with FI. Another option is to geographically assess proximity to food deserts as a marker for food resources that might be needed. Expanding the availability of nutritious and affordable food is also key, as limited access to affordable food choices can lead to increased FI. Food deserts need assistance in creating tax incentives, developing and equipping grocery stores, small retailers, corner markets, and farmer's markets, along with partnering with community allies that can enhance the ability to connect families to resources [36]. Driving economic growth in food deserts could lead to indirect health benefits. One example of a proposed intervention is the Healthy Food Financing Initiative implemented by the US government, which utilizes funding by the federal government to bring stores and food retailers to underserved communities [37]. Additionally, similar to farmer's markets, mobile food markets expand access to larger areas, but can cover more area in a shorter amount of time [37–39]. The Green Grocer mobile farmer's market, piloted by the Greater Pittsburgh Community Food Bank, is an example of a program that has the potential to reduce geographically based disparities in food deserts. It is owned and operated by a food bank and designed to be affordable and accept multiple forms of payment, including Electronic Benefits Transfer cards that have Supplemental Nutrition Assistance Program (SNAP) benefits [38,39]. Another successful program is the Just Harvest organizations, which targets FI at the individual, community, and national level by partnering with community organizations to develop grocery stores in food deserts. Additionally, schools and child nutrition programs are profoundly important in providing food access to children. Some of these programs include the School Breakfast Program, National School Lunch Program, and Afterschool Nutrition Program, all of which have been shown to have a positive effect on the health of food insecure children [40]. Social workers play an invaluable role in connecting patients and their families to local and federal food programs, such as the ones mentioned above. This is especially true for patients with chronic diseases, such as CF, because they are seeing the specialist on a more frequent basis, creating more opportunities for FI screening, resource distribution, and follow up. It is also important to acknowledge that technology can play a vital role in decreasing barriers to food access, as online grocery sales have become increasingly more widely available. Delivery services can be of great benefit for patients and families, who do not have a vehicle and are reliant on public transportation or ride sharing. However, this technology can be hindered by delivery and service fees, making this service inaccessible to an individual already facing food insecurity [41]. The Baltimore Ride Share Project is a program that was piloted in Baltimore in 2019, which provides transportation services to and from supermarkets at a discounted rate via a ride share service for residents living in food deserts. This is a program that can be replicated in other cities and decrease barriers to food access for individuals living in food deserts [42]. The CF community and advocacy organizations should consider partnering with similar organizations that are near CF care centers where a high proportion of their patients are either food insecure and/or live in or near a food desert. Additionally, government agencies can adopt the Healthy Food Priority Area (HFPA) designation, an initiative developed by Baltimore in 2018 that prioritizes certain areas for action against food insecurity. Each census tract is assigned a score for food availability, food access, and food utilization, and those with higher scores are designated HFPAs. This initiative helps identify high-risk communities and prioritize a city's action against food insecurity [41].

This study has several limitations. First, it examines a small sample from a single center in the Mid-Atlantic/Northeast US and, thus, may not be generalizable to other regions. Additionally, it does not capture everyone who experiences FI, as there are people with CF who live outside of food deserts who are food insecure. Future work may consider utilizing similar methods to analyze the health impacts of food deserts on people with CF nationally. Moreover, exploring the perspectives of people with CF and their families who experience FI and/or reside in food deserts may help target future interventions to increase access to healthy and affordable food.

5. Conclusions

Food deserts and their surrounding regions impact pediatric CF outcomes, particularly BMI. CF teams should routinely screen for FI and proximity to food deserts, and development of novel interventions are essential to increase access to healthy and affordable food. The methods used in this study can serve as a model to assess the association between food deserts and health outcomes more broadly in CF and in other chronic gastrointestinal diseases. Furthermore, as prior research has shown that FI may be strongly predictive of chronic illnesses, these results may have widespread applicability, as nutrition is one of the mainstays of therapy for many chronic diseases [43].

Author Contributions: Conceptualization, M.A.C.-H., K.S.K., D.J.W., S.C.K., T.M.K.; methodology, M.A.C.-H., K.S.K., A.F., T.M.K.; validation, M.A.C.-H., K.S.K., A.F., T.M.K.; formal analysis, M.A.C.-H., K.S.K., A.F., M.R.H., T.M.K.; investigation, M.A.C.-H., K.S.K., A.F., M.R.H., T.M.K.; resources, M.A.C.-H., K.S.K., A.F., M.R.H., T.M.K.; data curation, M.A.C.-H., T.M.K.; software, M.A.C.-H., A.F.; writing—original draft preparation, M.A.C.-H., T.M.K.; writing—review and editing, M.A.C.-H., K.S.K., M.R.H., A.F., D.J.W., S.C.K., T.M.K.; visualization, M.A.C.-H., K.S.K., T.M.K.; supervision, T.M.K.; project administration, M.A.C.-H.; funding acquisition, M.A.C.-H., K.S.K., D.J.W., S.C.K., T.M.K. All authors have read and agreed to the published version of the manuscript.

Funding: This research was funded by the Cystic Fibrosis Foundation (grant ID: CORBER20B0 and CORBER21D0).

Institutional Review Board Statement: This study was approved by the Institutional Review Board (IRB) of the University of Pittsburgh (protocol number: 20050263).

Informed Consent Statement: Patient consent was waived due to secondary research on data or specimens.

Data Availability Statement: The data presented in this study are available on request from the corresponding author.

Acknowledgments: We would like to thank the patients and families who consented to participate in our center's CF registry.

Conflicts of Interest: The authors declare no conflict of interest.

References

1. USDA Economic Research Service Definitions of Food Insecurity. Available online: https://www.ers.usda.gov/topics/food-nutrition-assistance/food-security-in-the-us/definitions-of-food-security.aspx (accessed on 1 April 2021).

2. Guide to Measuring Household Food Security, Revised 2000. Available online: https://naldc.nal.usda.gov/download/38369/PDF (accessed on 1 April 2021).
3. Centers for Disease Control and Prevention Geographic Information Systems. Available online: www.cdc.gov/gis/index.html (accessed on 1 April 2021).
4. USDA Economic Research Service Mapping Food Deserts in the United States. Available online: https://www.ers.usda.gov/amber-waves/2011/december/data-feature-mapping-food-deserts-in-the-us/ (accessed on 1 April 2021).
5. Addressing Food Insecurity in CF Care. Available online: https://www.cff.org/Care/Clinician-Resources/Network-News/January-2019/Addressing-Food-Insecurity-in-CF-Care/ (accessed on 1 May 2021).
6. Yen, E.; Quniton, B.; Borowitz, D. Better Nutritional Status in Early Childhood is Associated with Improved Clinical Outcomes and Survival in Patients with Cystic Fibrosis. *J. Pediatrics* **2013**, *162*, 530–535. [CrossRef]
7. Konstan, M.W.; Butler, S.M.; Wohl, M.E.; Stoddard, M.; Matousek, R.; Wagener, J.S.; Johnson, C.A.; Morgan, W.J. Growth and nutritional indexes in early life predict pulmonary function in cystic fibrosis. *J. Pediatrics* **2003**, *142*, 624–630. [CrossRef] [PubMed]
8. Brown, P.S.; Durham, D.; Tivis, R.D.; Stamper, S.; Waldren, C.; Toevs, S.E.; Gordon, B.; Robb, T.A. Evaluation of Food Insecurity in Adults and Children with Cystic Fibrosis: Community Case Study. *Front. Public Health* **2018**, *6*, 1–8. [CrossRef] [PubMed]
9. Borowitz, D.; Baker, R.D.; Stallings, V. Consensus Report on Nutrition for Pediatric Patients with Cystic Fibrosis. *J. Ped Gastroenterol. Nutr.* **2002**, *35*, 246–259. [CrossRef] [PubMed]
10. American Academy of Pediatrics and the Food Research & Action Center Screen and Intervene: A Toolkit for Pediatricians to Address Food Insecurity. Available online: https://frac.org/wp-content/uploads/FRAC_AAP_Toolkit_2021.pdf (accessed on 1 May 2021).
11. Banerjee, S.; Radak, T.; Khubchandani, J.; Dunn, P. Food Insecurity and Mortality in American Adults: Results From the NHANES-Linked Mortality Study. *Health Promot. Pract.* **2021**, *22*, 204–214. [CrossRef] [PubMed]
12. Clemm, C. Force Field Analysis of Food Insecurity Screening and Treatment in 15 CF Care Centers. In Proceedings of the 33rd Annual North American Cystic Fibrosis Conference, Nashville, TN, USA, 31 October–2 November 2019. Abstract Number 836.
13. USDA Economic Research Service Food Access Research Atlas. Available online: https://www.ers.usda.gov/data-products/food-access-research-atlas/ (accessed on 1 March 2021).
14. Noelke, C.; McArdle, N.; Baek, M.; Huntington, N.; Huber, R.; Hardy, E.; Acevedo-Garcia, D. Child Opportunity Index 2.0 Technical Documentation. Available online: http://new.diversitydatakids.org/sites/default/files/2020-01/ddk_coi2.0_technical_documentation_20200115_1.pdf (accessed on 1 April 2021).
15. Stallings, V.A.; Stark, L.J.; Robinson, K.; Feranchak, A.; Quinton, H. Evidence based practive recommendations for nutrition-related management of children and adults with cystic fibrosis and pancreatic insufficiency: Results of a systematic review. *J. Am. Diet. Assoc.* **2008**, *108*, 832–839. [CrossRef]
16. Flume, P.A.; O'Sullivan, B.P.; Robinson, K.A.; Goss, C.H.; Mogayzel, P.J.; Willey-Courand, D.B.; Bujan, J.; Finder, J.; Lester, M.; Quittell, L.; et al. Cystic Fibrosis Pulmonary Guidelines: Chronic Medications for Maintenance of Lung Health. *Am. J. Respir. Crit. Care Med.* **2007**, *176*, 957–969. [CrossRef]
17. Cystic Fibrosis Foundation Patient Registry 2019 Annual Data Report. Available online: https://www.cff.org/Research/Researcher-Resources/Patient-Registry/2019-Patient-Registry-Annual-Data-Report.pdf (accessed on 1 August 2021).
18. Beaulac, J.; Kristjansson, E.; Cummins, S. A Systematic Review of Food Deserts, 1966–2007. *Prev. Chronic Dis.* **2009**, *6*, 1–10.
19. Stronks, K.; Van De Mheen, D.; Looman, C.W.N.; Mackenbach, J. Behavioural and structural factors in the explanation of socio-economic inequalities in health: An empirical analysis. *Sociol. Health Illn.* **1996**, *18*, 653–674. [CrossRef]
20. Centers for Disease Control and Prevention Social Determinants of Health: Know What Affects Health. Available online: https://www.cdc.gov/socialdeterminants/index.htm (accessed on 1 July 2021).
21. Berkowitz, S.A.; Basu, S.; Venkataramani, A.; Reznor, G.; Fleegler, E.W.; Atlas, S.J. Association between access to social service resources and cardiometabolic risk factors: A machine learning and multilevel modeling analysis. *BMJ* **2019**, *9*, 1–8. [CrossRef]
22. Cook, J.T.; Frank, D.A.; Berkowitz, C.; Black, M.M.; Casey, P.H.; Cutts, D.B.; Meyers, A.F.; Zaldivar, N.; Skalicky, A.; Levenson, S.; et al. Food Insecurity Is Associated with Adverse Health Outcomes among Human Infants and Toddlers. *J. Nutr.* **2004**, *134*, 1432–1438. [CrossRef]
23. Diez Roux, A.V. Investigating Neighborhood and Area Effects on Health. *Am. J. Public Health* **2001**, *91*, 1783–1789. [CrossRef] [PubMed]
24. Walker, R.E.; Keane, C.R.; Burke, J.G. Disparities and access to healthy food in the United States: A review of food deserts literature. *Health Place* **2010**, *16*, 876–884. [CrossRef] [PubMed]
25. Food Insecurity and Chronic Disease. Available online: https://www.ifm.org/news-insights/food-insecurity-chronic-disease/ (accessed on 1 August 2021).
26. Kelli, H.M.; Hammadah, M.; Ahmed, H.; Ko, Y.A.; Topel, M.; Samman-Tahhan, A.; Awad, M.; Patel, K.; Mohammed, K.; Sperling, L.S.; et al. Association Between Living in Food Deserts and Cardiovascular Risk. *Circ. Cardiovas. Qual. Outcomes* **2017**, *10*, 1–22. [CrossRef] [PubMed]
27. Sturm, R.; Datar, A. Body mass index in elementary school children, metropolitan area food prices and food outlet density. *Public Health* **2005**, *119*, 1059–1068. [CrossRef]
28. Larson, N.I.; Story, M.T.; Nelson, M.C. Neighborhood environments: Disparities in access to healthy foods in the U.S. *Am. J. Prev. Med.* **2009**, *36*, 74–81. [CrossRef]

29. Leung, C.W.; Laraia, B.A.; Kelly, M.; Nickleach, D.; Adler, N.E.; Kushi, L.H.; Yen, I.H. The influence of neighborhood food stores on change in young girls' body mass index. *Am. J. Prev. Med.* **2011**, *41*, 43–51. [CrossRef] [PubMed]
30. Merchant, A.T.; Dehghan, M.; Behnke-Cook, D.; Anand, S.S. Diet, physical activity, and adiposity in children in poor and rich neighbourhoods: A cross-sectional comparison. *Nutr. J.* **2007**, *6*, 1–7. [CrossRef] [PubMed]
31. Oates, G.; Rutland, S.; Juarez, L.; Friedman, A.; Schechter, M.S. The association of area deprivation and state child health with respiratory outcomes of pediatric patients with cystic fibrosis in the United States. *Pediatric Pulmonol.* **2021**, *56*, 883–890. [CrossRef]
32. Schechter, M.S.; Shelton, B.J.; Margolis, P.A.; Fitzsimmons, S.C. The association of socioeconomic status with outcomes in cystic fibrosis patients in the United States. *Am. J. Respir. Crit. Care Med.* **2001**, *163*, 1331–1337. [CrossRef]
33. Bauer, S.R.; Monuteaux, M.C.; Fleegler, E.W. Geographic Disparities in Access to Agencies Providing Income-Related Social Services. *J. Urban Health* **2015**, *92*, 853–863. [CrossRef] [PubMed]
34. Berkowitz, S.A.; Karter, A.J.; Corbie-Smith, G.; Seligman, H.K.; Ackroyd, S.A.; Barnard, L.S.; Atlas, S.J.; Wexler, D.J. Food Insecurity, Food "Deserts", and Glycemic Control in Patients with Diabetes: A Longitudinal Analysis. *Diabetes Care* **2018**, *41*, 1188–1195. [CrossRef] [PubMed]
35. Durfey, S.; Kind, A.; Gutman, R.; Monteiro, K.; Buckingham, W.R.; DuGoff, E.H.; Trivedi, A.N. Impact of Risk Adjustment For Socioeconomic Status On Medicare Advantage Plan Quality Rankings. *Health Aff.* **2018**, *37*, 1065–1072. [CrossRef] [PubMed]
36. Incentivizing the Sale of Healthy and Local Food. Available online: http://growingfoodconnections.org/wp-content/uploads/sites/3/2015/11/GFCHealthyFoodIncentivesPlanningPolicyBrief_2016Feb-1.pdf (accessed on 15 September 2021).
37. CED Healthy Food Financing Initiative FY 2016. Available online: https://www.acf.hhs.gov/archive/ocs/programs/community-economic-development/healthy-food-financing (accessed on 23 September 2021).
38. Gary-Webb, T.L.; Bear, T.M.; Mendez, D.D.; Schiff, M.D.; Keenan, E.; Fabio, A. Evaluation of a Mobile Farmer's Market Aimed at Increasing Fruit and Vegetable Consumption in Food Deserts: A Pilot Study to Determine Evaluation Feasibility. *Health Equity* **2018**, *2*, 375–383. [CrossRef]
39. Mendez, D.D.; Fabio, A.; Robinson, T.; Bear, T.; Keenan, E.; Schiff, M.D.; Gary-Webb, T. Green Grocer: Using Spatial Analysis to Identify Locations for a Mobile Food Market. *Prog. Community Health Partnersh.* **2020**, *14*, 109–115. [CrossRef]
40. The Role of the Federal Child Nutrition Programs in Improving Health and Well-Being. Available online: https://frac.org/wp-content/uploads/hunger-health-role-federal-child-nutrition-programs-improving-health-well-being.pdf (accessed on 2 November 2021).
41. Understanding Food Insecurity in the City of Pittsburgh. Available online: https://apps.pittsburghpa.gov/redtail/images/11738_FeedPGH_2020.10.16.pdf (accessed on 2 November 2021).
42. Baltimore Ride Share Project to Support Health Food Priority Areas. Available online: https://phnci.org/uploads/resource-files/PHNCI-Case-Study-Baltimore.pdf (accessed on 3 November 2021).
43. USDA Economic Research Service Food Insecurity, Chronic Disease, and Health among Working-Age Adults. Available online: https://www.ers.usda.gov/webdocs/publications/84467/err-235.pdf?v=7518.3 (accessed on 15 June 2021).

Article

Vitamin D Status and Risk of Cystic Fibrosis-Related Diabetes: A Retrospective Single Center Cohort Study

Yiqing Peng [1], Malinda Wu [2], Jessica A. Alvarez [3] and Vin Tangpricha [3,4,*]

1. Emory College, Emory University, Atlanta, GA 30322, USA; whitney.peng@emory.edu
2. Division of Endocrinology, Department of Pediatrics, Emory University School of Medicine, Atlanta, GA 30322, USA; malinda.wu@alumni.emory.edu
3. Division of Endocrinology, Metabolism & Lipids, Department of Medicine, Emory University School of Medicine, Atlanta, GA 30322, USA; jessica.alvarez@emoryed.edu
4. Atlanta VA Medical Center, Decatur, GA 30300, USA
* Correspondence: vin.tangpricha@emory.edu

Abstract: Objective: Cystic fibrosis-related diabetes (CFRD) affects up to half of the people with cystic fibrosis (CF) by adulthood. CFRD is primarily caused by pancreatic dysfunction that leads to insufficient insulin release and/or insulin resistance. Exocrine pancreatic insufficiency in people with CF is associated with fat-soluble vitamin malabsorption, including vitamins A, D, E, and K. This study examined the relationship between vitamin D status, assessed by serum 25-hydroxyvitamin D (25(OH)D), and the development of CF-related diabetes (CFRD) in adults with CF. Methods: This was a retrospective cohort study of adults seen at a single CF center. The data were extracted from the electronic medical records and the Emory Clinical Data Warehouse, a data repository of health information from patients seen at Emory Healthcare. We collected age, race, the first recorded serum 25-hydroxyvitamin D (25(OH)D) concentration, body mass index (BMI), and onset of diabetes diagnosis. Log-rank (Mantel–Cox) tests were used to compare the relative risk of CFRD onset in the subjects with stratified vitamin D status and weight status. A sub-group analysis using chi-square tests assessed the independence between vitamin D deficiency and CFRD risk factors, including gender and CF mutation types (homozygous or heterozygous for F508del, or others). Unpaired t-tests were also used to compare the BMI values and serum 25(OH)D between the CF adults based on the CFRD development. Results: This study included 253 subjects with a mean age of 27.1 years (± 9.0), a mean follow-up time period of 1917.1 (± 1394.5) days, and a mean serum 25(OH)D concentration of 31.8 ng/mL (± 14.0). The majority (52.6%) of the subjects developed CFRD during the study period. Vitamin D deficiency (defined as 25(OH)D < 20 ng/mL) was present in 25.3% of the subjects. Close to two thirds (64.1%) of the subjects with vitamin D deficiency developed CFRD during the study. Vitamin D deficiency increased the risk of developing CFRD (chi-square, $p = 0.03$) during the course of the study. The time to the onset of CFRD stratified by vitamin D status was also significant (25(OH)D < 20 ng/mL vs. 25(OH)D $\geq$ 20 ng/mL) (95% CI: 1.2, 2.7, $p < 0.0078$). Conclusion: Our findings support the hypothesis that adults with CF and vitamin D deficiency are at a higher risk of developing CFRD and are at risk for earlier CFRD onset. The maintenance of a serum 25(OH)D concentration above 20 ng/mL may decrease the risk of progression to CFRD.

Keywords: cystic fibrosis; vitamin D; vitamin D deficiency; diabetes; epidemiology

Citation: Peng, Y.; Wu, M.; Alvarez, J.A.; Tangpricha, V. Vitamin D Status and Risk of Cystic Fibrosis-Related Diabetes: A Retrospective Single Center Cohort Study. *Nutrients* **2021**, *13*, 4048. https://doi.org/10.3390/nu13114048

Academic Editor: Genevieve Mailhot

Received: 22 September 2021
Accepted: 8 November 2021
Published: 12 November 2021

Publisher's Note: MDPI stays neutral with regard to jurisdictional claims in published maps and institutional affiliations.

Copyright: © 2021 by the authors. Licensee MDPI, Basel, Switzerland. This article is an open access article distributed under the terms and conditions of the Creative Commons Attribution (CC BY) license (https://creativecommons.org/licenses/by/4.0/).

1. Introduction

Cystic fibrosis (CF) is an autosomal recessive disorder caused by a mutation of the cystic fibrosis transmembrane conductance regulator (CFTR) gene. The complications of CF include chronic bacterial infection in the lungs, CF-related diabetes, pancreatic insufficiency, and a decline in lung function [1,2]. Exocrine pancreatic insufficiency leads to fat malabsorption and a deficiency in fat-soluble vitamins, including vitamin A, D, E, and K [3]. Vitamin D deficiency can subsequently lead to CF bone disease and an increased risk of fractures [4,5]. Endocrine pancreatic insufficiency, along with other contributing risk factors, such as body weight, diet, and physical activity, can lead to CF-related diabetes (CFRD). Close to half of the adults with CF will develop CFRD, which is characterized by insufficient insulin production, reduced pancreatic beta cell activity, and insulin resistance [2].

It is well established that vitamin D plays important roles in maintaining bone health, calcium homeostasis in blood, and preventing osteoporosis [6]. Vitamin D insufficiency (25(OH)D < 30 ng/mL), which is commonly observed in people with CF, has been associated with decreased insulin sensitivity and secretion in both animal and human studies [7,8]. The best marker of vitamin D status is the total serum 25-hydroxyvitamin D (25(OH)D), which accounts for both the endogenous production of vitamin D from the skin and dietary intake of vitamin D-containing foods and supplements [9]. A serum 25(OH)D level of less than 30 ng/mL is considered as insufficient, whereas a 25(OH)D of less than 20 ng/mL is considered as vitamin D deficient [10,11]. A recent study suggested that vitamin D status is associated with glucose intolerance and CFRD [12], but it did not address whether vitamin D status influences the risk of CFRD development or time to CFRD onset. The mechanisms by which vitamin D protects against the development of diabetes are uncertain. However, pre-clinical studies have indicated that vitamin D may regulate insulin secretion and improve beta-cell function [13].

This current study focuses on the relationship between vitamin D deficiency and CFRD, two common endocrine co-morbidities found in CF. Given the association between vitamin D status and the risk of CFRD, this study aimed to examine the impact of vitamin D status on the onset of CFRD in adults with CF. We hypothesized that low vitamin D status is one of the risk factors for the development of CFRD in adults. The design of the study was a retrospective longitudinal cohort study of adults with CF receiving care at a single center. The data on the subject demographics, serum vitamin D (25(OH)D) measurements, and diabetes status were extracted from the medical records to examine the influence of vitamin D status on CFRD risk.

2. Materials and Methods

2.1. Study Design

This was a retrospective chart review examining the relationship between vitamin D status, assessed by serum 25-hydroxyvitamin D (25(OH)D), and development of CFRD. The research study was approved by the Emory Institutional Review Board (IRB). Subjects were adults, age greater than 18, with CF treated by the Emory CF Center and receiving care by the Emory Clinic and Emory Hospital from 2002–2012. Inclusion criteria included a confirmed diagnosis of CF and at least one serum 25(OH)D measurement taken between 1 January 2002 and 31 December 2012. Exclusion criteria included a diagnosis of CFRD at the time of first serum 25(OH)D measurement. Development of CFRD was defined as a diagnosis of CFRD in the medical record, initiation with diabetes medication, fasting glucose $\geq$ 126 mg/dL, 2hr oral glucose tolerance test (OGTT) glucose $\geq$ 200 mg/dL, hemoglobin A1C (HgbA1c) $\geq$ 6.5%, or classical symptoms of diabetes in the presence of a casual glucose $\geq$ 200 mg/dL. Data from the electronic medical record or the CF Foundation Patient Registry (PortCF) were used to conduct the analysis. We collected data on each subject's serum 25(OH)D level, body mass index (BMI), date of diabetes diagnosis, presence of pancreatic insufficiency as defined as prescription of pancreatic enzymes, and the number of days since the start of study (1 January 2002) to diagnosis of CFRD, or, if not, to the end of study on 31 December 2012.

2.2. Database and Categorization

Data were obtained from the Emory Clinical Data Warehouse, an electronic database of all clinical laboratory data for the Emory Clinic, Emory University Hospital, and Emory CF Center. Data on pancreatic insufficiency by use of pancreatic enzymes and CFTR mutation were obtained from PortCF. The date of the onset of CFRD was determined by the first date recorded that the subject developed CFRD during the study period. Subjects who did not develop CFRD by the end of the study period were deemed not to have developed CFRD. The first serum 25(OH)D recorded in the electronic medical record during the study period was used to define a subject's vitamin D status. Vitamin D status deficiency was defined as a 25(OH)D < 20 ng/mL, and insufficient vitamin D level was defined as 25(OH)D < 30 ng/mL. Weight statuses were categorized based on subjects' BMI. A BMI of more than 25 kg/m^2 was considered as overweight, and BMI cut-offs greater than 22 kg/m^2 for female and 23 kg/m^2 for male subjects were considered as the recommended BMI level for adults with CF [14].

2.3. Statistical Analysis

Prism 9.0.1 was used to perform statistical analyses. The odds ratios between potential factors contributing to CFRD and vitamin D status were analyzed with 95% confidence interval. Freedom from CFRD was compared using log-rank (Mantel–Cox) tests, and all study subjects were free from CFRD at the beginning of the analysis. Log-rank tests compared the relative risk of CFRD onset in subjects with stratified vitamin D status and weight status. Sub-group analysis using chi-square tests assessed the independence between vitamin D deficiency and CFRD risk factors, including gender and CF mutation types (homozygous for F508del, heterozygous for F508del, or others). Unpaired t-test was also used to determine if BMI or first 25(OH)D measures differed between adults with CF who developed CFRD and those that did not.

3. Results

3.1. Subject Demographics

A total of 267 adults without CFRD were potentially eligible for the study. One patient was excluded due to an implausible BMI value of 43, and 13 subjects were excluded due to missing BMI values. Eventually, 253 subjects met the inclusion criteria, and their baseline demographics are presented in Table 1, stratified by vitamin D deficiency status. During the course of the study, 52.6% of the subjects with CF developed CFRD. During the course of the study, 53.1% of the subjects with insufficient vitamin D (25(OH)D levels < 30 ng/mL) developed CFRD. Using a cut-off value of 25(OH)D < 20 ng/mL, 64.1% of vitamin D deficient subjects (25(OH)D < 20 ng/mL) developed CFRD.

3.2. Time to Onset of Cystic Fibrosis-Related Diabetes

We examined the time to the onset of CFRD by vitamin D status (deficient or insufficient). The time to the onset of CFRD refers to the first instance that a subject had documentation of a CFRD diagnosis as defined in our methods. The log-rank (Mantel–Cox) test was used for the time of CFRD onset by insufficient vitamin D status (25(OH)D < 30 ng/mL) and deficient vitamin D status (25(OH)D < 20 ng/mL). No significant difference was observed in the time to the onset of CFRD in the subjects with CF compared to those with and without insufficient vitamin D status (hazard ratio: 0.84, 95% CI: 0.60, 1.18, $p = 0.31$). However, there was a statistically significant hazard ratio in the time to the onset of CFRD in the subjects with CF compared to those with 25(OH)D < 20 ng/mL versus those with 25(OH)D $\geq$ 20 ng/mL (Figure 1, hazard ratio: 1.76, 95% CI: 1.2, 2.7, $p = 0.0078$). Based on the survival curve in Figure 1, 50% of the subjects with a deficiency in vitamin D were diagnosed with CFRD at day 931, while, only after 3070 days, 50% of the subjects without deficient vitamin D were diagnosed with CFRD.

Table 1. Baseline subject demographics stratified by vitamin D deficiency at 20 ng/mL.

	All Subjects	Vitamin D Deficient (25(OH)D < 20 ng/mL)	Vitamin D Not Deficient (25(OH)D ≥ 20 ng/mL)
Subjects n (%)	253	64 (25.3)	189 (74.7)
Age at Entry, y	27.1 (±9.0)	26.9 (±8.3)	27.1 (±9.2)
Gender, male	132 (52.2)	39 (60.1)	93 (49.2)
Gender, female	121 (47.8)	25 (39.1)	96 (50.8)
Race, Caucasian or White [a]	231 (91.3)	47 (73.4)	186 (97.4)
Race, African American or Black [a]	18 (7.1)	16 (25)	2 (1.1)
Days without Diabetes Mellitus [b]	1917.1 (±1394.5)	2161.2 (±1627.6)	2410.7 (±1667.6)
BMI, kg/m^2	21.8 (±3.4)	21.8 (±3.8)	21.7 (±3.3)
BMI at goal [c]	165 (65.2)	45 (70.3)	120 (63.5)
BMI [d], <25 kg/m^2	211 (83.4)	53 (82.8)	158 (83.6)
Developed CFRD [e]	133 (52.6)	41 (64.1)	92 (48.7)
25(OH)D, ng/mL	31.8 (±14.0)	12.5 (±4.4)	36.9 (±15.5)

Entries are means (±standard deviation), n (%), or % (n). [a] Six missing values of race not reported. [b] Days, since study start on 1 January 2002. End date was the day when the subject was diagnosed with diabetes, or, if the subject never developed diabetes, 31 December 2012. [c] The BMI goal for adults with CF is greater than 22 kg/m^2 for females and 23 kg/m^2 for males. [d] BMI at or above 25 kg/m^2 is considered overweight. [e] Developed diabetes between 2002–2012.

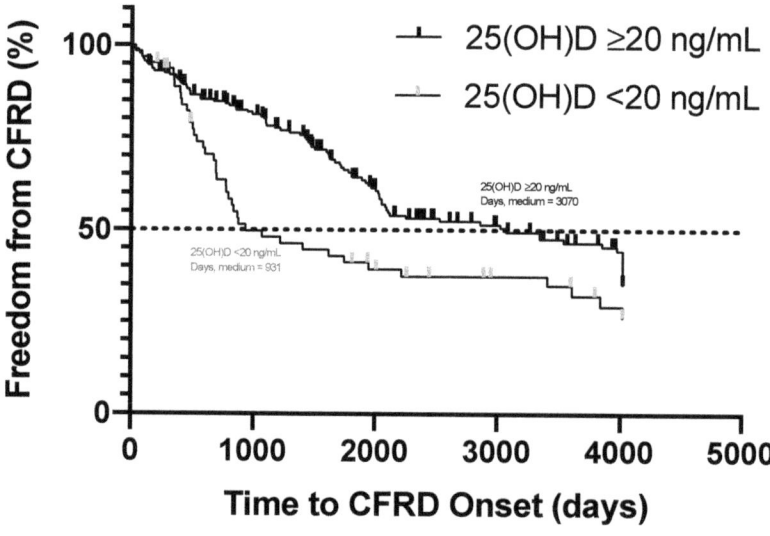

Figure 1. Time in days to cystic fibrosis-related diabetes (CFRD) onset by vitamin D status (25(OH)D < and ≥20 ng/mL). Hazard ratio of the vitamin D deficient subjects to not vitamin D deficient subjects is 1.76 (95% CI: 1.2, 2.7, p = 0.0078). The medium for days of CFRD onset was 931 for subjects with vitamin D deficiency and 3070 for those without.

3.3. Influence of Other Risk Factors, Including Vitamin D Status, on the Development of CFRD

We evaluated other potential risk factors that may be associated with the development of CFRD. The sub-group analyses using chi-square tests assessed the independence between vitamin D deficiency and CFRD risk factors, including sex, CFTR mutation status (F508del homozygous, F508del heterozygous, or no F508del mutation), and pancreatic enzyme usage. Vitamin D deficiency (25(OH)D < 20 ng/mL) was significantly associated with the development of CFRD (chi square, $p = 0.03$) (Table 2). Insufficient vitamin D status (25(OH)D < 30 ng/mL) was not associated with the risk of CFRD ($p = 0.87$) (Table 2). There were no significant associations between vitamin D deficiency and gender ($p = 0.10$), CF mutation type ($p = 0.88$), or pancreatic enzyme usage ($p = 0.42$). The unpaired t-test indicated that the first serum 25(OH)D measure at the start of the study ($p = 0.83$) and average BMI ($p = 0.08$) were not significantly different between the subjects who did and did not develop CFRD (Table 3).

Table 2. Sub-group analysis of vitamin D deficiency/insufficiency with selected demographic factors.

	Compared Value	X^2, Df	p-Value
Vitamin D Deficiency [a]	CF Mutation Type [b]	0.27, 2	0.88
Vitamin D Deficiency	Develop CFRD	4.54, 1	0.03 *
Vitamin D Deficiency	Gender, Male or Female	2.63, 1	0.10
Vitamin D Deficiency	Pancreatic Enzyme Usage [c]	0.66, 1	0.42
Develop CFRD [d]	Gender, Male or Female	0.43, 1	0.51
Vitamin D Insufficiency [e]	Develop CFRD	0.03, 1	0.87

Results from sub-group analysis based on chi-square test analysis. [a] Subjects stratified by vitamin D deficiency at 25(OH)D < 20 ng/mL. [b] CF mutation types categorized by F508del homozygous, F508del heterozygous, or no F508del mutation [c] Pancreatic enzyme usage, yes or no, recorded at the start of study [d] Subjects stratified by whether they were diagnosed with CFRD [e] Subjects stratified by vitamin D insufficiency at 25(OH)D < 30 ng/mL. An asterisk (*) indicates statistical significance at $p < 0.05$.

Table 3. Sub-group analysis of subjects developing CFRD during the course of the study.

	Compared Value	t-Score, df	p-Value
Develop CFRD [a]	First 25(OH)D, ng/mL [b]	0.22, 251	0.83
Develop CFRD	BMI, kg/m^2	1.74, 251	0.08

Results from sub-group analysis based on unpaired t-test. [a] Subjects stratified by whether they were diagnosed with CFRD [b] First serum vitamin D measure obtained from subjects at the start of the study.

4. Discussion

This retrospective study aimed to investigate the relationship between vitamin D status, assessed by serum 25(OH)D, and the development of CF-related diabetes (CFRD) in a cohort of subjects without CFRD followed over a 10-year period. As expected, we found that more than half of the subjects (52.6%) developed CFRD over the course of the study, which is consistent with previous reports indicating the prevalence of CFRD in the CF population [2]. The results of our statistical analysis demonstrate that subjects with CF had a higher risk of CFRD development, with earlier onset in those with a deficient serum 25(OH)D level at less than 20 ng/mL compared to those with a serum 25(OH)D greater than 20 ng/mL. Our sub-group analysis also indicated that, aside from the susceptibility to CFRD onset, the subjects with and without vitamin D deficiency were not significantly different in other CFRD-related markers, which included gender, BMI, first vitamin D measure, CF mutation types, and pancreatic enzyme usage.

Our findings support the findings that 25(OH)D > 20 ng/mL may be protective against the development of CFRD. Pincikova et al. found that a serum 25(OH)D less than 20 ng/mL was associated with higher hemoglobin A1c (HbA1c) in a cross-sectional cohort composed of about 900 children and adults with CF living in Scandinavia, suggesting a relationship between vitamin D status and glucose concentrations [12]. The investigators found the association between vitamin D status and HbA1c in children, but such an association was not as strong in adults with CF [12]. In contrast, a similar study performed by Coriati et al. found no relationship between the serum 25(OH)D and measures of glucose status performed by an oral glucose tolerance test in 270 adults with CF in Canada [15].

Studies examining vitamin D status and the risk of diabetes are also mixed in populations without CF. A systematic review by Mitri et al. found that, in eight observational studies, a greater than 500 IU of vitamin D intake per day was associated with a 13% decreased risk of type II diabetes compared to a vitamin D intake of less than 200 IU per day [16]. However, there were no associations found in the development of type II diabetes in the post hoc analyses of randomized controlled trials [16]. A meta-analysis of vitamin D and the risk of type I diabetes found, in four case control studies, that vitamin D supplementation in early childhood reduced the risk of type I diabetes compared to children not supplemented with vitamin D [17]. A recent large multi-center randomized controlled trial found that vitamin D supplementation did not protect against the progression from prediabetes to type II diabetes in adults [18]. Therefore, the current literature is mixed in regard to the potential effect of vitamin D on the development of diabetes, with some potential protection of vitamin D in children with type I diabetes but no benefits seen in adults with type II diabetes and prediabetes.

Vitamin D is thought to be protective against the development of type I diabetes by modulating the immune system by dampening the autoimmune response, decreasing self-destructive islet cell auto-antibodies produced by beta-cells, and, thus, preserving the beta-cell mass of the pancreas. In murine models of type I diabetes mellitus, supplementation with the active form of vitamin D, calcitriol, prevented the onset of diabetes, which further supported the role of vitamin D in the prevention of type I diabetes [19,20]. For type II diabetes, vitamin D was thought to enhance insulin sensitivity; however, a recent meta-analysis of 18 randomized controlled trials found no benefit of vitamin D supplementation on the markers of insulin sensitivity [21]. Finally, a recent meta-analysis of vitamin D on the markers of inflammation and oxidative stress in people with diabetes found that vitamin D reduced C-reactive protein and malondialdehyde levels, both markers of inflammation. The supplementation of vitamin D also increased the nitric oxide release, serum antioxidant capacity, and total glutathione concentrations, favorable changes in oxidative stress [22]. Since CFRD shares some features of type I and II DM, vitamin D may have many roles in the prevention of diabetes by reducing inflammation, oxidative stress, and, potentially, by preserving beta-cell mass.

Optimal vitamin D status has been associated with the decreased risk of diabetes and prediabetes in populations without CF [23]. In addition, better vitamin D status among patients with diabetes and prediabetes is associated with a reduced risk for all-cause mortality [24,25]. However, in a randomized controlled trial of subjects with prediabetes and without CF, intervention with 4000 IU of vitamin D for 2.5 years did not reduce the progression to diabetes compared to a placebo [18]. Despite the protective mechanisms of vitamin D, as outlined above, and the strong association between vitamin D status and the risk of diabetes, clinical trials have been negative in establishing a role of vitamin D supplementation for prevention of progression to diabetes. Thus, vitamin D intervention may need to be provided much earlier in life before beta cell dysfunction and/or vitamin D deficiency represents poor general health that is not modifiable by supplementation with vitamin D.

Our study found that a serum 25(OH)D > 20 ng/mL was protective against the risk of CFRD. However, larger studies in the general population using the NHANES data have suggested that a cut-off of 30 ng/mL was protective against diabetes and metabolic syndrome [26]. The CF Foundation recommends a 25(OH)D level of greater than 30 ng/mL to protect against CF bone disease [27]. The 25(OH)D cut-off of 30 ng/mL to establish vitamin D sufficiency is based on recommendations in people without CF, including a study showing no osteomalacia found on bone biopsy in adults with 25(OH)D concentrations greater than 30 ng/mL [28]. Some investigators have proposed even higher concentrations of 25(OH)D to prevent elevations in PTH concentrations to prevent bone loss [29]. While there is a general consensus that a 25(OH)D concentration of above 30 ng/mL is optimal for bone health in CF, it is not known if different thresholds are protective against conditions such as CFRD, pulmonary exacerbations, or infections. Furthermore, different organ systems may need different thresholds for optimal health. Finally, the duration that 25(OH)D concentrations need to be sufficient is not known either. Studies indicate that the short-term correction of vitamin D status in CF, such as at the time of pulmonary exacerbation, does not change the course of the disease [30]. Until data are available to determine the optimal 25(OH)D concentration for all the health conditions in CF, clinicians should still address vitamin D status by supplementation or encouraging outdoor sunlight exposure in people with CF [9,31].

This is the first study that analyzed the long-term impact of vitamin D status on the onset of diabetes specifically in adults with CF. Limitations to our study include the limited reports on the use of any pancreatic enzymes, the interaction of vitamin D and CFTR modulator therapy, and the types of CFTR mutations in the entire subject population. Another limitation is the lack of data on other potential risk factors for the development of CFRD. Since this was a retrospective study, we lacked information on diet, physical activity, medications, and other health conditions. Only 68 of the 253 subjects had records of CFTR mutation and whether they were prescribed with pancreatic enzymes during their treatment regimes. Although CF mutation types and pancreatic enzyme usage were used as indicators for pancreatic insufficiency, we were unable to draw substantial conclusion due to the sample size of less than 20 when stratified based on vitamin D status. Understanding the CFTR mutation type and the usage of pancreatic enzyme is important to adjust for equal sampling distribution because both factors can potentially affect vitamin D deficiency and diabetes onset. Having an F508 del homozygous mutation is associated with more severe CF complications and a higher risk of pancreatic defects [1]. Therefore, adults with an F508del homozygous mutation might have a higher risk of diabetes development despite the vitamin D status.

Our findings support the hypothesis that adults with CF and vitamin D deficiency are at a higher risk of developing CFRD and are at risk for earlier CFRD onset. We found that a serum 25(OH)D concentration above 20 ng/mL may decrease the risk of the progression to CFRD. However, what is not clear and cannot be answered in this retrospective study is whether providing vitamin D supplements will actually decrease the progression to CFRD. Future randomized prospective studies should evaluate whether vitamin D supplementation with vitamin D in adults and/or children can decrease the progression to CFRD.

Author Contributions: V.T., J.A.A., Y.P. were responsible for the conceptualization, design, and methodology of the study, data collection, formal analysis and interpretation of the data. V.T. and J.A.A. were responsible for the resources, project administration and supervision. V.T., J.A.A., Y.P., M.W. were responsible for the writing-original draft, writing-review and editing. All authors have read and agreed to the published version of the manuscript.

Funding: Supported by the National Institutes of Health (NIH) Center for Advancing Translational Sciences under Award Number UL1TR002378, National Institute of Diabetes and Digestive and Kidney Diseases under awards numbers P30DK125013 and R03DK117246, and the Cystic Fibrosis Foundation (V.T., M.W., J.A.A.).

Institutional Review Board Statement: The study was approved by Emory University IRB Study #IRB00002092.

Informed Consent Statement: Patient consent was waived due to the study design of a retrospective chart review.

Conflicts of Interest: The authors report no financial conflict of interest. The funders had no role in the design of the study; in the collection, analyses, or interpretation of data; in the writing of the manuscript, or in the decision to publish the results.

References

1. Strausbaugh, S.D.; Davis, P.B. Cystic fibrosis: A review of epidemiology and pathobiology. *Clin. Chest Med.* **2007**, *28*, 279–288. [CrossRef]
2. Alvarez, J.A.; Ashraf, A. Role of vitamin d in insulin secretion and insulin sensitivity for glucose homeostasis. *Int. J. Endocrinol.* **2010**, *2010*, 351385. [CrossRef] [PubMed]
3. Bertolaso, C.; Groleau, V.; Schall, J.I.; Maqbool, A.; Mascarenhas, M.; Latham, N.E.; Dougherty, K.A.; Stallings, V.A. Fat-soluble vitamins in cystic fibrosis and pancreatic insufficiency: Efficacy of a nutrition intervention. *J. Pediatr. Gastroenterol. Nutr.* **2014**, *58*, 443–448. [CrossRef]
4. Donovan, D.S., Jr.; Papadopoulos, A.; Staron, R.B.; Addesso, V.; Schulman, L.; McGREGOR, C.; Cosman, F.; Lindsay, R.L.; Shane, E. Bone mass and vitamin D deficiency in adults with advanced cystic fibrosis lung disease. *Am. J. Respir. Crit. Care Med.* **1998**, *157*, 1892–1899. [CrossRef] [PubMed]
5. Aris, R.M.; Renner, J.B.; Winders, A.D.; Buell, H.E.; Riggs, D.B.; Lester, G.E.; Ontjes, D.A. Increased rate of fractures and severe kyphosis: Sequelae of living into adulthood with cystic fibrosis. *Ann. Intern. Med.* **1998**, *128*, 186–193. [CrossRef]
6. Khazai, N.; Judd, S.E.; Tangpricha, V. Calcium and vitamin D: Skeletal and extraskeletal health. *Curr. Rheumatol. Rep.* **2008**, *10*, 110–117. [CrossRef]
7. Palomer, X.; González-Clemente, J.M.; Blanco-Vaca, F.; Mauricio, D. Role of vitamin D in the pathogenesis of type 2 diabetes mellitus. *Diabetes Obes. Metab.* **2008**, *10*, 185–197. [CrossRef]
8. Pittas, A.G.; Lau, J.; Hu, F.B.; Dawson-Hughes, B. The role of vitamin D and calcium in type 2 diabetes. A systematic review and meta-analysis. *J. Clin. Endocrinol. Metab.* **2007**, *92*, 2017–2029. [CrossRef] [PubMed]
9. Tangpricha, V.; Kelly, A.; Stephenson, A.; Maguiness, K.; Enders, J.; Robinson, K.A.; Marshall, B.C.; Borowitz, D.; Cystic Fibrosis Foundation Vitamin D Evidence-Based Review Committee. An update on the screening, diagnosis, management, and treatment of vitamin D deficiency in individuals with cystic fibrosis: Evidence-based recommendations from the Cystic Fibrosis Foundation. *J. Clin. Endocrinol. Metab.* **2012**, *97*, 1082–1093. [CrossRef]
10. Holick, M.F. Vitamin D deficiency. *N. Engl. J. Med.* **2007**, *357*, 266–281. [CrossRef] [PubMed]
11. Holick, M.F.; Binkley, N.C.; Bischoff-Ferrari, H.A.; Gordon, C.M.; Hanley, D.A.; Heaney, R.P.; Murad, M.H.; Weaver, C.M. Evaluation, treatment, and prevention of vitamin D deficiency: An Endocrine Society clinical practice guideline. *J. Clin. Endocrinol. Metab.* **2011**, *96*, 1911–1930. [CrossRef]
12. Pincikova, T.; Nilsson, K.; Moen, I.E.; Fluge, G.; Hollsing, A.; Knudsen, P.K.; Lindblad, A.; Mared, L.; Pressler, T.; Hjelte, L. Vitamin D deficiency as a risk factor for cystic fibrosis-related diabetes in the Scandinavian Cystic Fibrosis Nutritional Study. *Diabetologia* **2011**, *54*, 3007–3015. [CrossRef]
13. Daley, T.; Hughan, K.; Rayas, M.; Kelly, A.; Tangpricha, V. Vitamin D deficiency and its treatment in cystic fibrosis. *J. Cyst. Fibros.* **2019**, *18* (Suppl. 2), S66–S73. [CrossRef]
14. Stallings, V.A.; Stark, L.J.; Robinson, K.A.; Feranchak, A.P.; Quinton, H.; Clinical Practice Guidelines on Growth and Nutrition Subcommittee; Ad Hoc Working Group. Evidence-based practice recommendations for nutrition-related management of children and adults with cystic fibrosis and pancreatic insufficiency: Results of a systematic review. *J. Am. Diet. Assoc.* **2008**, *108*, 832–839. [CrossRef] [PubMed]
15. Coriati, A.; Dubois, C.L.; Phaneuf, M.; Mailhot, M.; Lavoie, A.; Berthiaume, Y.; Rabasa-Lhoret, R. Relationship between vitamin D levels and glucose tolerance in an adult population with cystic fibrosis. *Diabetes Metab.* **2016**, *42*, 135–138. [CrossRef] [PubMed]
16. Mitri, J.; Muraru, M.D.; Pittas, A.G. Vitamin D and type 2 diabetes: A systematic review. *Eur. J. Clin. Nutr.* **2011**, *65*, 1005–1015. [CrossRef]
17. Zipitis, C.S.; Akobeng, A.K. Vitamin D supplementation in early childhood and risk of type 1 diabetes: A systematic review and meta-analysis. *Arch. Dis. Child.* **2008**, *93*, 512–517. [CrossRef] [PubMed]
18. Pittas, A.G.; Dawson-Hughes, B.; Sheehan, P.; Ware, J.H.; Knowler, W.C.; Aroda, V.R.; Brodsky, I.; Ceglia, L.; Chadha, C.; Chatterjee, R.; et al. Vitamin D supplementation and prevention of type 2 diabetes. *N. Engl. J. Med.* **2019**, *381*, 520–530. [CrossRef]
19. Cristelo, C.; Machado, A.; Sarmento, B.; Gama, F.M. The roles of vitamin D and cathelicidin in type 1 diabetes susceptibility. *Endocr. Connect.* **2021**, *10*, R1–R12. [CrossRef]
20. Zella, J.B.; McCary, L.C.; DeLuca, H.F. Oral administration of 1, 25-dihydroxyvitamin D3 completely protects NOD mice from insulin-dependent diabetes mellitus. *Arch. Biochem. Biophys.* **2003**, *417*, 77–80. [CrossRef]
21. Pramono, A.; Jocken, J.W.E.; Blaak, E.E.; van Baak, M.A. The Effect of Vitamin D Supplementation on Insulin Sensitivity: A Systematic Review and Meta-analysis. *Diabetes Care* **2020**, *43*, 1659–1669. [CrossRef] [PubMed]
22. Mansournia, M.A.; Ostadmohammadi, V.; Lankarani, K.B.; Tabrizi, R.; Kolahdooz, F.; Heydari, S.T.; Kavari, S.H.; Mirhosseini, M.; Mafi, A.; Dastorani, M. The effects of vitamin D supplementation on biomarkers of inflammation and oxidative stress in diabetic patients: A systematic review and meta-analysis of randomized controlled trials. *Horm. Metab. Res.* **2018**, *50*, 429–440. [CrossRef] [PubMed]
23. Martins, D.; Wolf, M.; Pan, D.; Zadshir, A.; Tareen, N.; Thadhani, R.; Felsenfeld, A.; Levine, B.; Mehrotra, R.; Norris, K. Prevalence of cardiovascular risk factors and the serum levels of 25-hydroxyvitamin D in the United States: Data from the Third National Health and Nutrition Examination Survey. *Arch. Intern. Med.* **2007**, *167*, 1159–1165. [CrossRef] [PubMed]
24. Wan, Z.; Guo, J.; Pan, A.; Chen, C.; Liu, L.; Liu, G. Association of Serum 25-Hydroxyvitamin D Concentrations with All-Cause and Cause-Specific Mortality among Individuals with Diabetes. *Diabetes Care* **2021**, *44*, 350–357. [CrossRef] [PubMed]

25. Lu, Q.; Wan, Z.; Guo, J.; Liu, L.; Pan, A.; Liu, G. Association between Serum 25-hydroxyvitamin D Concentrations and Mortality among Adults with Prediabetes. *J. Clin. Endocrinol. Metab.* **2021**, *106*, e4039–e4048. [CrossRef] [PubMed]
26. Ganji, V.; Tangpricha, V.; Zhang, X. Serum Vitamin D Concentration ≥75 nmol/L Is Related to Decreased Cardiometabolic and Inflammatory Biomarkers, Metabolic Syndrome, and Diabetes; and Increased Cardiorespiratory Fitness in US Adults. *Nutrients* **2020**, *12*, 730. [CrossRef]
27. Chesdachai, S.; Tangpricha, V. Treatment of vitamin D deficiency in cystic fibrosis. *J. Steroid Biochem. Mol. Biol.* **2016**, *164*, 36–39. [CrossRef] [PubMed]
28. Priemel, M.; von Domarus, C.; Klatte, T.O.; Kessler, S.; Schlie, J.; Meier, S.; Proksch, N.; Pastor, F.; Netter, C.; Streichert, T.; et al. Bone mineralization defects and vitamin D deficiency: Histomorphometric analysis of iliac crest bone biopsies and circulating 25-hydroxyvitamin D in 675 patients. *J. Bone Miner. Res.* **2010**, *25*, 305–312. [CrossRef]
29. West, N.E.; Lechtzin, N.; Merlo, C.A.; Turowski, J.B.; Davis, M.E.; Ramsay, M.Z.; Watts, S.L.; Stenner, S.P.; Boyle, M.P. Appropriate goal level for 25-hydroxyvitamin D in cystic fibrosis. *Chest* **2011**, *140*, 469–474. [CrossRef] [PubMed]
30. Tangpricha, V.; Lukemire, J.; Chen, Y.; Binongo, J.N.G.; Judd, S.E.; Michalski, E.S.; Lee, M.J.; Walker, S.; Ziegler, T.R.; Tirouvanziam, R.; et al. Vitamin D for the Immune System in Cystic Fibrosis (DISC): A double-blind, multicenter, randomized, placebo-controlled clinical trial. *Am. J. Clin. Nutr.* **2019**, *109*, 544–553. [CrossRef]
31. Bhimavarapu, A.; Deng, Q.; Bean, M.; Lee, N.; Ziegler, T.R.; Alvarez, J.; Tangpricha, V. Factors Contributing to Vitamin D Status at Hospital Admission for Pulmonary Exacerbation in Adults with Cystic Fibrosis. *Am. J. Med. Sci.* **2021**, *361*, 75–82. [CrossRef] [PubMed]

Communication

Most Short Children with Cystic Fibrosis Do Not Catch Up by Adulthood

Margaret P. Marks [1,2], Sonya L. Heltshe [3,4], Arthur Baines [4], Bonnie W. Ramsey [3], Lucas R. Hoffman [3,5] and Michael S. Stalvey [1,2,*]

1. Department of Pediatrics, University of Alabama, Birmingham, AL 35233, USA; mpmarks@uabmc.edu
2. Cystic Fibrosis Research Center, University of Alabama, Birmingham, AL 35233, USA
3. Department of Pediatrics, University of Washington, Seattle, WA 98105, USA; sonya.heltshe@seattlechildrens.org (S.L.H.); bonnie.ramsey@seattlechildrens.org (B.W.R.); lhoffm@uw.edu (L.R.H.)
4. CFF TDNCC, Seattle Children's Research Institute, Seattle, WA 98121, USA; arthur.baines@seattlechildrens.org
5. Department of Microbiology, University of Washington, Seattle, WA 98105, USA
* Correspondence: mstalvey@uabmc.edu

Citation: Marks, M.P.; Heltshe, S.L.; Baines, A.; Ramsey, B.W.; Hoffman, L.R.; Stalvey, M.S. Most Short Children with Cystic Fibrosis Do Not Catch Up by Adulthood. *Nutrients* **2021**, *13*, 4414. https://doi.org/10.3390/nu13124414

Academic Editors: Jessica Alvarez and Maria R. Mascarenhas

Received: 29 October 2021
Accepted: 5 December 2021
Published: 10 December 2021

Publisher's Note: MDPI stays neutral with regard to jurisdictional claims in published maps and institutional affiliations.

Copyright: © 2021 by the authors. Licensee MDPI, Basel, Switzerland. This article is an open access article distributed under the terms and conditions of the Creative Commons Attribution (CC BY) license (https://creativecommons.org/licenses/by/4.0/).

Abstract: Poor linear growth is common in children with cystic fibrosis (CF) and predicts pulmonary status and mortality. Growth impairment develops in infancy, prior to pulmonary decline and despite aggressive nutritional measures. We hypothesized that growth restriction during early childhood in CF is associated with reduced adult height. We used the Cystic Fibrosis Foundation (CFF) patient registry to identify CF adults between 2011 and 2015 (ages 18–19 y, n = 3655) and had height for age (HFA) records between ages 2 and 4 y. We found that only 26% CF adults were ≥median HFA and 25% were <10th percentile. Between 2 and 4 years, those with height < 10th percentile had increased odds of being <10th percentile in adulthood compared to children ≥ 10th percentile (OR = 7.7). Of HFA measured between the 10th and 25th percentiles at ages 2–4, 58% were <25th percentile as adults. Only 13% between the 10th and 25th percentile HFA at age 2–4 years were >50th percentile as adults. Maximum height between ages 2 and 4 highly correlated with adult height. These results demonstrate that low early childhood CF height correlates with height in adulthood. Since linear growth correlates with lung growth, identifying both risk factors and interventions for growth failure (nutritional support, confounders of clinical care, and potential endocrine involvement) could lead to improved overall health.

Keywords: cystic fibrosis; growth restriction; final height

1. Introduction

Cystic fibrosis (CF) is an autosomal recessive disease caused by mutations in the cystic fibrosis transmembrane conductance regulator (CFTR) gene. CFTR encodes a chloride channel expressed in a variety of epithelial tissues, and mutations in this gene impair protein activity and/or expression [1]. CFTR protein dysfunction affects multiple organ systems and unequivocally contributes to growth deficits and progressive lung disease from birth and throughout infancy and childhood [2]. The association between nutrition and pulmonary outcomes in CF is well established, particularly when using weight-for-length (WFL) and body mass index (BMI) as key nutritional indices [3,4].

The CF Foundation recommends that children ages 0 to 2 years maintain their WFL at or above the 50th percentile, and that children ages 2 to 19 years maintain their BMI at or above the 50th percentile [4]. The advent of universal newborn screening for CF has allowed for earlier diagnosis and earlier interventions to improve nutritional status. For example, the Wisconsin CF Neonatal Screening Project was a randomized controlled trial initiated in 1984 that examined longitudinal nutritional outcomes in patients diagnosed with CF via newborn screening compared to control patients diagnosed with CF later in infancy. Weight-for-age (WFA) and height-for-age (HFA) Z-scores were significantly better throughout childhood and adolescence in patients diagnosed by newborn screening compared to controls. Although growth metrics in control CF participants improved following diagnosis, they never reached levels comparable to those diagnosed by newborn screening. This discrepancy was particularly notable in HFA Z-scores, indicating that early-life growth stunting associated with delayed CF diagnosis may have lasting effects on stature [5].

Strong correlations between WFA percentiles in early life and subsequent pulmonary status have been reported in cross-sectional studies [6,7]. Recently, a longitudinal study by Sanders et al. examining early life growth trajectories showed that infants and children with CF who maintained their WFL and BMI consistently above the 50th percentile had the best percent-predicted forced expiratory volume in 1 s (ppFEV1) at ages 6–7. Furthermore, infants who entered the study with a WFL < 50th percentile, but who eventually achieved a WFL ≥ 50th percentile before age 2, had higher ppFEV1 at age 6 years compared to children who did not achieve BMI ≥ 50th percentile until after age 2 [3]. Similar to WFL and BMI, linear growth is increasingly recognized as an important prognosticator of respiratory morbidity and mortality in CF. A longitudinal study by Assael et al. reported that people with CF who developed severe respiratory compromise had reduced early-life growth velocity, even before any appreciable decline in lung function. Thus, impaired linear growth is an early indicator of pulmonary disease severity that manifests before spirometry can be reliably performed [8]. The importance of early-life growth trajectories was also demonstrated in another study by Sanders et al., who reported that children who maintain HFA above the 50th percentile between diagnosis of CF and age 6–7 years have the highest pulmonary function at age 6–7 years, irrespective of their BMI percentile [9]. Furthermore, stunting, defined as a height below the 5th percentile, is an independent predictor of mortality [10].

These data suggest that improving nutrition early in life may also improve CF lung disease outcomes, and perhaps overall survival, underscoring the importance of stature in the prognosis and perhaps pathogenesis of CF lung disease. Based on data from the US CF Foundation Patient Registry (CFFPR), meticulous attention to nutritional interventions in both early life and throughout childhood and adolescence is successful in achieving the goals of WFL and BMI targets of above the 50th percentile. Over the past 15 years, the median BMI percentile in individuals 2–19 years has improved from 45.1% to 58.3%. Further, the number of children with WFA <10th percentile has improved from 20.9% to 10.1%. Additionally, use of supplemental feeding tubes has increased from 8.8% to 10.3%. However, HFA less than the fifth percentile has only improved from 14.6% to 9.5%. Presently, the median BMI for adults with CF over the age of 20 is 23.0. For CF individuals between 2 and 19 years, the median weight percentile is 49.3%, but the height percentile is only 38.4%. In children less than 24 months old, the median WFL percentile is 63.7%, the weight percentile is 42.8%, and—consistent with the above—the length percentile is 30.2% [11]. Thus, risk factors for short stature itself remain to be defined. We hypothesized that impaired early linear growth would correlate with stunting as an adult among people with CF.

To test this hypothesis, we used CFFPR data to determine whether linear growth impairment in early life in CF is associated with a reduced final adult height. Given the importance of short stature as a prognosticator of morbidity and mortality in CF, the early identification of children at risk for stunting and appropriate interventions (nutritional support, potentially mediating confounders from clinical care, and endocrine involvement) could improve overall health outcomes in patients with CF.

2. Materials and Methods

The United States CFFPR was used for this retrospective case–control study. People with CF who were age 18–19 years between 2011 and 2015 with any height measurements were selected. Then, the height measurements for those young adults were retrospectively ascertained from the CFFPR from when they were between 2 and 4 years of age. CFFPR provides one annualized height measurement per person per year calculated by averaging the maximum value per quarter. Heights were standardized to US Centers for Disease Control (CDC) height for age percentiles (HFA%) [12]. For a given individual, the lowest height percentile of the two recorded at 18 and 19 years of age in the CFFPR defined their categorization into groups: <10th, 10th–25th, 25th–50th, or ≥50th HFA%. Similarly, the lowest annualized HFA%, up to three recorded for each person between 2 and 4 years of age, defined their inclusion into early childhood HFA groups. As a sensitivity analysis, children who's annualized HFA% < 10 in all three years between 2 and 4 years of age were examined. Summary statistics, odds ratios (OR), and Pearson correlation coefficients with 95% confidence intervals (CI) are reported. The Seattle Children's Hospital (Seattle, WA, USA) Institutional Review Board granted approval for this research. SAS version 9.2 (Cary, NC, USA) or R version 3.3.3 (R Foundation for Statistical Computing, Vienna, Austria) were used for all analyses.

3. Results

Early Childhood Height for Age Is Associated with Adult Height

Among the 3655 individuals with CF aged 18–19 between 2011 and 2015 in the CFFPR, only 26% (*n* = 939) were at or above the median height for their age (50th HFA%) and 25% (*n* = 915) had an HFA% < 10 in adulthood (Table 1). Between the ages of 2 and 4 years, 28% (1034/3566) of those same children had at least one annualized HFA% < 10, and 54% of those short children went on to be <10 HFA% in early adulthood (increased odds of 7.7 (95% CI = 6.5–9.1) compared to children with CF between 2 and 4 years with HFA% ≥ 10). Among the children with HFA% < 10 in all three years between 2 and 4 years, 67% were short as adults (<10 HFA%) and had much higher odds of being so compared to children between 2 and 4 years of age who were not always <10 HFA% in early childhood (OR = 9.0 (95% CI = 7.3–11.0)). Only 13% of children with HFA% between 10 and 25 at between 2 and 4 years achieved or exceeded 50 HFA% as adults. The maximum height measurements between 2 and 4 years were highly correlated with maximum height at age 18–19 (r = 0.64, 95% CI = (0.62, 0.66)) (Figure 1).

Table 1. Height percentile grouping at 2–4 years of age in CFFPR versus height percentile grouping at age 18–19 years of age. The individuals lowest annual HFA% in the 2- or 3-year windows, adult and childhood, respectively, defined their inclusion into categories.

Age 2–4 Height Percentile	Age 18–19 Height Percentile *n* % of Row				Total *n* % of Column
	<10th	10th–25th	25th–50th	≥50th	
<10th	563 54.4%	283 27.4%	150 14.5%	38 3.7%	1034 28.3%
10th–25th	231 23.8%	327 33.7%	285 29.4%	126 13%	969 26.5%
25th–50th	97 10.8%	207 22.9%	323 35.8%	275 30.5%	902 24.7%
≥50th	24 3.2%	63 8.4%	163 21.7%	500 66.7%	750 20.5%
Total	915 25%	880 24.1%	921 25.2%	939 25.7%	3655 100%

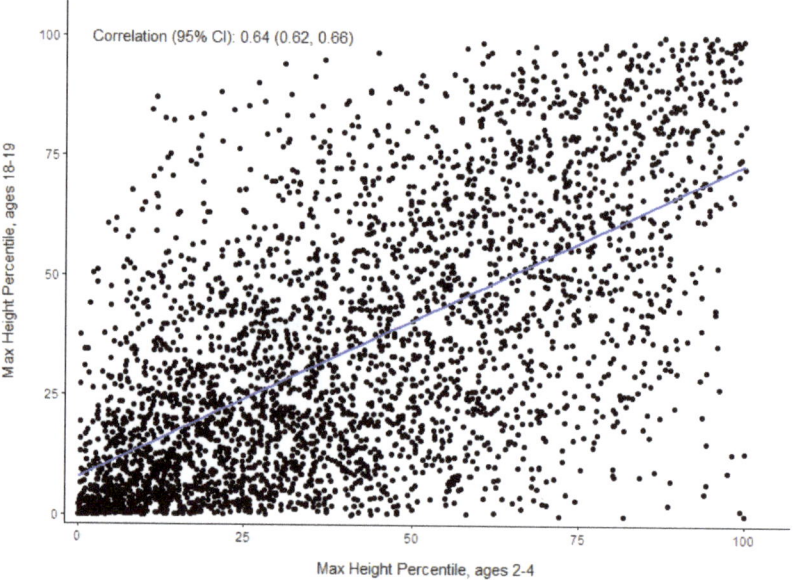

Figure 1. Maximum annualized HFA% in CFFPR between ages of 18 and 19 versus maximum annualized HFA% between ages 2 and 4 years (n = 3566).

4. Discussion

We found that early-life growth impairments of preschool children with CF correlated with stunting in early adulthood. Growth impairment in people with CF remains common and challenging to identify and address, although improvements in nutritional metrics have been observed in all ages in CF over the past several decades. According to the CF Foundation's 2019 Patient Registry Annual Data Report, the median WFL among children with CF in the US ages 0 to 2 years was above the recommended 50th percentile, and the median BMI for children ages 2 to 19 years was above the recommended 50th percentile. However, relying solely on WFL and BMI as markers of nutritional status fails to identify children with suboptimal nutrition based on WFA and HFA percentiles, as children with stunting may maintain relatively normal WFL and BMI due to proportionally poor linear and weight growth. For example, despite the improvements in median WFL and BMI, both LFA and HFA percentiles in children with CF remain well below those of the general population [11]. A 2017 study by Konstan et al. examining growth parameters of 11,669 children with CF ages 2 to 18 years reported that 20.5% of participants whose BMI were at or above the recommended 50th percentile had HFA below the 10th percentile [13]. Growth trajectories of patients with CF diagnosed by newborn screening suggest that the stunting frequently observed in CF begins in early infancy and persists despite weight normalization. The Baby Observational and Nutrition Study (BONUS) examined the growth patterns of 231 infants diagnosed with CF via newborn screening compared to healthy cohorts. BONUS infants achieved normal weight by 12 months of age; however, their length lagged behind those of healthy peers [14]. As an aside, the consideration of BMI as the sole growth measurement would not reflect such stunting, and we therefore considered height and weight separately in this study.

In addition to early stunting, a "second hit" to the growth of children with CF may occur in puberty. Historically, studies have suggested that children with CF also have reduced height velocities (HV) throughout childhood that are particularly compromised during puberty. These studies used Tanner–Davies growth curves developed in the USA in 1985 to compare HV in children with CF to children without CF [15,16]. The Tanner–Davies curves were generated using longitudinal growth data from a restricted population of European descent with superimposed cross-sectional data from the US National Center for Health Statistics to establish ages of peak HV for children who matured early, average, and late [17]. More recently, reference data based on longitudinal growth metrics in a more diverse population of healthy youths in the USA allowed for the calculation of HV percentiles and Z-scores, which facilitates comparison between children with CF and healthy peers [18]. Using these contemporary data, Zysman-Colman et al. demonstrated that the HV percentiles of children with CF ages 5–17 years fall within the 25th to 75th percentiles of healthy children. Furthermore, shorter children with CF tended to have lower HV Z-scores than taller children with CF. This difference was most notable in pre-pubertal children, which suggests that final height is determined early in life in CF. The study cohort had below average lengths from birth that persisted into adulthood despite normal childhood and pubertal growth velocities and adequate nutrition [19].

The mechanism underlying growth restriction in CF remains unclear. Impaired CFTR activity has detrimental effects on multiple organ systems that could contribute to poor growth. Malnutrition can undoubtedly lead to poor growth; however, as BONUS demonstrated, stunting in CF begins in early infancy despite nutritional interventions that generally normalized weight achievement [14]. Infants included in the BONUS study with pancreatic insufficiency initiated pancreatic enzyme replacement therapy at a mean age of 2 months, with doses consistent with the CF Foundation guidelines. In the infants that were exclusively formula-fed, 40.2% received high calorie formula—greater than or equal to 24 kcal/oz—at 3 months, 52.4% at 6 months, and 49.0% at 12 months. Infants that were exclusively breastfed could not be assessed for total caloric intake; however, they weighed more than formula-fed or a combination of the two, at 3 months. This finding did not persist at 6 or 12 months of age. Interestingly, the feeding type (whether breast, formula-fed, or a combination of the two) was not associated with infant length during the BONUS study. The summation findings of BONUS demonstrated normalization of weight in CF infants by 12 months; however, despite these nutritional improvements, length did not normalize. Consistent with the studies previously mentioned, only 13.6% were less than the 10th percentile WFA, but 23.9% were less than the 10th percentile LFA.

Historically, providers (both pulmonary and endocrine) have attributed poor nutritional absorption and chronic disease as the primary influence on linear growth in CF. Similarly, lung disease could diminish nutritional outcomes. However, CF animal models have demonstrated growth restriction in the absence of these confounding variables. CF mice do not develop lung disease—and lung disease only manifests after 6 months of age in CF rats—although both exhibit impaired growth [20–22]. Data from these animal models combined with clinical data offer clues regarding other potential contributors to poor growth in CF. For example, both animals and people with CF have reduced anabolic drive secondary to lower tissue concentrations of insulin-like growth factor-1 (IGF-1) and insulin deficiency associated with CF-related diabetes [23–26].

One of the most remarkable findings suggestive of a primary defect in the growth axis unrelated to nutritional intake is the decreased IGF-1 concentrations seen in many animal models of CF, including newborn CF piglets (compared to non-CF littermates) [23]. CFTR-deficient mice and rats have reduced total body length and femur length as well as reduced serum IGF-1 concentrations. These findings suggest that CFTR dysfunction intrinsically affects the endocrine growth axis. Furthermore, growth-restricted CF rats exhibit growth plate alterations compared to controls, demonstrated by reduced hypertrophic chondrocyte volume as well as an overall reduction in growth plate thickness [22]. Similarly low IGF-1 was found in newborn blood spots of CF infants, which also suggests intrinsic defects in the growth axis and argues against intestinal malabsorption as the primary etiology of poor growth [23]. Further addressing the implication that nutritional intake is or is not the causative agent for decreased IGF-1 concentrations, a study by Hardin et al. investigated enteral nutritional supplementation versus enteral nutritional supplementation combined with treatment with human growth hormone (hGH). The authors evaluated outcomes in both growth and contribution to serum IGF-1 levels. Despite one year of enteral nutrition supplementation, subjects not on hGH therapy exhibited no change in serum IGF-1 concentrations. However, after the addition of hGH to the enteral feeding group—at year 2—there was an improvement [27].

Additional evidence of dysfunctional CFTR protein's intrinsic effect on growth, as well as a potential source for improvement, is demonstrated by improved growth with CFTR modulator therapy, which increases the activity of the defective chloride channel in vivo. Stalvey et al. demonstrated this in a post hoc analysis of linear growth observed longitudinally in pre-pubertal children with CF and at least one copy of the G551D CFTR mutation who were enrolled in the G551D Observational Study (GOAL) of children initiating the CFTR modulator ivacaftor. At baseline, participants were below average in HFA. Six months following the initiation of ivacaftor, both HFA and HV significantly increased. In the placebo-controlled, randomized Evaluation of Efficacy and Safety of VX-770 in Children Six to Eleven Years Old with CF (ENVISION) study, there was continued improvement in HFA beyond 6 months in children treated with ivacaftor compared to controls [28]. Newer, highly effective modulators are now available for people with the more common F508del CFTR mutation, and ongoing studies are collecting growth data on pubertal and pre-pubertal children with CF following the initiation of treatment. As part of the outcomes studied in children and adults, these studies are evaluating changes in body composition. Other studies are underway to assess essential growth changes in early infancy and childhood (when growth rates tend to be relatively fast), as well as growth factors (such as IGF-1), and to monitor the influence of CFTR therapies as they are initiated. Investigators hope to provide insights into the biological–pathologic process of growth restriction, which begins in infancy and carries through to adulthood.

There are certain limitations in utilizing the CFFPR for our retrospective case–control analysis. The first would be how our population compares to healthy children growing at the 10th or below percentile at age 2–4 years. The authors do not argue that children—with or without CF—growing along that percentile may be growing as intended for their genetic potential. We are in fact attempting to illustrate and discredit the common misconception by providers, that poor growth in CF children early in life has time to "catch up" by adulthood. In our comparisons, we did not have heights from the parents to calculate mid-parental height targets for the subjects, but given the data utilized is obtained from the national registry, one would anticipate a near mean average adult height of the parents. Additionally, the recent report by Zysman-Colman et al. demonstrates that when parental heights are known, there is a reduction in obtainment of mid-parental linear growth targets in children with CF [19]. Our analysis would further suggest that CF children with evidence of poor growth early in life warrant closer attention. It has also been observed that the age of the onset of puberty was historically often delayed in children with CF, raising the possibility that some in our study population were still growing at the end of the study period [29–31]. While we did not analyze the effects of some potential confounders, including gender/sex, CFTR genotype, or differences in therapy (including CFTR modulator use, which was likely rare in this study population), future studies could address the effects of these covariates. Similarly, registry data could be used to construct predictive models based on early childhood growth characteristics of later growth outcomes, an analysis we did not perform here.

In summary, our study demonstrates that height during preschool ages correlates with subsequent adult height achievement in people with CF. Recognition of impairment in early life may be essential to improving outcomes. A team-based multi-disciplinary approach to these high-risk children by their pulmonologists, registered dietitians, social services, and consultative services (endocrine and gastroenterology) will achieve the best outcomes. Additionally, the incorporation of highly effective CFTR modulator therapies in early life may be one such way to alter both long-term outcomes and the relationship between early-life and adult height achievement.

Linear growth correlates closely with lung growth in people with CF; therefore, identifying both risk factors and effective preventative treatments for linear growth failure could lead to improved overall CF patient health. If new CFTR modulator therapies are beneficial to overall growth (weight and height), nutritional guidelines may need modification to address overall health outcomes. Encouraging calorie intake and high-fat diets may not continue to be advantageous, given the balance with obesity, fat mass versus lean body mass, and the potential for the development of CFRD.

Author Contributions: Conceptualization, M.S.S. and S.L.H.; methodology, M.S.S. and S.L.H.; SAS version 9.2; formal analysis, S.L.H. and A.B.; data curation, S.L.H. and A.B.; writing—original draft preparation, M.P.M. and M.S.S.; writing—review and editing, M.P.M., S.L.H., A.B., B.W.R., L.R.H., M.S.S.; supervision, M.S.S. All authors have read and agreed to the published version of the manuscript.

Funding: This research was supported by NIH P30DK072482, P30DK089507, R01DK095738 and CFF.

Institutional Review Board Statement: The study was conducted according to the guidelines of the Declaration of Helsinki, and approved by the Institutional Review Board of Seattle Children's Hospital and Advarra (Seattle Children's Hospital IRB: BONUS (22 August 2011–29 June 2021) and BEGIN (28 October 2020); Advarra IRB: BEGIN (23 April 2020).

Informed Consent Statement: Informed consent was obtained from all subjects involved in the study through the CFFPR.

Data Availability Statement: Data were acquired though the CFF Patient Registry and are available to investigators upon request and scientific approval (https://www.cff.org/researchers/patient-registry-data-requests).

Conflicts of Interest: The authors declare no conflict of interest.

References

1. Rommens, J.M.; Iannuzzi, M.C.; Kerem, B.; Drumm, M.L.; Melmer, G.; Dean, M.; Rozmahel, R.; Cole, J.; Kennedy, D.; Hidaka, N.; et al. Identification of the cystic fibrosis gene: Chromosome walking and jumping. *Science* **1989**, *245*, 1059–1065. [CrossRef] [PubMed]
2. VanDevanter, D.R.; Kahle, J.; O'Sullivan, A.K.; Sikirica, S.; Hodgkins, P.S. Cystic fibrosis in young children: A review of disease manifestation, progression, and response to early treatment. *J. Cyst. Fibros.* **2016**, *15*, 147–157. [CrossRef]
3. Sanders, D.B.; Fink, A.; Mayer-Hamblett, N.; Schechter, M.S.; Sawicki, G.S.; Rosenfeld, M.; Flume, P.A.; Morgan, W.J. Early Life Growth Trajectories in Cystic Fibrosis are Associated with Pulmonary Function at Age 6 Years. *J. Pediatr.* **2015**, *167*, 1081–1088.e1. [CrossRef]
4. Stallings, V.A.; Stark, L.J.; Robinson, K.A.; Feranchak, A.P.; Quinton, H. Evidence-Based Practice Recommendations for Nutrition-Related Management of Children and Adults with Cystic Fibrosis and Pancreatic Insufficiency: Results of a Systematic Review. *J. Am. Diet. Assoc.* **2008**, *108*, 832–839. [CrossRef]
5. Farrell, P.M.; Lai, H.J.; Li, Z.; Kosorok, M.R.; Laxova, A.; Green, C.G.; Collins, J.; Hoffman, G.; Laessig, R.; Rock, M.J.; et al. Evidence on improved outcomes with early diagnosis of cystic fibrosis through neonatal screening: Enough is enough! *J. Pediatr.* **2005**, *147*, S30–S36. [CrossRef] [PubMed]
6. Konstan, M.W.; Butler, S.M.; Wohl, M.E.B.; Stoddard, M.; Matousek, R.; Wagener, J.S.; Johnson, C.A.; Morgan, W.J. Growth and nutritional indexes in early life predict pulmonary function in cystic fibrosis. *J. Pediatr.* **2003**, *142*, 624–630. [CrossRef] [PubMed]
7. Yen, E.H.; Quinton, H.; Borowitz, D. Better Nutritional Status in Early Childhood Is Associated with Improved Clinical Outcomes and Survival in Patients with Cystic Fibrosis. *J. Pediatr.* **2013**, *162*, 530–535.e1. [CrossRef]
8. Assael, B.M.; Casazza, G.; Iansa, P.; Volpi, S.; Milani, S. Growth and long-term lung function in cystic fibrosis: A longitudinal study of patients diagnosed by neonatal screening. *Pediatr. Pulmonol.* **2009**, *44*, 209–215. [CrossRef] [PubMed]
9. Sanders, D.B.; E Slaven, J.; Maguiness, K.; Chmiel, J.F.; Ren, C.L. Early Life Height Attainment in Cystic Fibrosis Is Associated with Pulmonary Function at Age 6 Years. *Ann. Am. Thorac. Soc.* **2021**, *18*, 1335–1342. [CrossRef]
10. Vieni, G.; Faraci, S.; Collura, M.; Lombardo, M.; Traverso, G.; Cristadoro, S.; Termini, L.; Lucanto, M.C.; Furnari, M.L.; Trimarchi, G.; et al. Stunting is an independent predictor of mortality in patients with cystic fibrosis. *Clin. Nutr.* **2013**, *32*, 382–385. [CrossRef]
11. Foundation, C.F. *Cystic Fibrosis Foundation Patient Registry 2019 Annual Data Report*; Cystic Fibrosis Foundation: Bethesda, MD, USA, 2020.
12. Centers for Disease Control and Prevention, National Center for Health Statistics: CDC Growth Charts: United States. 2000. Available online: http://www.cdc.gov/growthcharts/ (accessed on 15 September 2017).
13. Konstan, M.W.; Pasta, D.J.; Wagener, J.S.; VanDevanter, D.R.; Morgan, W.J. BMI fails to identify poor nutritional status in stunted children with CF. *J. Cyst. Fibros.* **2017**, *16*, 158–160. [CrossRef]

14. Leung, D.H.; Heltshe, S.L.; Borowitz, D.; Gelfond, D.; Kloster, M.; Heubi, J.E.; Stalvey, M.; Ramsey, B.W.; for the Baby Observational and Nutrition Study (BONUS) Investigators of the Cystic Fibrosis Foundation Therapeutics Development Network. Effects of Diagnosis by Newborn Screening for Cystic Fibrosis on Weight and Length in the First Year of Life. *JAMA Pediatr.* **2017**, *171*, 546–554. [CrossRef] [PubMed]
15. Zhang, Z.; Lindstrom, M.J.; Lai, H.J. Pubertal Height Velocity and Associations with Prepubertal and Adult Heights in Cystic Fibrosis. *J. Pediatr.* **2013**, *163*, 376–382.e1. [CrossRef] [PubMed]
16. Byard, P.J. The adolescent growth spurt in children with cystic fibrosis. *Ann. Hum. Biol.* **1994**, *21*, 229–240. [CrossRef] [PubMed]
17. Tanner, J.; Davies, P.S. Clinical longitudinal standards for height and height velocity for North American children. *J. Pediatr.* **1985**, *107*, 317–329. [CrossRef]
18. Kelly, A.; Winer, K.K.; Kalkwarf, H.; Oberfield, S.E.; Lappe, J.; Gilsanz, V.; Zemel, B.S. Age-Based Reference Ranges for Annual Height Velocity in US Children. *J. Clin. Endocrinol. Metab.* **2014**, *99*, 2104–2112. [CrossRef] [PubMed]
19. Zysman-Colman, Z.N.; Kilberg, M.J.; Harrison, V.S.; Chesi, A.; Grant, S.F.A.; Mitchell, J.; Sheikh, S.; Hadjiliadis, D.; Rickels, M.R.; Rubenstein, R.C.; et al. Genetic potential and height velocity during childhood and adolescence do not fully account for shorter stature in cystic fibrosis. *Pediatr. Res.* **2020**, *89*, 653–659. [CrossRef]
20. A Rosenberg, L.; Schluchter, M.D.; Parlow, A.F.; Drumm, M.L. Mouse as a Model of Growth Retardation in Cystic Fibrosis. *Pediatr. Res.* **2006**, *59*, 191–195. [CrossRef]
21. Darrah, R.; Bederman, I.; Vitko, M.; Valerio, D.M.; Drumm, M.L.; Hodges, C.A. Growth deficits in cystic fibrosis mice begin in utero prior to IGF-1 reduction. *PLoS ONE* **2017**, *12*, e0175467. [CrossRef] [PubMed]
22. Stalvey, M.S.; Havasi, V.; Tuggle, K.L.; Wang, D.; Birket, S.; Rowe, S.M.; Sorscher, E.J. Reduced bone length, growth plate thickness, bone content, and IGF-I as a model for poor growth in the CFTR-deficient rat. *PLoS ONE* **2017**, *12*, e0188497. [CrossRef]
23. Rogan, M.P.; Reznikov, L.; Pezzulo, A.; Gansemer, N.D.; Samuel, M.; Prather, R.; Zabner, J.; Fredericks, D.C.; McCray, P.; Welsh, M.J.; et al. Pigs and humans with cystic fibrosis have reduced insulin-like growth factor 1 (IGF1) levels at birth. *Proc. Natl. Acad. Sci. USA* **2010**, *107*, 20571–20575. [CrossRef]
24. Gifford, A.; Nymon, A.; Ashare, A. Serum insulin-like growth factor-1 (IGF-1) during CF pulmonary exacerbation: Trends and biomarker correlations. *Pediatr. Pulmonol.* **2013**, *49*, 335–341. [CrossRef] [PubMed]
25. Terliesner, N.; Vogel, M.; Steighardt, A.; Gausche, R.; Henn, C.; Hentschel, J.; Kapellen, T.; Klamt, S.; Gebhardt, J.; Kiess, W.; et al. Cystic-fibrosis related-diabetes (CFRD) is preceded by and associated with growth failure and deteriorating lung function. *J. Pediatr. Endocrinol. Metab.* **2017**, *30*, 815–821. [CrossRef]
26. Cheung, M.; Bridges, N.; Prasad, S.; Francis, J.; Carr, S.; Suri, R.; Balfour-Lynn, I. Growth in children with cystic fibrosis-related diabetes. *Pediatr. Pulmonol.* **2009**, *44*, 1223–1225. [CrossRef]
27. Hardin, D.S.; Rice, J.; Ahn, C.; Ferkol, T.; Howenstine, M.; Spears, S.; Prestidge, C.; Seilheimer, D.K.; Shepherd, R. Growth hormone treatment enhances nutrition and growth in children with cystic fibrosis receiving enteral nutrition. *J. Pediatr.* **2005**, *146*, 324–328. [CrossRef] [PubMed]
28. Stalvey, M.S.; Pace, J.; Niknian, M.; Higgins, M.N.; Tarn, V.; Davis, J.; Heltshe, S.L.; Rowe, S.M. Growth in Prepubertal Children With Cystic Fibrosis Treated With Ivacaftor. *Pediatrics* **2017**, *139*. [CrossRef] [PubMed]
29. Mitchell-Heggs, P.; Mearns, M.; Batten, J.C. Cystic Fibrosis in Adolescents and Adults. *QJM Int. J. Med.* **1976**, *45*. [CrossRef]
30. Reiter, E.O.; Stern, R.C.; Root, A.W. The Reproductive Endocrine System in Cystic Fibrosis. *Am. J. Dis. Child.* **1981**, *135*, 422–426. [CrossRef] [PubMed]
31. Buntain, H.M.; Greer, R.M.; Wong, J.C.; Schluter, P.J.; Batch, J.; Lewindon, P.; Bell, S.C.; E Wainwright, C. Pubertal development and its influences on bone mineral density in Australian children and adolescents with cystic fibrosis. *J. Paediatr. Child Health* **2005**, *41*, 317–322. [CrossRef]

Article

The Relationship between Body Composition, Dietary Intake, Physical Activity, and Pulmonary Status in Adolescents and Adults with Cystic Fibrosis

Kevin J. Scully [1,2], Laura T. Jay [3], Steven Freedman [2,4], Gregory S. Sawicki [2,5], Ahmet Uluer [2,5,6], Joel S. Finkelstein [2,7] and Melissa S. Putman [2,7,*]

1. Division of Endocrinology, Boston Children's Hospital, Boston, MA 02115, USA; kevin.scully@childrens.harvard.edu
2. Harvard Medical School, Boston, MA 02115, USA; sfreedma@bidmc.harvard.edu (S.F.); gregory.sawicki@childrens.harvard.edu (G.S.S.); Ahmet.Uluer@childrens.harvard.edu (A.U.); finkelstein.joel@mgh.harvard.edu (J.S.F.)
3. Division of Gastroenterology, Hepatology and Nutrition, Boston Children's Hospital, Boston, MA 02115, USA; laura.jay@childrens.harvard.edu
4. Division of Gastroenterology, Beth Israel Deaconess Hospital, Boston, MA 02115, USA
5. Division of Pulmonary Medicine, Boston Children's Hospital, Boston, MA 02115, USA
6. Division of Pulmonary and Critical Care Medicine, Brigham and Women's Hospital, Boston, MA 02115, USA
7. Division of Endocrinology, Massachusetts General Hospital, Boston, MA 02115, USA
* Correspondence: msputman@partners.org; Tel.: +1-857-218-5017; Fax: +1-617-730-0194

Abstract: Measures of body fat and lean mass may better predict important clinical outcomes in patients with cystic fibrosis (CF) than body mass index (BMI). Little is known about how diet quality and exercise may impact body composition in these patients. Dual X-ray absorptiometry (DXA) body composition, 24-h dietary recall, and physical activity were assessed in a cross-sectional analysis of 38 adolescents and adults with CF and 19 age-, race-, and gender-matched healthy volunteers. Compared with the healthy volunteers, participants with CF had a lower appendicular lean mass index (ALMI), despite no observed difference in BMI, and their diets consisted of higher glycemic index foods with a greater proportion of calories from fat and a lower proportion of calories from protein. In participants with CF, pulmonary function positively correlated with measures of lean mass, particularly ALMI, and negatively correlated with multiple measures of body fat after controlling for age, gender, and BMI. Higher physical activity levels were associated with greater ALMI and lower body fat. In conclusion, body composition measures, particularly ALMI, may better predict key clinical outcomes in individuals with CF than BMI. Future longitudinal studies analyzing the effect of dietary intake and exercise on body composition and CF-specific clinical outcomes are needed.

Keywords: body composition; cystic fibrosis; dual-energy X-ray absorptiometry; lean body mass; appendicular lean mass index; fat mass index; dietary intake

1. Introduction

Nutritional optimization has long been a focus of care in patients with cystic fibrosis (CF), with body mass index (BMI) being utilized as the primary marker of health and survival [1–3]. Undernutrition in CF has been associated with worsening pulmonary status, decreased exercise tolerance, immunologic impairment, impaired growth, decreased quality of life, and a shorter life expectancy [4–6]. Conversely, optimized nutritional status is associated with improved lung function, clinical outcomes and survival [7,8]. As a result of this, nutritional guidelines recommend maintaining a body mass index (BMI) at or above the 50th percentile of age for children and adolescents, and a level of at least 22 kg/m² in adult females and 23 kg/m² in adult males aged 18 and over. In the past, this concern for malnutrition has often led to physicians recommending high-calorie diets without concern for diet quality [9]. This has led to a tendency for individuals with CF to overconsume energy-dense, nutrient-poor foods, particularly foods high in added sugars and refined carbohydrates that have a high glycemic index [9–12].

Life expectancy and clinical outcomes for patients with CF have significantly improved with widespread use of highly effective cystic fibrosis transmembrane conductance regulator (CFTR) modulators, including the risk for undernutrition [6]. However, advancements in CF care have also led to significantly increased rates of overweight and obesity [13–17]. Additionally, the prevalence of non-pulmonary complications such as CF-related diabetes (CFRD) continues to increase, particularly as the CF population ages [18,19]. At present, there are few published studies investigating the impact of these high-calorie, lower-quality diets on body composition and the development of CF-related metabolic comorbidities [9,20].

While BMI has classically been the primary measure of nutritional outcomes in patients with CF, there is interest in evaluating other potentially more meaningful predictors of health status [9,20]. There is growing evidence that BMI may not accurately reflect body composition, particularly the distinction between fat mass and fat-free mass [21–23]. As a result, the use of BMI as the primary marker of nutritional status in CF may have significant drawbacks in routine clinical care.

Dual-energy X-ray absorptiometry (DXA) body composition analysis has become increasingly utilized in individuals with CF, particularly given its ability to provide more detailed information regarding the distribution of fat mass, lean mass, and bone density [20]. Several groups have described an increased prevalence of normal weight obesity (NWO) and decreased fat-free mass distribution (FFMD) in individuals with CF, as well as a link between this type of body habitus and poorer lung function [20,24,25]. There may exist specific DXA body composition variables that better predict long-term CF-specific clinical outcomes than BMI [26]. If identified, these variables could predict the risk of pulmonary decline and metabolic abnormalities in patients with CF, and help guide individualized advice regarding dietary composition and physical activity.

We performed a cross-sectional analysis comparing the dietary intake, physical activity, and DXA body composition measures in adolescents and adults with CF and age-, race- and gender-matched healthy volunteers. We also investigated how body composition correlated with pulmonary status and dietary intake in participants with CF. We hypothesized that adults with CF would have higher carbohydrate intake, greater fat mass, and lower lean mass than healthy volunteers, and that measures of lean mass would correlate more strongly with percent predicted forced expiratory volume in 1 second (FEV1) than BMI.

2. Materials and Methods
2.1. Participants and Eligibility Criteria

Cross-sectional data were analyzed from the baseline visit of a prospective observational study investigating the effect of ivacaftor on bone density and microarchitecture in individuals with CF [27]. Participants with CF were recruited from the Massachusetts General Hospital and Boston Children's Hospital Cystic Fibrosis Center. Exclusion criteria for participants with CF included history of solid organ transplantation, current pregnancy, and *Burkholderia dolosa* infection (due to institutional infection control issues). Matched healthy volunteers were recruited from the community. Exclusion criteria for healthy volunteers included current pregnancy, a history of medications or disorders known to affect bone metabolism, cumulative use of oral glucocorticoids for greater than two months, or BMI < 18.5 or >30 kg/m^2 (or <5th percentile or >95th percentile for pediatric participants) at the time of screening.

The parent study included children and adults with CF and at least one copy of the G551D mutation, who were matched by age (± 2 years and by Tanner stage in pediatric participants), race, and gender to a cohort of participants with CF and other CFTR mutations, and to a cohort of healthy volunteers. For the present analysis, only post-pubertal (Tanner stage V) participants aged 15 years and above were included, due to the rapid and variable changes in diet and body composition occurring during growth. The protocol was approved by the Mass General Brigham Institutional Review Board (IRB) with ceded review by the Boston Children's Hospital IRB and was registered on clinicaltrials.gov (NCT01549314). Written informed consent was obtained from all participants.

2.2. Clinical Assessments

All participants were queried regarding medical history, medication use including oral and inhaled glucocorticoids, alcohol and tobacco use, and pubertal and reproductive history. Tanner staging in pediatric participants was performed by a board-certified pediatric endocrinologist. Additional data obtained from participants with CF included the number of CF exacerbations in the past year, defined as treatment with intravenous antibiotics and/or hospitalization. CFTR genotype and percent predicted forced expiratory volume in 1 s (FEV1) in the most recent pulmonary function testing were obtained in participants with CF by chart review. In all participants, height was measured on a wall-mounted stadiometer and weight on an electronic scale. Race and ethnicity were self-reported. Serum glucose and insulin levels were obtained after fasting at least 8 h overnight; these data were used to calculate a homeostatic model assessment for insulin resistance (HOMA-IR), with a level >2 considered consistent with insulin resistance [28].

2.3. Dietary Intake and Physical Activity Assessments

A registered dietician assessed nutritional intake with a 24-h dietary recall. Dietary composition of fat (g), protein (g), and carbohydrates (g), percentage of calories from each macronutrient, total energy (kcal), added sugar (g), glycemic index (GI) and glycemic load as defined by the International Carbohydrate Quality Consortium (ICQC) [29], were quantified using the validated Nutrient Data System for Research (NDSR) [30]. Physical activity was also assessed by a registered dietician using the Modifiable Activity Questionnaire, a self-reported tool that assesses each individual's degree of physical activity over the last 1 year, based on 40 leisure and occupational activity items, ranging in intensity [31].

2.4. Assessment of Body Composition

Whole and regional body composition analyses were obtained from whole body DXA scans (Discovery A, Hologic Inc., Bedford, MA, USA). To account for variants in stature, height-normalized indexes were determined for fat-mass and lean mass (mass in kg/height in m^2). DXA quality control included daily measurement of a Hologic DXA anthropomorphic spine phantom and visual review of all images by an experienced investigator.

Participants were classified as normal weight (BMI 18–24.9 kg/m^2), overweight (BMI 25–29.9 kg/m^2), or obese (BMI $\geq$ 30 kg/m^2). Normal weight obesity (NWO) was defined as those with a normal BMI < 25 kg/m^2 but with a body fat percentage of >30% in women and >23% in men [20,32].

2.5. Statistical Analyses

Statistical analyses were performed using STATA (version 16, StataCorp LLC, College Station, TX, USA). Normality was assessed for all variables using the Shapiro–Wilk test. Clinical characteristics, dietary intake, and body composition measures were compared between participants with CF and healthy volunteers using independent *t*-tests or Wilcoxon rank sum tests for normally and non-normally distributed data, respectively. Categorical variables were compared using chi square tests. In participants with CF, the relationship between body composition measures and FEV1 or dietary intake variables was determined via Pearson or Spearman correlation analysis, for normally and non-normally distributed data, respectively. Multivariable regression analysis was used to assess the relationship between various body composition measures and FEV1 Two regression models were used to adjust for potential confounding effects: Model 1, adjusting for age and gender; Model

2, adjusting for age, gender and BMI. A *p*-value of <0.05 was considered statistically significant.

3. Results

3.1. Clinical Characteristics

Thirty-eight adolescents and adults with CF and 19 healthy volunteers were included in the analysis. Clinical characteristics are summarized in Table 1. Participants ranged in age from 15 to 56 years and included eight adolescents with CF and four adolescent healthy volunteers aged 15–17 years. There were no significant differences between age, gender, race/ethnicity, anthropometric measures, HOMA-IR, or physical activity between the participants with CF and the healthy volunteers. Of the participants with CF, 30 (78.9%) had a history of pancreatic insufficiency and 5 (13.2%) had a history of CFRD. Five of the participants with CF without CFRD had fasting glucose levels consistent with impaired fasting glucose (100–125 mg/dL). Approximately half of these individuals (n = 20, 52.6%) were heterozygous and seven (18.4%) were homozygous for the F508del mutation. As the initial CF cohort was recruited to study the effects of ivacaftor on BMD, half (n = 19) of these participants had the G551D mutation. Ten participants (26.3%), all of whom had the G551D mutation, were taking ivacaftor, which was initiated within three months of the study visit. No other modulators were available for clinical use at the time of study enrollment. In the 12 months prior to the study visit, thirteen (34%) of the participants with CF experienced one or more CF exacerbations and 15 (40%) reported treatment with systemic glucocorticoids. Three of the participants with CF were overweight (BMI $\geq$ 25 kg/m^2), and two were obese (BMI $\geq$ 30 kg/m^2). Of the remaining 32 participants, approximately one-third (n = 10, 31.25%) met criteria for NWO, including five women (33.3%) and five men (29.4%).

3.2. Body Composition and Dietary Intake in Participants with CF and Healthy Volunteers

Comparisons of DXA body composition measures between participants with CF and healthy volunteers are presented in Table 2. Participants with CF had a lower appendicular lean mass/height^2 (appendicular lean mass index, ALMI) compared with the healthy volunteers. All other body composition measures were similar between the two cohorts. As shown in Table 2, participants with CF reported a significantly higher total fat intake, greater % calories from fat, higher glycemic index, and lower % calories from protein than the healthy volunteers.

3.3. Relationship between Body Composition and Pulmonary Function in Participants with CF

Table 3 presents the results of correlation analyses and multiple linear regression models for FEV1 and DXA body composition measures in the participants with CF (n = 38). On univariable correlation analysis, only ALMI was significantly correlated with FEV1. When controlling for age and gender (regression Model 1), FEV1 showed significant positive correlations with lean mass, lean mass/height^2 (lean mass index, LMI), and ALMI. Subsequent analysis adjusting for age, gender, and BMI (regression Model 2) displayed an even stronger relationship between FEV1 and ALMI (Figure 1), but the relationship with LMI was no longer significant. In addition, a significant negative correlation between FEV1 and fat measures (% fat, trunk % fat, and fat mass/height^2 (fat mass index, FMI)) was noted when age, gender, and BMI were included in the model. BMI was significantly correlated with FEV1; however, this significance was lost after adjusting for age and gender.

Table 1. Clinical Characteristics.

	CF (n = 38)	Healthy Volunteers (n = 19)	p-Value
Age (years)	27.9 ± 2.0	28.8 ± 2.7	0.796
Female, n (%)	20 (52.6%)	10 (52.6%)	
Race, n (%)			
White	38 (100%)	19 (100%)	
Black			
Asian			
Native Haqaiian or Pacific Islander			
American Indian or Alaskan Native			
Ethnicity, n (%)			
Hispanic	0 (0%)	0 (0%)	
Non-Hispanic	38 (100%)	38 (100%)	
Height (cm)	166.9 ± 1.5	170.3 ± 1.9	0.185
Weight (kg)	59.6 ± 1.9	64.4 ± 2.6	0.154
BMI (kg/m^2)	21.4 ± 0.6	22.1 ± 0.5	0.447
HOMA-IR	1.2 ± 0.1	1.0 ± 0.1	0.455
Physical Activity Score	22.3 ± 2.8	16.4 ± 2.6	0.311
Genotype, n (%)			
F508del homozygous	7 (18.4%)		
F508del heterozygous	20 (52.6%)		
Other	11 (28.9%)		
Pancreatic insufficiency, n (%)	30 (78.9%)		
FEV1 (% predicted)	73 ± 5		
CFRD, n (%)	5 (13.2%)		
CF Exacerbations in the past prior year (Y/N)	25 (65.8%)		
Number of CF Exacerbations in the prior year	1.5 ± 0.25		
Glucocorticoid Use in the prior year	15 (39.5%)		
Ivacaftor Use, n (%)	10 (26.3%)		

Data displayed as mean ± standard error (SE) or n (%) unless otherwise indicated. BMI, body mass index kg/m^2; FEV1, forced expiratory volume; CFRD, cystic fibrosis related diabetes; HOMA-IR, homeostatic model assessment for insulin resistance.

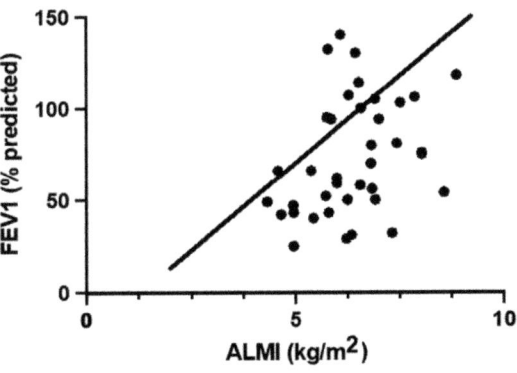

Figure 1. Multivariable regression of FEV1 vs. ALMI in participants with CF. Figure 1 displays the relationship between individuals' FEV1 and ALMI values (circles) as well as the regression line of best fit across the whole dataset when controlling for age, gender and BMI (Model 2). FEV1, % predicted forced expiratory volume in 1 s; ALMI, appendicular lean mass index.

Table 2. Body composition and dietary intake in participants with CF and healthy volunteers.

	CF (*n* = 38)	Healthy Volunteers (*n* = 19)	*p*-Value
Body Composition			
Fat mass (g)	15,740 ± 1119	16,687 ± 898	0.097
Lean mass (g)	42,438 ± 1301	46,109 ± 2593	0.352
Total mass (g)	60,274 ± 1940	65,163 ± 2611	0.07
% fat	25.6 ± 1.2	26.2 ± 1.7	0.778
Trunk fat mass (g)	7051 ± 646	6709 ± 426	0.294
Trunk total mass (g)	30,267 ± 1111	30,330 ± 1299	0.618
Trunk % fat	22.3 ± 1.3	22.4 ± 1.3	0.52
Fat mass/height^2 (kg/m^2)	5.8 ± 0.4	5.9 ± 0.4	0.275
Lean/height^2 (kg/m^2)	15.9 ± 0.4	16.4 ± 0.6	0.481
Appen lean/height^2 (kg/m^2)	6.4 ± 0.2	7.2 ± 0.4	0.029
Dietary intake			
Energy (kcal)	2880 ± 258	2203 ± 196	0.137
Total Fat (g)	118.8 ± 11.7	74.8 ± 8	0.016
Total Carb (g)	362.6 ± 34.4	292 ± 22.3	0.418
Total Protein (g)	104 ± 10.2	96.5 ± 11.4	0.667
Cholesterol (mg)	313.9 ± 42.9	248.3 ± 26.9	0.569
% Calories from Fat	36.5 ± 1.6	29.3 ± 1.3	0.002
% Calories from Carbohydrates	48.5 ± 1.8	52.8 ± 1.4	0.092
% Calories from Protein	14.9 ± 0.8	17.2 ± 0.8	0.015
Added Sugar (g)	120.6 ± 17.5	66.7 ± 6.2	0.142
Glycemic Index	61.4 ± 1.2	57.7 ± 1.1	0.036
Glycemic Load	213.1 ± 22	154.1 ± 11.6	0.131

Data displayed as mean ±SE; CF, cystic fibrosis.

Table 3. Correlation analyses and multiple linear regression of FEV1 vs. body composition measures in participants with CF.

	% Predicted FEV1					
	r	*p*-Value	Model 1 Beta Coefficient	*p*-Value	Model 2 Beta Coefficient	*p*-Value
Fat mass (g)	0.174	0.296	0.0004 ± 0.0009	0.656	−0.003 ± 0.002	0.051
Lean mass (g)	0.194	0.243	**0.003 ± 0.001**	**0.001**	**0.003 ± 0.001**	**0.003**
% fat	0.065	0.7	−1.12 ± 0.992	0.263	**−4.066 ± 0.1.121**	**0.001**
Trunk fat mass (g)	0.157	0.348	0.001 ± 0.002	0.513	−0.005 ± 0.003	0.111
Trunk total mass (g)	0.265	0.108	0.002 ± 0.001	0.072	0.001 ± 0.002	0.451
Trunk % fat	0.081	0.629	−0.259 ± 0.87	0.768	**−3.206 ± 1.184**	**0.011**
FMI (kg/m^2)	0.171	0.305	−0.141 ± 2.35	0.953	**−14.684 ± 3.931**	**0.001**
LMI (kg/m^2)	0.264	0.11	**5.242 ± 2.325**	**0.031**	4.839 ± 3.459	0.171
ALMI (kg/m^2)	**0.332**	**0.042**	**15.021 ± 4.45**	**0.002**	**18.972 ± 6.542**	**0.007**
BMI (kg/m^2)	**0.405**	**0.012**	2.574 ± 1.505	0.096		

Data displayed as correlation coefficient/Spearman's rho for correlation analyses or Beta coefficient ± SE for regression models. Model 1: adjusted for age and gender; model 2: adjusted for age, gender and BMI. Significant results (*p* < 0.05) are in bold. FEV1, % predicted forced expiratory volume in 1 s; ALMI, appendicular lean mass index; LMI, lean mass index; FMI, fat mass index.

3.4. Relationship of Body Composition with Dietary Intake and Physical Activity in Participants with CF

Table 4 outlines the results of the correlation analysis between DXA body composition measures, dietary intake components, and physical activity scores. Multiple measures of body fat composition, particularly fat mass, % fat, and FMI, negatively correlated with total amount of macronutrients (energy, fat, carbohydrate and protein), though the relative proportion of each macronutrient intake did not correlate with body composition. Added sugar and glycemic load both negatively correlated with multiple measures of body fat, including % fat and FMI. Lean mass, LMI, and ALMI were not significantly associated with dietary composition. HOMA-IR did not significantly correlate with any DXA body composition measures or dietary variables (data not shown).

Table 4. Correlation analyses between dietary intake and body composition in participants with CF.

	Energy (kcal)	Total Fat (g)	Total Carb (g)	Total Protein (g)	% Cal Fat	% Cal Carb	% Cal Protein	Added Sugars (g)	GI (Glucose)	GL (Glucose)	Physical Activity
Fat mass (g)	−0.533 (0.002)	−0.498 (0.004)	−0.445 (0.012)	−0.508 (0.004)	0.091 (0.626)	−0.104 (0.576)	0.096 (0.608)	−0.354 (0.051)	−0.036 (0.846)	−0.445 (0.012)	−0.469 (0.003)
Lean mass (g)	0.181 (0.331)	0.194 (0.296)	0.073 (0.696)	0.277 (0.131)	0.248 (0.179)	−0.26 (0.158)	0.161 (0.389)	0.005 (0.979)	−0.024 (0.9)	0.072 (0.70)	0.304 (0.068)
% fat	−0.679 (<0.0001)	−0.621 (0.0002)	−0.512 (0.003)	−0.705 (<0.0001)	−0.041 (0.827)	0.084 (0.653)	0.015 (0.936)	−0.409 (0.022)	−0.024 (0.898)	−0.481 (0.006)	−0.55 (0.0004)
Trunk fat mass (g)	−0.54 (0.002)	−0.504 (0.004)	−0.443 (0.013)	−0.536 (0.002)	0.084 (0.655)	−0.08 (0.668)	0.017 (0.928)	−0.334 (0.066)	−0.057 (0.763)	−0.451 (0.011)	−0.457 (0.005)
Trunk % fat	−0.65 (0.0001)	−0.618 (0.0002)	−0.512 (0.003)	−0.684 (<0.0001)	0.016 (0.931)	−0.009 (0.962)	−0.02 (0.916)	−0.349 (0.054)	−0.035 (0.853)	−0.508 (0.004)	−0.532 (0.0007)
FMI (kg/m^2)	−0.626 (0.0002)	−0.567 (0.0009)	−0.525 (0.002)	−0.623 (0.0002)	0.097 (0.605)	−0.097 (0.604)	0.049 (0.794)	−0.409 (0.022)	−0.051 (0.784)	−0.517 (0.003)	−0.51 (0.001)
LMI (kg/m^2)	−0.135 (0.469)	−0.099 (0.595)	−0.223 (0.228)	0.012 (0.949)	0.192 (0.301)	−0.271 (0.14)	0.259 (0.159)	−0.196 (0.292)	−0.156 (0.403)	−0.228 (0.219)	0.225 (0.181)
ALMI (kg/m^2)	0.004 (0.984)	0.063 (0.737)	−0.029 (0.878)	0.153 (0.41)	0.206 (0.266)	−0.288 (0.116)	0.232 (0.209)	−0.062 (0.742)	−0.113 (0.546)	−0.043 (0.819)	0.367 (0.025)

Data displayed as correlation coefficient/Spearman's rho (p-value). Significant results ($p < 0.05$) are in bold. CF, cystic fibrosis, ALMI, appendicular lean mass index; LMI, lean mass index; FMI, fat mass index.

Physical activity score was negatively correlated with multiple measures of body fat, including fat mass, % fat, trunk fat mass, trunk % fat, and FMI. In contrast, ALMI was positively correlated with physical activity, with no other relationship noted between physical activity and other lean mass measures.

4. Discussion

In this cross-sectional study, individuals with CF had significantly lower ALMI compared with the healthy volunteers, despite no observed differences in BMI. In participants with CF, pulmonary function was positively associated with measures of lean mass but negatively associated with measures of fat mass when accounting for age, gender, and BMI, with ALMI having the strongest correlation. Higher physical activity levels were also correlated with greater ALMI and lower body fat measures. Participants with CF consumed significantly more fat, had higher glycemic index diets, and had a lower proportion of calories from protein than their healthy peers. Interestingly, measures of lean mass were not associated with key dietary intake variables in participants with CF; however, those with the lowest body fat had the greatest caloric intake, without a significant relationship to the relative macronutrient composition of their diet.

BMI has been the primary measure of nutritional status in patients with CF for many years, due to its established strong correlation with pulmonary function and mortality [1,3,6]. However, BMI does not distinguish between fat mass and lean mass, and may be an insensitive marker of both fat-free mass deficits and excess adiposity in patients with CF [20,23,24]. For example, in a cross-sectional study of 86 adults with CF, fat-free mass depletion was found in 14% of participants, but was undetected by BMI in 58% of cases [23]. A study by Alvarez et al., of 32 adults with CF reported that 31% had NWO, defined as % fat > 30% in women and >23% in men in the setting of a normal BMI, and that these subjects had a lower fat free mass index and pulmonary function than overweight subjects, suggesting that excess adiposity may impact clinical outcomes even in the setting of a normal BMI [20]. In our study, we found a similar proportion of participants with NWO (31%). In addition, we found no difference in BMI between participants with CF and healthy volunteers; however, those with CF had a significantly lower ALMI, which has been identified as an important marker of sarcopenia and low muscle mass [33]. Lower

ALMI has been associated with increased mortality in the healthy older population [33], but this measure has not previously been reported in CF.

ALMI was also significantly correlated with pulmonary function in participants with CF in our study, with an even stronger relationship after adjusting for age, gender, and BMI. In contrast, BMI was correlated with FEV1 in the univariate analysis, but lost significance after adjustment for age and gender, suggesting that ALMI may be a superior measure than BMI in predicting lung function. After multivariable adjustment, lean mass was also positively associated with FEV1, whereas measures of body fat (fat mass, % fat, and FMI) were negatively correlated with FEV1. Similar to our findings, other studies in the CF population have reported associations of lower fat-free mass or lean body mass with lower pulmonary function [20,21,24]. Although ALMI correlations with clinical outcomes have not previously been reported in CF, appendicular lean mass (ALM) was found to be associated with FEV1 in one study of 69 adolescents with CF [5], and another study reported a greater number of CF exacerbations in those with lower appendicular fat-free mass [34]. In addition, our findings of a negative correlation between body fat measures and pulmonary function are consistent with the previously cited study by Alvarez et al., in which FMI was negatively associated with FEV1 after adjusting for age, gender and BMI [20]. Altogether, these results build a strong case supporting the importance of lean mass, particularly ALMI, in promoting lung function while implicating excess adiposity in pulmonary decline.

To achieve and maintain an adequate BMI, patients with CF are encouraged to consume a caloric intake of 120–150% of the dietary reference intake (DRI) for the typical healthy adult [1,35–38]. The CF Foundation, American Diabetes Association (ADA), and European Society for Clinical Nutrition and Metabolism (ESPEN) recommend a similar caloric composition for children and adults with CF, comprised of 20% of calories from protein, 35–50% from fat, and 40–50% from carbohydrate, though these recommendations are built on a general consensus rather than evidence-based data [1,36]. Current guidelines do not specify the composition of carbohydrate intake apart from avoiding artificial sweeteners and closely monitoring carbohydrate intake to maintain glycemic control [8,36–38].

Nutritional interventions in patients with CF often target increasing or maintaining BMI with high-calorie, high-carbohydrate and high-fat diets. However, in contrast with the trends observed in the general population, the impact of such diets on body composition in adults with CF is less clear [10,20]. In our study, we found that participants with CF consumed a higher calorie diet driven predominately by the intake of fat, and higher glycemic foods with a lower proportion of calories from protein, as compared with healthy volunteers. Similar to our findings, a cross-sectional study of 80 children with CF aged 2–18 years (mean age 9.3 years), with age- and gender-matched controls, found that children with CF consumed significantly more energy-dense, nutrient-poor foods than the controls. In addition, another study noted significantly higher added sugar intake, lower Healthy Eating Index scores, and higher visceral adipose tissue (VAT) in 24 adults with CF compared with the matched controls [9]. Interestingly, VAT was associated with higher added sugar intake and fasting glucose levels in that study. In contrast, we found a negative correlation between body fat measures and both added sugar and glycemic load, and a negative correlation was also noted with total energy intake as well as the absolute value of all macronutrients, irrespective of macronutrient intake as a proportion of total calories. No correlations were noted between diet and any measures of lean mass. Although unexpected, these results suggest that the relative macronutrient composition of the diet may not directly impact body composition, and that those patients with the lowest body fat were consuming the greatest number of calories from all sources, perhaps related to increased metabolic needs.

Not surprisingly, physical activity levels were positively correlated with ALMI and negatively correlated with multiple measures of body fat, supporting the beneficial role of exercise on body composition and muscle health. Several prior studies have shown a strong correlation between fat-free mass (FFM) content and exercise capacity [39–41]. In a prospective pilot observational study of 28 adults with CF participating in an 8-week exercise training (ET) program, with 15 CF controls with no ET, Prevotat et al., found that ET resulted in an increased FFM compared with the controls [39]. Another study in 18 adolescents with CF reported that FFM measured by bioimpedance correlated with FVC z-score, maximal inspiratory pressure, and exercise tolerance. Interestingly, BMI did not significantly correlate with pulmonary or respiratory muscle function in this study.

Strengths of this study included the comprehensive clinical, DXA, and dietary measures prospectively collected in a relatively large number of patients with CF, as well as the inclusion of an age-, race-, and gender-matched healthy control group. However, important limitations of this study should be noted. The cross-sectional study design limited the conclusions on causality that could be drawn, and further prospective longitudinal studies investigating body composition in CF are needed. Participants were recruited from a single study center, potentially limiting the heterogeneity of the study population. Nutrition data were limited to a single 24-h diet recall, as opposed to 3-day food diary; repeated prospective data collection within participants may have provided more accurate dietary information. Similarly, physical activity was measured by survey and not by a wearable activity tracker or fitness monitor. Although there were multiple definitions of NWO in the literature, we utilized a definition that has previously been studied in CF. Given that the primary outcome for the parent study was to assess the effect of ivacaftor therapy on bone density and microarchitecture, half of the participants had the G551D mutation. A genotypically more diverse CF population may have impacted our outcomes. In addition, the small number of adolescent participants limited our ability to investigate differences between adolescents and adults. Lastly, few of the participants were on CFTR modulators at the time of the study, which could limit the applicability of these results to patients on highly effective CFTR modulator therapy.

5. Conclusions

In conclusion, body composition measures may more accurately predict key clinical outcomes in individuals with CF than BMI. In particular, ALMI was significantly lower in individuals with CF than healthy volunteers, despite no differences in BMI, which may have important clinical implications given the observed correlation between ALMI and pulmonary function. In contrast, FMI and other measures of body fat were negatively correlated with FEV1 when accounting for age, gender, and BMI, suggesting a detrimental effect of adiposity in CF. Our data also support the beneficial impact of physical activity on body composition, including increased ALMI and decreased body fat measures. Future prospective longitudinal studies analyzing the effect of body composition, dietary intake, and physical activity on CF-specific clinical outcomes are greatly needed, particularly in the post-modulator era.

Author Contributions: Conceptualization, K.J.S., L.T.J., S.F., G.S.S., A.U., J.S.F. and M.S.P.; data curation, M.S.P.; formal analysis, K.J.S. and M.S.P.; funding acquisition, M.S.P.; investigation, K.J.S. and J.S.F.; supervision, M.S.P.; writing—original draft, K.J.S.; writing—review and editing, L.T.J., S.F., G.S.S., A.U., J.S.F. and M.S.P. All authors have read and agreed to the published version of the manuscript.

Funding: This study was supported by NIH K23DK102600 and a Vertex Pharmaceuticals Investigator Initiated Studies Grant. Resources utilized for this study were provided by the Massachusetts General Hospital Clinical Research Center funded by the Harvard Catalyst 1UL1 TR001102.

Institutional Review Board Statement: The study approved by the Institutional Review Board of Mass General Brigham (IRB 2012P000269) with ceded review by the Boston Children's Hospital IRB and was registered on clinicaltrials.gov (NCT01549314).

Informed Consent Statement: All participants provided written informed consent.

Data Availability Statement: Some or all datasets generated during and/or analyzed during the current study are not publicly available but are available from the corresponding author on reasonable request.

Acknowledgments: The authors would like to gratefully acknowledge the support of the dedicated staff of the MGH Clinical Research Center and the MGH and BCH Cystic Fibrosis Centers. We thank the study volunteers for their participation.

Conflicts of Interest: Putman reports grants and speaking fees from Vertex Pharmaceuticals as well as grants from the Cystic Fibrosis Foundation, outside the submitted work. Sawicki reports personal fees from Vertex Pharmaceuticals, outside the submitted work. Uluer reports grants from the Cystic Fibrosis Foundation and serves an advisory board for Vertex Pharmaceuticals and as an unpaid board member for the Cystic Fibrosis Research Institute. Freedman reports grants from the Cystic Fibrosis Foundation. The other authors have nothing to disclose.

References

1. Moran, A.; Brunzell, C.; Cohen, R.C.; Katz, M.; Marshall, B.C.; Onady, G.; Robinson, K.A.; Sabadosa, K.A.; Stecenko, A.; Slovis, B.; et al. Clinical care guidelines for cystic fibrosis-related diabetes: A position statement of the American Diabetes Association and a clinical practice guideline of the Cystic Fibrosis Foundation, endorsed by the Pediatric Endocrine Society. *Diabetes Care* **2010**, *33*, 2697–2708. [CrossRef]
2. Yen, E.H.; Quinton, H.; Borowitz, D. Better nutritional status in early childhood is associated with improved clinical outcomes and survival in patients with cystic fibrosis. *J. Pediatr.* **2013**, *162*, 530–535.e1. [CrossRef]
3. Borowitz, D.; Baker, R.D.; Stallings, V. Consensus report on nutrition for pediatric patients with cystic fibrosis. *J. Pediatr. Gastroenterol. Nutr.* **2002**, *35*, 246–259. [CrossRef] [PubMed]
4. Morton, A.M. Symposium 6: Young people, artificial nutrition and transitional care the nutritional challenges of the young adult with cystic fibrosis: Transition. *Proc. Nutr. Soc.* **2009**, *68*, 430–440. [CrossRef]
5. Calella, P.; Valerio, G.; Thomas, M.; McCabe, H.; Taylor, J.; Brodlie, M.; Siervo, M. Association between body composition and pulmonary function in children and young people with cystic fibrosis. *Nutrition* **2018**, *48*, 73–76. [CrossRef] [PubMed]
6. McDonald, C.M.; Alvarez, J.A.; Bailey, J.; Bowser, E.K.; Farnham, K.; Mangus, M.; Padula, L.; Porco, K.; Rozga, M. Academy of Nutrition and Dietetics: 2020 Cystic Fibrosis Evidence Analysis Center Evidence-Based Nutrition Practice Guideline. *J. Acad. Nutr. Diet.* **2021**, *121*, 1591–1636.e3. [CrossRef] [PubMed]
7. Culhane, S.; George, C.; Pearo, B.; Spoede, E. Malnutrition in cystic fibrosis: A review. *Nutr. Clin. Pract.* **2013**, *28*, 676–683. [CrossRef] [PubMed]
8. Engelen, M.P.K.J.; Com, G.; Deutz, N.E.P. Protein is an important but undervalued macronutrient in the nutritional care of patients with cystic fibrosis. *Curr. Opin. Clin. Nutr. Metab. Care* **2014**, *17*, 515–520. [CrossRef] [PubMed]
9. Bellissimo, M.P.; Zhang, I.; Ivie, E.A.; Tran, P.H.; Tangpricha, V.; Hunt, W.R.; Stecenko, A.A.; Ziegler, T.R.; Alvarez, J.A. Visceral adipose tissue is associated with poor diet quality and higher fasting glucose in adults with cystic fibrosis. *J. Cyst. Fibros.* **2019**, *18*, 430–435. [CrossRef]
10. Sutherland, R.; Katz, T.; Liu, V.; Quintano, J.; Brunner, R.; Tong, C.W.; Collins, C.E.; Ooi, C.Y. Dietary intake of energy-dense, nutrient-poor and nutrient-dense food sources in children with cystic fibrosis. *J. Cyst. Fibros.* **2018**, *17*, 804–810. [CrossRef]
11. Calvo-Lerma, J.; Boon, M.; Hulst, J.; Colombo, C.; Asseiceira, I.; Garriga, M.; Masip, E.; Claes, I.; Bulfamante, A.; Janssens, H.M.; et al. Change in Nutrient and Dietary Intake in European Children with Cystic Fibrosis after a 6-Month Intervention with a Self-Management mHealth Tool. *Nutrients* **2021**, *13*, 1801. [CrossRef] [PubMed]
12. Woestenenk, J.W.; Castelijns, S.J.A.M.; van der Ent, C.K.; Houwen, R.H.J. Dietary intake in children and adolescents with cystic fibrosis. *Clin. Nutr.* **2014**, *33*, 528–532. [CrossRef] [PubMed]
13. Panagopoulou, P.; Fotoulaki, M.; Nikolaou, A.; Nousia-Arvanitakis, S. Prevalence of malnutrition and obesity among cystic fibrosis patients. *Pediatr. Int.* **2014**, *56*, 89–94. [CrossRef] [PubMed]
14. Hanna, R.M.; Weiner, D.J. Overweight and obesity in patients with cystic fibrosis: A center-based analysis. *Pediatr. Pulmonol.* **2015**, *50*, 35–41. [CrossRef] [PubMed]
15. Stephenson, A.L.; Mannik, L.A.; Walsh, S.; Brotherwood, M.; Robert, R.; Darling, P.B.; Nisenbaum, R.; Moerman, J.; Stanojevic, S. Longitudinal trends in nutritional status and the relation between lung function and BMI in cystic fibrosis: A population-based cohort study. *Am. J. Clin. Nutr.* **2013**, *97*, 872–877. [CrossRef]
16. Harindhanavudhi, T.; Wang, Q.; Dunitz, J.; Moran, A.; Moheet, A. Prevalence and factors associated with overweight and obesity in adults with cystic fibrosis: A single-center analysis. *J. Cyst. Fibros.* **2020**, *19*, 139–145. [CrossRef]
17. Guimbellot, J.S.; Baines, A.; Paynter, A.; Heltshe, S.L.; VanDalfsen, J.; Jain, M.; Rowe, S.M.; Sagel, S.D. Long term clinical effectiveness of ivacaftor in people with the G551D CFTR mutation. *J. Cyst. Fibros.* **2021**, *20*, 213–219. [CrossRef]
18. Cystic Fibrosis Foundation. *2019 Patient Registry Annual Data Report*; Cystic Fibrosis Foundation: Bethesda, MD, USA, 2019.
19. Moran, A.; Pekow, P.; Grover, P.; Zorn, M.; Slovis, B.; Pilewski, J.; Tullis, E.; Liou, T.G.; Allen, H. Cystic Fibrosis Related Diabetes Therapy Study Group. Insulin therapy to improve BMI in cystic fibrosis-related diabetes without fasting hyperglycemia: Results of the cystic fibrosis related diabetes therapy trial. *Diabetes Care* **2009**, *32*, 1783–1788. [CrossRef]

20. Alvarez, J.A.; Ziegler, T.R.; Millson, E.C.; Stecenko, A.A. Body composition and lung function in cystic fibrosis and their association with adiposity and normal-weight obesity. *Nutrition* **2016**, *32*, 447–452. [CrossRef]
21. Ritchie, H.; Nahikian-Nelms, M.; Roberts, K.; Gemma, S.; Shaikhkhalil, A. The prevalence of aberrations in body composition in pediatric cystic fibrosis patients and relationships with pulmonary function, bone mineral density, and hospitalizations. *J. Cyst. Fibros.* **2021**, *20*, 837–842. [CrossRef]
22. Nevill, A.M.; Stewart, A.D.; Olds, T.; Holder, R. Relationship between adiposity and body size reveals limitations of BMI. *Am. J. Phys. Anthropol.* **2006**, *129*, 151–156. [CrossRef]
23. King, S.J.; Nyulasi, I.B.; Strauss, B.J.G.; Kotsimbos, T.; Bailey, M.; Wilson, J.W. Fat-free mass depletion in cystic fibrosis: Associated with lung disease severity but poorly detected by body mass index. *Nutrition* **2010**, *26*, 753–759. [CrossRef]
24. Sheikh, S.; Zemel, B.S.; Stallings, V.A.; Rubenstein, R.C.; Kelly, A. Body composition and pulmonary function in cystic fibrosis. *Front. Pediatr.* **2014**, *2*, 33. [CrossRef]
25. Baker, J.F.; Putman, M.S.; Herlyn, K.; Tillotson, A.P.; Finkelstein, J.S.; Merkel, P.A. Body composition, lung function, and prevalent and progressive bone deficits among adults with cystic fibrosis. *Jt. Bone Spine* **2016**, *83*, 207–211. [CrossRef]
26. King, S.J.; Tierney, A.C.; Edgeworth, D.; Keating, D.; Williams, E.; Kotsimbos, T.; Button, B.M.; Wilson, J.W. Body composition and weight changes after ivacaftor treatment in adults with cystic fibrosis carrying the G551 D cystic fibrosis transmembrane conductance regulator mutation: A double-blind, placebo-controlled, randomized, crossover study with open-label extension. *Nutrition* **2021**, *85*, 111124.
27. Putman, M.S.; Greenblatt, L.B.; Bruce, M.; Joseph, T.; Lee, H.; Sawicki, G.; Uluer, A.; Sicilian, L.; Neuringer, I.; Gordon, C.M.; et al. The Effects of Ivacaftor on Bone Density and Microarchitecture in Children and Adults with Cystic Fibrosis. *J. Clin. Endocrinol. Metab.* **2021**, *106*, E1248–E1261. [CrossRef]
28. Matthews, D.R.; Hosker, J.P.; Rudenski, A.S.; Naylor, B.A.; Treacher, D.F.; Turner, R.C. Homeostasis model assessment: Insulin resistance and beta-cell function from fasting plasma glucose and insulin concentrations in man. *Diabetologia* **1985**, *28*, 412–419. [CrossRef] [PubMed]
29. Augustin, L.S.A.; Kendall, C.W.C.; Jenkins, D.J.A.; Willett, W.C.; Astrup, A.; Barclay, A.W.; Björck, I.; Brand-Miller, J.C.; Brighenti, F.; Buyken, A.E.; et al. Glycemic index, glycemic load and glycemic response: An International Scientific Consensus Summit from the International Carbohydrate Quality Consortium (ICQC). *Nutr. Metab. Cardiovasc. Dis.* **2015**, *25*, 795–815. [CrossRef] [PubMed]
30. Sievert, Y.A.; Schakel, S.F.; Buzzard, I.M. Maintenance of a nutrient database for clinical trials. *Control. Clin. Trials* **1989**, *10*, 416–425. [CrossRef]
31. Kriska, A.M.; Bennett, P.H. An epidemiological perspective of the relationship between physical activity and NIDDM: From activity assessment to intervention. *Diabetes. Metab. Rev.* **1992**, *8*, 355–372. [CrossRef]
32. Madeira, F.B.; Silva, A.A.; Veloso, H.F.; Goldani, M.Z.; Kac, G.; Cardoso, V.C.; Bettiol, H.; Barbieri, M.A. Normal weight obesity is associated with metabolic syndrome and insulin resistance in young adults from a middle-income country. *PLoS ONE* **2013**, *8*, e60673. [CrossRef] [PubMed]
33. Bunout, D.; de la Maza, M.P.; Barrera, G.; Leiva, L.; Hirsch, S. Association between sarcopenia and mortality in healthy older people. *Australas. J. Ageing* **2011**, *30*, 89–92. [CrossRef] [PubMed]
34. Alicandro, G.; Bisogno, A.; Battezzati, A.; Bianchi, M.L.; Corti, F.; Colombo, C. Recurrent pulmonary exacerbations are associated with low fat free mass and low bone mineral density in young adults with cystic fibrosis. *J. Cyst. Fibros.* **2014**, *13*, 328–334. [CrossRef] [PubMed]
35. Vaisman, N.; Pencharz, P.B.; Corey, M.; Canny, G.J.; Hahn, E. Energy expenditure of patients with cystic fibrosis. *J. Pediatr.* **1987**, *111*, 496–500. [CrossRef]
36. Turck, D.; Braegger, C.P.; Colombo, C.; Declercq, D.; Morton, A.; Pancheva, R.; Robberecht, E.; Stern, M.; Strandvik, B.; Wolfe, S.; et al. ESPEN-ESPGHAN-ECFS guidelines on nutrition care for infants, children, and adults with cystic fibrosis. *Clin. Nutr.* **2016**, *35*, 557–577. [CrossRef]
37. Matel, J.L. Nutritional Management of Cystic Fibrosis. *J. Parenter. Enter. Nutr.* **2012**, *36* (Suppl. 1), 60S–67S. [CrossRef]
38. Gaskin, K.J. Nutritional care in children with cystic fibrosis: Are our patients becoming better? *Eur. J. Clin. Nutr.* **2013**, *67*, 558–564. [CrossRef] [PubMed]
39. Prévotat, A.; Godin, J.; Bernard, H.; Perez, T.; Le Rouzic, O.; Wallaert, B. Improvement in body composition following a supervised exercise-training program of adult patients with cystic fibrosis. *Respir. Med. Res.* **2019**, *75*, 5–9. [CrossRef]
40. Papalexopoulou, N.; Dassios, T.G.; Lunt, A.; Bartlett, F.; Perrin, F.; Bossley, C.J.; Wyatt, H.A.; Greenough, A. Nutritional status and pulmonary outcome in children and young people with cystic fibrosis. *Respir. Med.* **2018**, *142*, 60–65. [CrossRef]
41. Fielding, J.; Brantley, L.; Seigler, N.; McKie, K.T.; Davison, G.W.; Harris, R.A. Oxygen uptake kinetics and exercise capacity in children with cystic fibrosis. *Pediatr. Pulmonol.* **2015**, *50*, 647–654. [CrossRef]

Review

What Do We Know about the Microbiome in Cystic Fibrosis? Is There a Role for Probiotics and Prebiotics?

Josie M. van Dorst [1], Rachel Y. Tam [1] and Chee Y. Ooi [1,2,3,*]

1. Discipline of Paediatrics & Child Health, Randwick Clinical Campus, School of Clinical Medicine, UNSW Medicine & Health, UNSW, Sydney 2031, Australia; j.vandorst@unsw.edu.au (J.M.v.D.); yantungrachel.tam@unsw.edu.au (R.Y.T.)
2. Molecular and Integrative Cystic Fibrosis (miCF) Research Centre, Sydney 2031, Australia
3. Department of Gastroenterology, Sydney Children's Hospital Randwick, Sydney 2031, Australia
* Correspondence: keith.ooi@unsw.edu.au

Abstract: Cystic fibrosis (CF) is a life-shortening genetic disorder that affects the cystic fibrosis transmembrane conductance regulator (CFTR) protein. In the gastrointestinal (GI) tract, CFTR dysfunction results in low intestinal pH, thick and inspissated mucus, a lack of endogenous pancreatic enzymes, and reduced motility. These mechanisms, combined with antibiotic therapies, drive GI inflammation and significant alteration of the GI microbiota (dysbiosis). Dysbiosis and inflammation are key factors in systemic inflammation and GI complications including malignancy. The following review examines the potential for probiotic and prebiotic therapies to provide clinical benefits through modulation of the microbiome. Evidence from randomised control trials suggest probiotics are likely to improve GI inflammation and reduce the incidence of CF pulmonary exacerbations. However, the highly variable, low-quality data is a barrier to the implementation of probiotics into routine CF care. Epidemiological studies and clinical trials support the potential of dietary fibre and prebiotic supplements to beneficially modulate the microbiome in gastrointestinal conditions. To date, limited evidence is available on their safety and efficacy in CF. Variable responses to probiotics and prebiotics highlight the need for personalised approaches that consider an individual's underlying microbiota, diet, and existing medications against the backdrop of the complex nutritional needs in CF.

Keywords: cystic fibrosis; dysbiosis; inflammation; nutrition; prebiotic; probiotic

1. Introduction

Cystic fibrosis (CF) is a genetic condition of autosomal recessive inheritance related to mutations in the gene coding for the cystic fibrosis transmembrane conductance regulator (CFTR) protein [1]. The CFTR protein affects the fluid secretion and mucus hydration of epithelial cells in the airway, pancreas, intestines, and hepatobiliary tracts [2]. Chronic suppurative respiratory disease arising due to impaired clearance of dehydrated airway secretions is typically the principal cause of morbidity and mortality. However, the majority (>90%) of patients with CF also suffer from gastrointestinal (GI) symptoms and complications [3,4]. Dysfunction of the CFTR protein in the GI system results in low intestinal pH, thick and inspissated mucus, a lack of endogenous pancreatic enzymes, reduced motility, and possibly an impaired innate immunity [5–7] (Figure 1). These mechanisms are proposed drivers of local GI inflammation and contribute to a range of intestinal morbidities, including an increased risk of early-onset adult GI cancer [8–11]. GI dysfunction combined with antibiotic therapies also drives significant alteration (dysbiosis) of the GI microbiota (Figure 1). Altered CF microbiota is likely to compound the proinflammatory effects of the underlying disease.

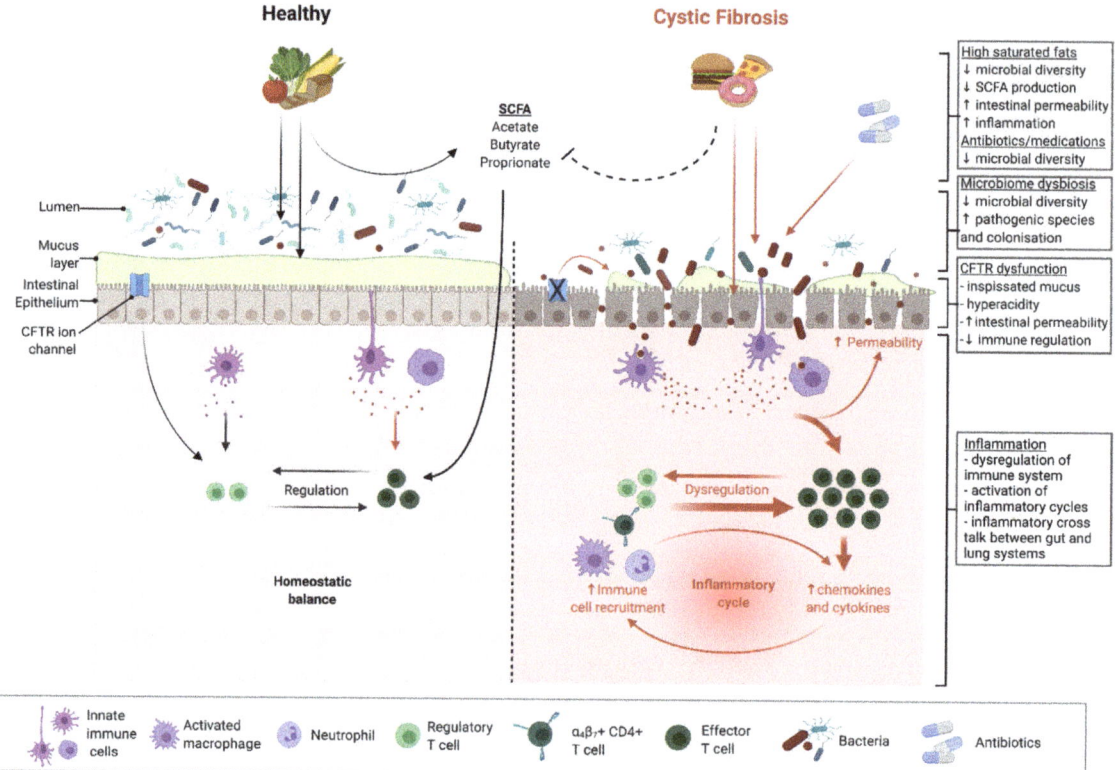

Figure 1. Microbiome- and CFTR-related dysfunction and inflammation in cystic fibrosis. Black arrows indicate direction of known homeostatic effects. Red arrows indicate direction of known inflammatory effects. Broken lines indicate proposed mechanisms of inhibition or dysfunction. Figure was created with Biorender.com.

Evidence is accumulating that GI bacterial strains, which occur differentially between CF and healthy controls (HC), are linked to inflammatory [12–14] and malignancy processes [7,14]. Supplementation with prebiotics and probiotics are thought to provide clinical benefit by promoting commensal bacteria and biosynthesis of immunomodulatory metabolites. As public awareness and acceptance of probiotics and prebiotics continue to expand, there is a growing interest in the potential clinical benefits of dietary prebiotics and probiotics in CF. A total of 17 probiotic trials, including 12 RCTs, have thus far investigated the safety and efficacy of individual probiotic strains and strain combinations in children and adults with CF [15]. Promising improvements in inflammation [16–18], nutritional status [19], and health outcomes [20,21] have been observed. However, due to selective reporting and incomplete outcome data, the certainty of evidence has been evaluated as low. Furthermore, large variations between protocols, probiotic formulas, dosage, and duration of treatments limit the potential for clinical application. Prebiotics are often included in probiotic preparations, but evidence surrounding safety and efficacy for the exclusive use of prebiotics is trailing. There is only one clinical trial investigating the prebiotic high-amylose maize starch (HAMS) in adults with CF [22]. The efficacy of prebiotic supplementation is based on the selective utilisation of substrates (usually indigestible carbohydrates) by beneficial bacteria. Critically, it is not yet known whether the altered CF intestinal microbiota retains the capacity to exploit prebiotic substrates.

This review describes the physiology of the GI tract in CF and the clinical relevance of GI microbiome dysbiosis and inflammation. We discuss the current understanding of probiotic and prebiotic mechanisms of action, provide important examples of clinical studies examining probiotic and prebiotic applications in CF, and discuss considerations for clinical translation.

2. Cystic Fibrosis in the Gastrointestinal Tract

CFTR is an important contributor to the normal physiology of the gastrointestinal (GI) tract; as such, its dysfunction in CF disease has profound impacts on GI homeostasis. CFTR is an epithelial cyclic adenosine monophosphate (cAMP)-dependent anion-selective channel. It primarily secretes bicarbonate and chloride, and therefore exerts great influence on the acidity and viscosity of secretions. In CF, the dysfunction of this ion channel is clearly manifested in the systemic production of hyperacidic and viscid mucus [23,24]. CFTR also plays a role in the maintenance of epithelial tight junctions, modulation of fluid flow, regulation of ion channels (such as sodium, potassium, calcium, and other chloride channels [25,26]), and coordination of gut motility [27,28]. Altogether, disruptions to these normal and vital functions of CFTR culminate in an abnormal GI tract (Figure 1).

The altered GI environment in CF results in various clinical sequelae that can be collectively referred to as "obstructive tubulopathies." The most common of these is pancreatic insufficiency, which affects as much as 90% of patients with CF [29]. From as early as in utero, the presence of concentrated pancreatic ductal secretions leads to luminal protein precipitation with resultant obstruction and dilation of the pancreatic ducts, culminating in progressive, irreversible destruction and fibrosis of the acinar tissue. The resultant pancreas is dysfunctional and severely impaired in its ability to secrete critical enzymes necessary for the digestion of carbohydrates, fats, and proteins [30,31]. Obstructive tubulopathies are also evident in the intestines in the form of meconium ileus (MI) and distal intestinal obstruction syndrome (DIOS). There are numerous other GI manifestations of CF, including gastroesophageal reflux disease, pancreatitis, and liver disease, which have been detailed elsewhere [32–36]. There are also less clinically obvious, but equally significant, manifestations that arise as a result of the altered GI milieu in CF; namely, alterations to the gut microbiota and intestinal inflammation. These are discussed in detail below.

3. The Human Gut Microbiome

The gut microbiome is a sophisticated, functional environment comprising an abundance of microbes along the GI tract. These microorganisms and their metabolites perform homeostatic functions, including the regulation of the gastrointestinal epithelial barrier, fermentation of dietary starches and fibres, synthesis of amino acids and essential vitamins, and modulation of the immune system locally and distally [37,38]. While a small proportion of the gut microbiota is heritable, it is largely influenced by nongenetic factors [39]. The early development of the gut microbiome in infancy is predominantly shaped by one's mode of birth and feeding, with the cessation of breastfeeding being the driver of functional maturation into an adultlike microbiota [40,41]. Subsequently, diet plays a primary role in shaping the gut microbiome, as organisms respond to selective pressures from dietary patterns throughout life [42,43]. Numerous other environmental factors can also affect the gut microbiota, but perhaps the most well-established are medications, including antibiotics [41,44–46].

One important aspect to spotlight when characterising the gut microbiota is microbial diversity. Microbial diversity refers to species richness (the number of species) and/or evenness (the relative distribution of species). Reduced microbial diversity is broadly associated with ill health, as it is hypothesised that species diversity confers the ability to withstand environmental threats and maintain homeostasis. This is attributable to compensatory functional redundancies enabled by a more robust ecological environment [47,48]. Disruption to the normal composition, physiology, and diversity of the gut microbiota is an increasingly recognised feature of numerous disease processes. Decreased microbial diversity has repeatedly been observed in patients with chronic conditions, including obesity, inflammatory bowel disease (IBD), type 1 and type 2 diabetes mellitus, and asthma. Dysbiosis, the collective term for alterations to the normal balance or composition of gut microbes, is also evident in many of those disease processes [48–51]. These observations point to the critical involvement of the gut microbiota in health and disease, and solidify the rationale for utilising microbial modulation as a therapeutic target.

4. The CF Gut Microbiome

Given the significant alterations to the intestinal environment resulting from CFTR dysfunction, it is unsurprising that the CF gut microbiome differs from that of the healthy gut from early life onwards. One key features of the CF gut microbiome is decreased species diversity [7,12,52–54] (Figure 1). In addition, paediatric studies have demonstrated that the CF gut microbiome diversifies and matures at a significantly slower rate than that of a healthy child [54–56]. Compositionally, the CF gut microbiome also differs from that of the healthy gut. Reductions in *Bacteroidetes, Ruminococcaceae, Bifidobacterium,* and *Roseburia* have consistently been observed. In contrast, abundances of *Enterococcus, Veillonella,* and *Enterobacter* have been shown to be relatively increased in the CF gut [7,12,52–55,57]. Use of the CFTR modulator ivacaftor is associated with arguably "healthier" microbiome profiles, reinforcing the concept that dysbiosis is driven by CFTR dysfunction [58]. Recent advancements in metagenomic methods have enhanced the ability to characterise the functionality of the gut microbiota, thereby elucidating the physiological consequences of dysbiosis. It has been demonstrated that the CF gut microbiome displays an increased capacity to metabolise nutrients, antioxidants, and short-chain fatty acids (SCFAs), as well as a relatively decreased propensity to synthesise fatty acids [7,56,59].

The key drivers of these changes to the gut microbiota involve the downstream effects of CFTR dysfunction. The production of dehydrated mucus, changes to intestinal pH, nutrient malabsorption, and prolonged intestinal transit secondary to intestinal dysmotility all have the potential to exert selective pressure on enteric microorganisms and ultimately alter the microbiome [60–62]. Notably, fat malabsorption following exocrine pancreatic insufficiency could also confer survival advantage to certain organisms that adapt well to high-fat intestinal environments [63]. These CFTR-related factors are further compounded by iatrogenic causes. Antibiotic exposure, which is prevalent in CF for the prophylaxis and treatment of respiratory tract infections, may contribute to changes in the gut microbiota. Studies in the CF population have consistently demonstrated an association between antibiotic use and decreased alpha diversity (within-sample species diversity) in the gut [12,53,64,65] (Figure 1). Multiple studies have also highlighted a correlation between antibiotic exposure and relative depletions of the bacterial genus *Bifidobacterium* [64–67]. The high-energy and high-fat diet prescribed in CF is another likely contributor (discussed below).

5. Intestinal Inflammation

Disruption to the gut microbiota is associated with intestinal inflammation in CF. Chronic inflammation is a well-recognised feature of the CF intestine, primarily evidenced by elevated faecal inflammatory markers in patients with CF in many studies [68–73] (Figure 1). The earliest evidence of GI inflammation was elevated concentrations of inflammatory markers such as interleukin-8, interleukin-1β, neutrophil elastase, and immunoglobulins on whole-gut lavage, reported by Smyth et al. [74]. Imaging techniques including endoscopy and capsule endoscopy have subsequently revealed a high prevalence of mucosal pathologies, including ulcerations and oedema in the CF GI tract [71,75,76].

Gut inflammation in CF is of a multifactorial aetiology. Mucus hyperviscosity and hyperacidity as a result of CFTR dysfunction likely promote gut inflammation [8,77,78]. CFTR itself is also involved in downregulating proinflammatory pathways, and hence its dysfunction in CF may contribute to the altered intestinal milieu [79] (Figure 1). Additionally, inflammation may be precipitated by intestinal dysmotility and the intraluminal pooling of inspissated contents [77,80]. The same iatrogenic factors that contribute to intestinal dysbiosis, namely antibiotic exposure and the high-fat CF diet, have also been shown to be correlated with intestinal inflammation in CF and other contexts [81–83] (Figure 1). The mechanisms by which antibiotics may induce inflammation are not well-known. However, it has been demonstrated in animal models that antibiotic administration promotes the translocation of microorganisms through goblet-cell-mediated pathways, subsequently increasing the release of inflammatory cytokines [84].

Notably, the aforementioned dysbiosis is a key contributor to intestinal inflammation in CF. Reductions in the abundances of bacteria with anti-inflammatory properties, including *Faecalibacterium prausnitzii,* has been widely observed in CF cohorts [7,52,53,66,70,85]. Many of these bacteria are known producers of short-chain fatty acids (SCFAs), the primary metabolites of anaerobic fermentation of dietary fibres and starches. SCFAs perform homeostatic functions, including intestinal epithelial maintenance, colonocyte nourishment, and immunomodulation (Figure 1). Accordingly, a relative depletion of SCFA-producing bacteria and subsequent reductions in SCFA levels may contribute to inflammation [60,86,87]. This is supported by numerous animal models in which SCFAs have been shown to improve epithelial integrity and ameliorate intestinal inflammation [88–93]. However, the interactions between the microbiota and inflammation are also bidirectional. Chronic inflammation results in the release of reactive oxygen and nitrogen species that supply terminal electron acceptors required for anaerobic respiration. This exerts selective pressure on gut microbes and may contribute to dysbiosis, as organisms with the ability to efficiently perform anaerobic respiration have a growth advantage [94]. For example, intestinal inflammation is associated with the proliferation of *Enterobacteriaceae*, a bacterial family that has high nitrate reductase activity and can undergo efficient nitrate respiration [95,96]. Indeed, organisms within the *Enterobacteriaceae* family (i.e., the *Enterobacter* genus) are relatively more abundant in the CF gut [7,12,64,85]. Many of the mechanistic aspects of the relationship between the gut microbiota and inflammation remain unknown, highlighting the intricacy of these complex interactions.

6. Nutritional Management in CF

In 1988, Corey et al. [97] published a landmark study that led to pivotal paradigm shifts in CF nutritional optimisation. It has since been established that energy requirements are increased in CF due to increased energy expenditure from chronic lung inflammation and increased work of breathing, as well as malabsorption secondary to exocrine pancreatic insufficiency and gastrointestinal disease [29,98,99]. Patients with CF are also at risk of deficiencies in fat-soluble vitamins due to fat and bile acid malabsorption, which often necessitates supplementation [100,101]. Good nutritional status beginning in childhood is now well-documented to be associated with better pulmonary function and survival in CF [102–106]. Body mass index (BMI) is positively correlated with forced expiratory volume in 1 second (FEV1) [107–110], and a low BMI at the age of 10 years is a risk factor for lung transplantation in adulthood [111]. Greater weight-for-age percentile at the age of 4 years is also associated with better pulmonary function and survival through to 18 years, as well as a reduced likelihood of subsequent pulmonary exacerbations, hospitalisations, or CF-related diabetes [106].

Today, patients with CF are recommended a high-energy diet (110–200% of the age- and sex-appropriate recommended daily energy intake) to maintain growth. While macronutrient targets are individual-specific, the current consensus generally advises that 15–20% of total energy intake be derived from protein, 40–45% from carbohydrates, and up to 35–40% from fat [29,100,112]. Despite the clear benefits of nutritional optimisation, it has become increasingly evident that patients with CF tend to overconsume "energy-dense, nutrient-poor" foods high in salt, sugar, and saturated fat (i.e., junk foods) in order to meet daily macronutrient requirements [113–117]. The proportion of patients with CF who are overweight or obese is increasing. While patients who are overweight or obese are reported to have better lung function than their normal weight or underweight counterparts in some studies, this finding may be confounded by the fact that these patients are also more likely to be pancreatic sufficient and have milder disease genotypes [118–122]. Additionally, weight gain and increased BMI, fat mass, and fat-free mass are reported outcomes of CFTR modulator therapies that need to be taken into account as modulator therapies gradually become the cornerstone of CF treatment [123–126]. Importantly, both high-fat diets and obesity may exacerbate existing alterations in gut microbial composition and chronic intestinal inflammation, with important clinical implications for individuals with CF [81–83,127,128] (Figure 1).

7. Clinical Significance of the CF Gut Microbiome

Intestinal dysbiosis and inflammation have been demonstrated to be significantly associated with clinical outcomes. In a recent study, Hayden et al. [55] observed a distinctly more marked dysbiosis in infants with CF who had low length compared to infants with CF who had normal length. Notably, the gut microbiome of infants with low length exhibited a reduced abundance of *Bacteroidetes* and relatively delayed maturation compared to that of infants with normal length [55]. Coffey et al. [7] had also previously reported a positive correlation between *Ruminococcaceae UCG 014* and BMI. Additionally, the CF intestinal microbiota contains a comparatively lower prevalence of proteins that facilitate carbohydrate transport, metabolism, and conversion, which may impact nutrient utilisation and thus adversely affect growth [85]. It has also been demonstrated that faecal calprotectin levels are inversely correlated with weight and height z-scores, and elevated calprotectin levels are associated with underweight BMI (<18.5 kg/m^2) [13,70,72].

Additionally, there is a growing body of evidence that suggests that the intestinal microbiome is related to lung function. It has been reported that patients with lower FEV1 have reduced intestinal microbial diversity compared to their counterparts with better pulmonary function [12]. Positive correlations between intestinal bacterial genera such as *Ruminococcaceae NK4A214* and FEV1 have also previously been documented [7]. Furthermore, one study reported an association between microbial diversity in the gut microbiota and pulmonary exacerbation events [57]. Some studies have also demonstrated associations between gut inflammation and lower FEV1, although this has not yet been widely validated [68,72]. It is postulated that these associations reflect a physiological phenomenon termed the "gut–lung axis." Along this axis, the intestinal and respiratory microbiota engage in cross-talk to regulate immunity and homeostasis in both the enteric and pulmonary environments [129]. In the intestinal compartment, this is achieved by gut-microbiota-derived metabolites, including SCFAs, which coordinate immune cell signaling cascades that ultimately involve the lungs through G-protein coupled receptor (GPCR)-mediated pathways and histone deacetylase inhibition [130–132]. In support of this, Hoen et al. [133] observed in a paediatric CF cohort that pulmonary colonisation with the pathogen *Pseudomonas aeruginosa*, a known contributor to declining lung function, was preceded by a reduction in the abundance of *Parabacteroides* in the gut. Notably, *Parabacteroides* is associated with immunomodulation and anti-inflammatory properties [134]. While there remain many unknowns with regard to the mechanistic aspects of the gut–lung axis, these findings suggested that the intestinal microbiome is a site of therapeutic potential that could be manipulated to optimise lung function.

Intestinal dysbiosis and inflammation have been linked to a number of serious morbidities. Firstly, while patients with CF do not typically present with overt GI symptoms similar to those of inflammatory bowel disease (IBD), elevated faecal calprotectin levels are correlated with a worse quality of life [4,135,136]. Elevated calprotectin has also been highlighted as a predictive factor of GI-related hospitalisations for infants with CF in their first year of life [136,137]. Importantly, intestinal dysbiosis and inflammation may be contributors to the increased risk of GI malignancies that is evident in the CF population [138]. While a clear causative mechanism has yet to be established in the context of CF, it is well recognised from studies pertaining to IBD that chronic inflammation poses a significant risk for the development of GI cancers [139,140]. This is largely due to oxidative stress and the resultant DNA damage, culminating in epigenetic disturbances to the expression of tumour-suppressive regulatory proteins, transcription factors, and signalling molecules [141,142]. Furthermore, the inflamed gut may confer a growth advantage to genotoxic organisms, especially *E. coli* [143]. Indeed, the relative abundance of *E. coli* is increased in CF, as well as in IBD and colorectal cancer [144,145]. The depletion of SCFA-producing organisms in the CF gut may also be a key factor, as SCFAs exhibit tumour-suppressive properties [146,147]. For example, in an animal model of colitis-associated colorectal cancer, SCFAs have been shown to mediate reductions in proinflammatory cytokine release and tumour size and incidence [148]. All in all, while intestinal dysbiosis and inflammation tend to be clinically silent in CF, they may be associated with serious complications. This emphasises the importance of optimising gut health in the management of CF.

8. Microbiome Modulation with Probiotics

Improving gut health through microbiome modulation is gaining traction in GI and respiratory diseases [149,150]. The microbiome can be modulated through administration of a single or combination of commensal strains (probiotics), indigestible carbohydrates to promote the expansion of commensal strains (prebiotics), or a combination of both (synbionts). Probiotics were first described in 1907, and have been utilised as a beneficial dietary supplement since. In 2002, a consensus was reached by a joint FAO/WHO working group on the definition of probiotics: "Live microorganisms that, when administered in adequate amounts, confer a health benefit on the host" [151]. Existing probiotic preparations are based primarily on strains from the genus lactobacilli, bifidobacterial, and other lactic acid-producing bacteria (LAB) isolated from fermented dairy products and faecal microbiome samples [152]. However, rapidly expanding research into host–microbe interactions is increasing the impetus for the development of next-generation probiotics from beneficial microbes including *Akkermansia, Eubacterium, Propionibacterium, Faecalibacterium,* and *Roseburia* species [153–155].

Probiotics have reported beneficial effects in diseases with links to a GI dysbiosis, inflammation, and respiratory function [15,152,156]. However, knowledge gaps exist related to robust evidence-based probiotic use as a result of the significant heterogeneity between studies and variability in the probiotic strains studied. Specific probiotic strains have been indicated in the reduction in necrotizing enterocolitis (NEC) incidence [157] and the management of *Clostridium difficile* [158,159], though the quality of evidence remains low [160]. The rise of in vitro, animal, and cell culture research has expanded our understanding of prosed mechanisms of action, and include direct interaction with commensal gut microbiota, modulation of the immune system, production of organic acids, colonization resistance, improved barrier function, production of hormones and other small molecules with systemic effects, and probiotic–host interactions mediated by cell surface structures [149].

8.1. Mechanisms of Action

Interaction with microbiome. The direct interaction with the microbiome is mediated through the increasing microbial stability [161–163], cross-feeding [164], substrate formation, and antagonistic action through competition and production of antimicrobials and bacteriocins [165–167]. Competitive exclusion and inhibition of pathogenic species is a primary function of probiotics. In 2007, Collado et al. [168] tested 12 probiotic strains against 8 pathogenic strains in a pig intestinal mucosa model, and found that all probiotic strains tested were able to inhibit and displace pathogenic species of *Bacterioides, Clostridium, Staphylococcus* and *Enterobacter*. Another in vitro assay demonstrated that *B. animalis* subsp. *lactis* BB-12 and *Lactobacillus reuteri* DSM 17938 inhibited the growth of pathogenic bacteria *E. coli* [169]. Likewise, *Lactobacillus paracasei* FJ861111.1 has demonstrated significant inhibition against several common intestinal pathogens including *Shigella dysenteriae, Escherichia coli,* and *Candida albicans* [159].

Modulate immune system. The interaction between microbiota and the immune system, reviewed in [170], has impacts systemwide. Probiotics have been shown to modulate immune function through an increase in anti-inflammatory cytokines [171,172], a reduction in proinflammatory cytokines [149,152,173,174], and augmentation of vaccines and antibody response [175–177]. The most common species to demonstrate immune modulation include *Lactobacillus*, *Bacillus*, and *Bifidobacterium*, and the genus *Saccharomyces* [178]. The modulation of the immune system through probiotics is not consistent across species or strains, and exhibits variability between hosts [149,172]. Yet recently, Sanders et al. [179] identified that some immune modulatory mechanisms related to cell surface infrastructure were conserved across species and even genera.

Production of organic acids. Probiotic species belonging to the Lactobacillus and Bifidobacterium genera produce lactic and acetic acids as end products of carbohydrate metabolism. These organic acids can reduce colonic pH, discouraging the growth of pathogens. Fredua-Agyeman et al. demonstrated that commercial cocultures of *Bifidobacterium* and *Lactobacillus* strains inhibited *Clostridioides difficile* growth in a pH-dependent manner [180]. Through the process of cross-feeding commensal bacterial species such as *Faecalibacterium*, *Lactobacillus*, and *Bifodobacterium*, probiotics can also increase levels of beneficial short-chain fatty acids (SCFA), including butyrate. SCFA have demonstrated anti-inflammatory and antitumour properties [86,181,182].

Improve barrier function. Tight junctions are critical to epithelial cell function, preventing translocation of microbial species and proinflammatory metabolites [183]. Probiotic *Lactobacillus* and *Bifidobacterium* strains have been shown to increase the expression of tight junction proteins [184,185] and reduce the severity of acute gastroenteritis in children through fortification of tight junctions [186]. Several *Lactobacillus* probiotic strains have also demonstrated regulatory effects on the epithelial mucus layer [153,187–189]. The demonstrated upregulation of mucin production genes and enhanced mucin secretion improves barrier function, inhibiting pathogen binding to epithelial cells [185].

Production of small molecules with systemic effects. Probiotic strains have been implicated in the production of a range of small molecules and hormones that influence systemic function. Interestingly, these include neurotransmitters such as cortisol, serotonin, tryptamine, noradrenaline gamma-aminobutyric acid (GABA), and dopamine, highlighting the potential of probiotics to modulate the gut–brain axis [190,191]. A range of satiety hormones and enzymes that can aid digestion are also produced by some probiotic strains. For example, *Streptococcus thermophilus* can facilitate lactose digestion through the production of microbial β-galactosidase [192].

Probiotic–host interactions mediated by cell surface structures. The cell surface architecture of probiotic strains is critical to probiotic–host cell interactions. Many Gram-positive probiotic strains share cell surface macromolecules that mediate these interactions, including surface layer associated proteins (SLAPS), mucin-binding proteins (MUBs), fibronectin binding proteins, and pili that interact directly with the intestinal epithelium, mucus, and gastrointestinal mucosa receptors. These demonstrated interactions reviewed in [179] can improve host barrier integrity, intestinal motility, and binding to intestinal and vaginal cells.

8.2. Probiotics in CF

The use of probiotics in CF has been investigated in 17 clinical trials. Of those 17 trials, 12 were RCTs, with 8 trials including children and 4 trials including both children and adults [15]. The number of subjects within the RCTs ranged from 22 to 81, and the trial duration ranged from 1 month to 12 months. The probiotic formulations varied in dosage from 10^8 CFU/day to 10^{11} CFU/day. Strain formulations with six of the trials utilised a single Lactobacillus strain *L rhamnosus GG* [16,19,193,194] or *L. reuteri* [18,20,195], two trials utilised multistrain formulations with fructooligosaccharides (FOS) [17,66], and three trials utilised a multistrain without FOS [18,21,196]. Results from individual trials have cited a reduction in in inflammation [16–18], nutritional status [19], and pulmonary health outcomes [20,21] (Table 1).

Table 1. Evidence for the use of probiotics in CF from RCTs.

Year	Probiotic Preparation (Dose)	Study Design	Duration	Probiotic Participants	Primary Results	Ref
1998	L. rhamnosus strain GG (6×10^9 CFU/day)	RCT (Cross-over)	6 months	28	Increased weight gain (placebo 2.7 ± 2.5%, probiotic 8.7 ± 8.1%, $p < 0.05$). Reduced risk of infections (infections requiring antibiotic treatment per child in 6 months, placebo 39 and 1.7 ± 0.3, probiotic 19 or 0.9 ± 0.6, ($p < 0.05$)). Reduction in abdominal pain (placebo = 6 patients with abdominal pain, probiotic = 1, $p < 0.05$).	[194]
2007	Lactobacillus GG (6×10^9 CFU/day)	RCT (Cross-over)	6 months	38	Reduction in pulmonary exacerbations (median 1 vs. 2, range 4 vs. 4, median difference 1, CI 95% 0.5–1.5; $p = 0.003$). Reduction in hospital admissions (median 0 vs. 1, range 3 vs. 2, median difference 1, CI 95% 1.0–1.5; $p = 0.001$). Increase in FEV1 (3.6% ± 5.2 vs. 0.9% ± 5; $p = 0.02$) and body weight (1.5 kg ± 1.8 vs. 0.7 kg ± 1.8; $p = 0.02$).	[19]
2009	CasenBiotic [a] (1×10^8 CFU/day) VLS3 [b] (9×10^{11}/day)	RCT (Cross-over)	6 months	40	Increased Quality of Life score from the PedsQL™ survey. (Probiotics group—parent-reported, 0.87 higher (SD 0.19 higher to 1.55 higher)), (Probiotics group—child-reported, 0.59 higher (SD 0.07 lower to 1.26 higher)).	[18]
2013	Protexin capsule [c] (2×10^9 CFU/day)	RCT (Parallel)	1 month	20	Rate of pulmonary exacerbation significantly reduced among probiotic group ($p < 0.01$). Parent-reported quality of life improved in probiotic group compared with placebo group at 3rd month ($p = 0.01$), not significant at 6th month of probiotic treatment.	[21]
2013	Protexin Restor sachet [d] (1×10^9 CFU/day)	RCT (Parallel)	1 month	24	Mean faecal calprotectin levels decreased with probiotics 56.2 µg/g, compared to placebo 182.1 µg/g ($p = 0.031$).	[17]
2014	L. reuteri DSM 17938 (1×10^8 CFU/day)	RCT (Cross-over)	6 months	30	Significant improvement in gastrointestinal health (GIQLY score placebo 11.2 ± 0.3, probiotic 11.4 ± 0.3, ($p = 0.0036$)). Decreased calprotectin (µg/ml) (placebo 33.8 ± 23.5, probiotic 20.3 ± 19.3, ($p = 0.003$)).	[195]
2014	L. reuteri ATCC55730 (10^{10} CFU/day)	RCT (Parallel)	6 months	30	Reduced pulmonary exacerbations (odds ratio 0.06 ([95% confidence interval (CI) 0–0.40); number needed to treat 3 (95% CI 2–7), $p < 0.01$. Reduced number of upper respiratory tract infections (odds ratio 0.14 ([95% CI 0–0.96); number needed to treat 6 (95% CI 3–102), $p < 0.05$).	[20]
2014	Lactobacillus GG (6×10^9 CFU/day)	RCT (Parallel)	1 month	10	Reduced calprotectin concentrations from baseline, compared to placebo (164 ± 70 vs. 78 ± 54 µg/g, $p < 0.05$; 251 ± 174 vs. 176 ± 125 µg/g, $p = 0.3$).	[16]
2018	Lactobacillus GG (6×10^9 CFU/day)	RCT (Parallel)	12 months	41	No significant difference in odds of pulmonary exacerbations (OR 0.83; 95% CI 0.38 to 1.82, $p = 0.643$). No significant difference in odds of hospitalisations (OR 1.67; 95% CI 0.75 to 3.72, $p = 0.211$). No significant difference was for body mass index and FEV1.	[193]
2018	FOS + multi strain powder [e] (10^8–10^9 CFU/day each strain)	RCT (Parallel)	90 days	22	No significance difference in FEV1 and nutritional status markers. Patients with Staphylococcus aureus + supplementation had reduced NOx ($p = 0.030$), IL-6 ($p = 0.033$), and IL-8 ($p = 0.009$).	[66]
2018	L. rhamnosus SP1 (DSM 21690) and B. animalis lactis spp. BLC1 (LMG 23512) (10^{10} CFU/day)	RCT (Cross-over)	4 months	31	No significant changes in the clinical parameters (BMI, FEV1%, abdominal pain, exacerbations). Normalization of gut permeability was observed in 13% of patients during probiotic treatment.	[196]

[a] CasenBiotic (CasenFleet) 100 million (108 CFU/day), L. reuteri Protectis (DSM 17938), sweeteners (isomaltose (E-953), xylitol (E-967)), calcium stearate, palmitic acid, citric acid, strawberry aroma as a capsule. [b] VLS3 (Faes Farma) 450 million, B. breve, B. longum, B. infantis, L. acidophilus, L. plantarum, L. paracasei, L. delbrueckii subsp. bulgaricus, S. thermophilus as a powder sachet. [c] Protexin capsule containing L. casei, L. rhamnosus, S. thermophilus, B. breve, L. acidophilus, B. infantis, and L. bulgaricus. [d] Protexin Restor sachet, FOS and a mixture of 1×10^9 CFU/sachet bacteria (L. casei, L. rhamnosus, S. thermophilus, B. breve, L. acidophilus, B. infantis, L. bulgaricus). [e] FOS + multistrain powder (5.5 g), L. paracasei, L. rhamnosus, L. acidophilus, and B. lactis.

From 2016 to 2021, six systematic reviews have attempted to synthesise the expanding evidence for probiotics in CF [15,197–201]. The first review in 2016 [201] examined a total of nine trials with a total of 275 subjects, and found that probiotics were likely to decrease gut dysbiosis and improve gut maturity and function. In 2017, three more reviews were published that were broadened to include evidence on pulmonary exacerbations and quality-of-life indicators [197,199,200]. The latest and most comprehensive systematic review was based on data from the 12 RCTs only [15]. Combined data from four trials (225 participants) found that probiotics may reduce pulmonary exacerbations when administered over a four-to-12-month period mean difference (MD) of −0.32 episodes per participant (95% confidence interval (CI) −0.68 to 0.03; $p = 0.07$)). The 95% confidence intervals included the possibility of both an increased and deceased number of exacerbations. The combined data from four trials (177 participants) also indicated that probiotics may reduce faecal calprotectin, MD −47.4 µg/g (95% CI −93.28 to −1.54; $p = 0.04$). Due to (i) a high risk of bias due to selective reporting; (ii) a high risk of bias due to incomplete outcome data; and (iii) a lack of generalisability, the evidence for these results was evaluated as low certainty [15].

The results from other biomarkers and health outcomes including lung function (forced expiratory volume at one second (FEV_1)% predicted) (five trials, 284 participants); duration of antibiotic therapy (two trials, 127 participants); hospitalisation rates (two trials, 115 participants); height, weight, or body mass index (two trials, 91 participants); and reported health-related quality of life scores (1 trial, 37 participants) did not demonstrate any difference between placebo and treatment groups, (all low-certainty evidence). Only two studies included a microbial analysis, and insufficient data was available to analyse in the systematic review. Likewise, there was insufficient evidence to evaluate gastrointestinal symptoms. The probiotics evaluated in the RCTs were associated with four adverse events, including vomiting, diarrhoea, and allergic reactions [15].

Results from individual trials and systematic reviews have consistently indicated that probiotics are likely to have beneficial effects in CF, especially for inflammation and pulmonary exacerbations. However, all systematic reviews cited a limited amount of low-quality data as a barrier to justifying the inclusion of probiotics in current CF treatment protocols [15,197–201]. Furthermore, a high variation between trial protocols, probiotic formulation, dose, duration of therapy, and clinical outcomes measured make predictions about effective strains and dosages and clinical translation difficult. To address data quality, Coffey et al. [15] recommended multicentre RCTs of at least 12 months duration to best assess the efficacy and safety of probiotics for children and adults with CF.

9. Microbiome Modulation with Prebiotics

Modulation of the microbiome can also be targeted through the administration of generally non-digestible compounds known as prebiotics. Prebiotics provide health benefits by promoting the proliferation of commensal gut species and subsequent production of beneficial metabolites [202]. Prebiotics were first defined in 1995, with an updated definition published in 2017 as "a substrate that is selectively utilized by host microorganisms conferring a health benefit" [203]. Prebiotics occur naturally in foods such as breads, cereals, onions, garlics, and artichokes [204] but are also available as dietary supplements. The most established and well documented prebiotics include inulin, fructo-oligosaccharides (FOS), oligofructose, galacto-oligosaccharides (GOS), and lactulose [205,206]. Other potential prebiotics with expanding evidence of effect include resistant starch, high amylose maize starch (HMAS), glucans, arabinoxylan oligosaccharides, xylooligosaccharides, soybean oligosaccharides, isomalto-oligosaccharides, and pectin [206,207].

9.1. Mechanisms of Action

The underlying hypothesis of prebiotics is that the additional fermentable substrates drive the proliferation of keystone commensal bacteria, and subsequently the production of beneficial metabolites such as SCFA [152]. Commensal bacteria and SCFA metabolites then directly and indirectly improve host health through colonisation inhibition, increased barrier integrity, and immune modulation [183]. Currently, the mechanisms of action postulated for prebiotics are primarily based on in vitro models. Validation of proposed mechanisms and demonstrated effects in human models is limited.

Interaction and modulation of microbiome. Prebiotics in the form of undigestible carbohydrates promote proliferation of beneficial bacteria. As mentioned above, subsequent health benefits are a direct result of increased beneficial bacteria such as colonisation inhibition, or an indirect result of increased beneficial metabolite production such as improved barrier function and immune modulation [183]. In a randomised, double-blind, placebo-controlled clinical trial, a prebiotic intervention with GOS reduced intestinal permeability in obese adults. The probiotic intervention of *Bifidobacterium adolescentis* also reduced intestinal permeability, but interestingly, no synergistic effect was observed when the two were combined [208].

Defence against pathogens. As with probiotics, the modulation of the microbiome through prebiotics results in the generation of organic acids, reducing luminal pH, which inhibits growth of pathogens. The increase in commensal species also reduces nutrient availability for invasive species as described above. There is also evidence to suggest that GOS prebiotics can directly interfere with E. coli adhesion to tissue culture cells [209].

Metabolic effects. The metabolic effects of prebiotics have been the subject of several meta-analyses [210–212]. The evidence suggests that GOS and inulin can reduce high sensitivity C-reactive protein, plasma cholesterol, triglycerides, and fasting plasma insulin associated with obesity and diabetes. The exact mechanisms of action, duration of effects, or results from long-term consumption has not yet been established [211].

Immune modulation. Prebiotic immune modulation is primarily activated through microbial fermentation and subsequent production of SCFA metabolites. However, some prebiotics have been demonstrated to bind directly to some immune cell receptors [183,213]. Immune modulation is not consistent between probiotic categories or conditions, or even within conditions. A RCT with 259 infants concluded that GOS and long-chain FOS administered in formula may regulate immune function in infants, with a 50% reduction in atopic dermatitis, wheezing, and urticaria to when compared to non-prebiotic formula-fed infants [214]. Yet, a subsequent multicentre RCT with 365 infants found that while GOS supplementation altered faecal frequency and consistency, there was no effect on incidence of infection or allergic manifestation during the first year of life [215]. Likewise, conflicting studies in elderly individuals have proposed prebiotic GOS supplementation may have either no effect on immune function [216], or may increase immune function through enhanced phagocytic activity and activity of natural killer cells [217,218].

9.2. Prebiotics in CF

Prebiotics have been combined with probiotics in synbiotic preparations for use in CF probiotic trials [17,66,219]. However, only one study has investigated the exclusive use of prebiotics in CF for GI microbiome modulation [22]. Effective clinical use of prebiotics assumes a selective utilisation of the supplemented substrate by the recipient's microbiota. It is not yet known if the disrupted microbiota in CF that is depleted in key SCFA-producing organisms, has the functional capacity to utilise prebiotic substrates. In a pilot study, Wang et al. [22] used a combination of metagenomic sequencing, invitro fermentation, amplicon sequencing, and metabolomics to investigate the HAMS fermentation capacity of 19 adults with CF and 16 non-CF controls. They demonstrated that despite low abundances of common taxa attributed to fermentation of HMAS (*Faecalibacterium*, *Roseburia* and *Coprococcus*), the production of butyrate and propionate was consistent with healthy control slurries, while the production of acetate was reduced. In the absence of *Faecalibacterium*, the CF SCFA biosynthesis was attributed to *Clostridium ss1*. Importantly, in a subset of CF patients, the presence of HAMS led to enterococcal overgrowth and the accumulation of lactate [22]. Likewise, a murine study found that supplementation with purified prebiotics inulin, fructooligosaccharides, or pectin may result in hepatocellular carcinoma in mice with pre-existing perturbed microbial communities [220]. These results demonstrated the potential for variable responses to prebiotics, dependent on the underlying microbiome [220].

In the absence of further CF-specific research, we examined evidence of prebiotics in GI disorders with an inflammatory link, including ulcerative colitis (UC), Crohn's disease, colorectal cancer (CRC), and chronic respiratory disease including *Psuedomonas aeruginosa* infections, asthma, and emphysema (Table 2). In colitis animal models, a range of prebiotics including 2-fructosyl lactose [221], barley leaf insoluble dietary fibre (BLIDF) [222], psyllium [223], wheat bran [224], and butyrate [225] have been shown to reduce colitis symptoms and inflammatory markers, while increasing bacterial diversity and SCFA levels. Specific inflammatory markers, bacterial species, and SCFA vary between studies and prebiotic type (Table 2). The effects of inulin-type fructins (ITF) at 7.5 g/day (n = 12) or 15 g/day (n = 13) were tested in a human trial with 25 patients with mild/moderate UC [226]. The high-ITF-dose group showed significantly reduced colitis and calprotectin concentrations, and increased butyrate levels. The bacterial species *Bifidobacteriaceae* and *Lachnospiraceae* also increased in the high-dose group, but their abundance was not correlated to improved disease scores. The lack of taxonomic correlation suggested that functional shifts may be more relevant than compositional shifts in UC (Table 2). A RCT involving 140 preoperative patients with CRC investigated the role of prebiotics (fructooligosaccharides, xylooligosaccharides, polydextrose, and resistant dextrin) on immune function and intestinal microbiota. They reported that probiotics led to improved serum immunological markers and abundances of commensal bacterial species before surgeries [227]. Prebiotics did not protect from surgery-related microbial stress observed in both postoperative groups (Table 2).

Table 2. Evidence of prebiotics and dietary effects on microbiome in chronic inflammatory and respiratory disease.

Dietary Component	Study Model	Disease Type	Effect on Disease	Effect on Gut Microbiome	Effect on Host Biomarkers	Ref
			Specific diet			
Low fat, high fibre	Human	Ulcerative colitis (IBD)	↑ QoL IBD questionnaire scores	↑ *Bacteroidetes*, *Faecalibacterium prausnitzii* ↓ *Actinobacteria*	↑ Acetate, tryptophan ↓ Lauric acid	[228]
			Monosaccharides			
High-sugar diet	Mouse	DSS-induced colitis (IBD)	↑ Colitis	↑ *Verruncomicrobiaceae*, *Porphyromonadaceae* ↓ α-Diversity, *Prevotellaceae*, *Lachnospiraceae*, *Anaeroplasmataceae*	↑ Intestinal permeability, proinflammatory cytokines, BMDM reactivity to LPS.	[229]
Artificial sweetener	Mouse	SAMP1/YitFc ileitis (Crohn's disease)	No change	↑ *Proteobacteria*	↑ Ileal myeloperoxidase reactivity	[230]
			Milk oligosaccharides			
GOS	Human crossover	NA	NA	↑ *Bifidobacterium* ↓ *Ruminococcus*, *Synergistes*, *Dehalobacterium*, *Holdemania*	↓ Butyrate (NS), *Bacteroides* predicts OGTT	[231]
pAOS	Mouse	*P. aeruginosa* infection	↑ Bacterial clearance	↑ *Bifidobacterium*, *Sutturella wadsworthia*, *Clostridium* cluster XI	↑ Butyrate, propionate ↑ IFN-γ, t-bet gene, M1 macrophage, IL10 ↓ TNF a, IL-4, gata 3 gene	[232]
2'-Fucosyl lactose	Mouse	IBD	↓ Colitis	↑ *Ruminococcus gnavus* ↓ *Bacteroides acidifaciens*, *Bacteroides vulgatus*	↑ Acetate, propionate, valerate, TGFβ ↓ iNOS, IL-1β, IL-6	[221]
			Plant polysaccharides			
Dietary fibre	Mouse	T-cell-transfer colitis (IBD)	↓ Colitis	No change in microbial load or *Clostridiales* abundance, metabolic changes between high-fibre and low-fibre diets presumed based on butyrate output	↑ Treg cells, caecal and luminal butyrate, Foxp3 histone H3 acetylation	[233]
Dietary fibre	Human, RCT meta-analysis	NA	NA	↑ *Bifidobacterium*, *Lactobacillus* No change in α-diversity	↑ Faecal butyrate FOS and GOS drove microbial shifts	[234]

Table 2. Cont.

Dietary Component	Study Model	Disease Type	Effect on Disease	Effect on Gut Microbiome	Effect on Host Biomarkers	Ref
Dietary fibre	Mouse	Emphysema	↓ Alveolar destruction and inflammation in BALF	↑ Bacteroidetes ↓ Lactobacillaceae, Defluviitaleaceae	↑ SCFA, bile acids, sphingolipids ↓ Macrophages and neutrophils in BALF ↓ mRNA expression of IFN-γ, IL-1β, IL-6, IL-8, IL-18, IRF-5, MMP-12, TNF-α, TGF-β, and cathepsin S	[235]
BLIDF	Mouse	DSS-induced acute colitis (IBD)	Reduced colitis symptoms	↓ Akkermansia ↑ Parasutterella, Alistipes, Erysipelatoclostridium	↑ SCFA, secondary bile acids, claudin-1 ↑ Occludin and mucin 2 expression	[222]
FOS	Human, crossover	NA	NA	↑ Bifidobacterium ↓ Phascolarctobacterium, Enterobacter, Turicibacter, Coprococcus, Salmonella	↓ Butyrate, Bacteroides predicts OGTT	[231]
FOS, XOS, polydextrose, resistant dextrin	Human, RCT	CRC	↓ Inflammation	(Preoperative) ↓ Bacteroides ↑ Bifidobacterium, Enterococcus (Postoperative) ↓ Bacteroides ↑ Enterococcus, Lactococcus, Streptococcus	(Preoperative) ↑ IgG, IgM, transferrin (Postoperative) ↑ IgG, IgA, suppressor/cytotoxic T cells CD3+CD8+, total B lymphocytes	[227]
ITF	Human, RCT	Ulcerative colitis	↑ Remission ↓ Colitis	↑ Bifidobacteriaceae, Lachnospiraceae (not correlated with colitis reduction)	↑ Total SCFA, butyrate ↓ Faecal calprotectin	[226]
Psyllium	Mouse	DSS-induced, T-cell-transfer colitis (IBD)	↓ Colitis	↑ α-Diversity ↓ Microbial density	↑ Butyrate, Treg cells ↓ IL-6, faecal LCN-2, intestinal permeability	[223]
Wheat bran	Pig	NA	↓ Inflammation pathways	↑ Bifidobacterium, Lactobacillis ↓ Escherichia coli	↓ TNF-α, IL-1β, IL-6 and TLRs/MyD88/NF-κB pathways	[224]
SCFA						
Butyrate	Mouse	IBD	↓ Colitis	↑ α-Diversity (NS), Lactobacillaceae, Erysipelotrichaceae ↓ IgA-coated bacteria, Prevotellaceae	↓ TNF, IL-6, infiltration of inflammatory cells in colonic mucosa, acetate	[225]
Dietary fats						
Saturated fats	Mouse	Il10−/−, DSS-induced colitis (IBD)	↑ Colitis	↑ Bacteroidetes, Bilophila wadsworthia ↓ α-Diversity, Firmicutes	↑ TH1 mucosal response due to change in bile acid production	[236]

↑, increased; ↓, decreased; BALF, bronchoalveolar lavage fluid; BMDM, bone-marrow-derived macrophages; Bregs, regulatory B cells; CRP, C-reactive protein; DSS, dextran–sulfate–sodium; FOS, fructooligosaccharides; GOS, galactooligosaccharides; IBD, inflammatory bowel disease; ILA, indole-3-lactic acid; iNOS, inducible nitric oxide synthase; ITF, inulin-type fructans; LCN-2, lipocalin-2; LPS, lipopolysaccharide; NA, not applicable; NS, not significant; RCT, randomized controlled trial; SCFA, short-chain fatty acids; TGFβ, transforming growth factor-β; TH, T helper; TLR, Toll-like receptor; TNF, tumour necrosis factor; QoL, Quality of life; OGTT, oral glucose tolerance test; Treg, regulatory.

In 2015, a mouse model study investigated the effects of dietary-pectin-derived acidic oligosaccharides (pAOS) on *Pseudomonas (P) aeruginosa*, and found that pAOS may limit the number and severity of pulmonary exacerbations chronically infected with *P. aeruginosa* [232] (Table 2). These results are highly relevant to individuals with CF, but have not been validated in human trials. In another small RCT study, either a placebo or a galactooligosaccharides (GOS) preparation highly selective to Bifidobacterium (B-GOS) was given to 10 adults with asthma-associated, hyperpnoea-induced bronchoconstriction (HIB) and 8 adult controls. Pulmonary function remained unchanged in the control group, but in the HIB group, FEV1 was attenuated by 40%, and the baseline chemokine CC ligand and TNF-a was reduced after B-GOS supplementation [237] (Table 2). As with probiotics, the large number and source of prebiotics limits the ability to effectively compare treatments across studies and conditions.

9.3. Microbiome Modulation with Diet

There is an increasingly apparent link between diet, microbiome modulation, and host health [238,239]. In the first few years of life, the interactions between diet and the microbiome are especially relevant. The oligosaccharides present in breast milk encourage the colonization of *Bifidobacterium* spp. The subsequent metabolites produced by *Bifidobacterium* spp. support the expansion of the microbiome through cross-feeding, and promote immune tolerance to other commensal bacteria [240,241]. The progression of the microbiome develops alongside increases in diet diversity, with the introduction of solid food triggering a rapid expansion in the bacterial community and the subsequent quantity and variety of metabolites. Metabolites associated with solid food ingestion, primarily butyrate, have been demonstrated to drive the maturation of the mucosal barrier in Caco-2 cells [242], which is critical to colonization inhibition of pathogens [243]. As mentioned earlier, iatrogenic factors such as antibiotic use can disrupt microbiome progression [41]. More broadly, early life microbiome disruption has been implicated in the development of autoimmune disease in mouse models [244], and has been associated with lasting metabolic and autoimmune disease consequences in observational studies [245–248].

The progression of the microbiome continues until it establishes near-adult levels of diversity at 3–5 years of age. Once established, the microbiome is more resistant to disruption [249]. However, diet remains a key modulator, with accumulating evidence of the role of dietary components in local inflammation, intestinal barrier function, and host immune dysregulation [220]. High-fat diets promote the translocation of certain bacteria by enhancing intestinal permeability and preferentially increasing the relative abundances of lipopolysaccharide-bearing bacteria [82,83,250–253]. Dietary fibres microbially fermented in the gut lead to the production of short-chain fatty acid metabolites, which have demonstrated roles in immune regulation [254–256], maintenance of epithelial barrier [223,257], and microbiome modification [258]. Low-fibre diets also have long-term implications for cancer [259] and metabolic and autoimmune diseases [260,261].

9.4. Diet in CF

Consistent with the general population, early-life microbiome development in CF is correlated to infant breastfeeding [262] and the initiation of solid foods [133]. Infant breastfeeding and solid food intake are further linked to the respiratory microbiome [262] and health outcome indicators, supporting the concept of a gut–lung axis in CF [133]. The progression of the CF microbiome is disrupted in early life [53,54,59], characterized by reduced taxonomic and functional profiles (dysbiosis) [7] that persist into adulthood [12]. No dietary fibre interventions have been trialled in CF, but epidemiological studies and preclinical and clinical trials support the potential of fibre to modulate the microbiome structure [234] and improve function in chronic gastrointestinal [263] and respiratory conditions [235,264].

A low-fat, high-fibre diet in individuals with ulcerative colitis improved the quality of life (QoL) as quantified through the QoL IBD survey. The intervention led to increased acetate and tryptophan levels and modulated the microbiome, increasing the abundance of *Bacteroidetes* and *Faecalibacterium* [228] (Table 2). A high-fibre diet also reduced inflammation and attenuated pathological changes associated with emphysema in an emphysema mouse model exposed to cigarette smoke. Alveolar destruction and inflammatory cytokines in bronchoalveolar lavage fluid (BALF) were reduced, while SCFA were increased [235] (Table 2). In contrast, saturated fat decreased microbial diversity and increased colitis severity and the TH1 mucosal response in a DSS-induced colitis mouse model [236] (Table 2). In 37 adults with asthma, high-fat intake was also demonstrated to increase inflammation and attenuate both the duration and magnitude of recovery from an aerosol-administered bronchodilator. However, the potential role of the microbiome was not investigated [265]. A range of micronutrients and antioxidant supplements have been trialled in CF to ameliorate fat-soluble vitamin deficiencies, altered fatty-acid synthesis, and increased oxidative stress [266]. In 2020, a meta-analysis concluded that the benefits sometimes observed across 8 antioxidant and 15 essential-fatty-acid supplementation studies was not consistent enough to recommend their routine use in CF [266]. With the exception of an exploratory study correlating vitamin D insufficiency with increases in potential pathogenic species [267], the effect of these supplementations on the microbiome have not been widely explored.

10. Considerations for Clinical Application and Future Studies

A survey from a CF clinic in the USA found that 60% of CF patients currently use some type of probiotic [268], despite the majority of probiotics available for sale having little to no evidence to support effectiveness, dose, or disease specificity [269]. Conditional use of specific probiotic strains in (non-CF) gastrointestinal disorders has been recommended for prevention of antibiotic-associated *C. difficile* infections, pouchitis, and prevention of necrotizing enterocolitis [160]. While these conditions may impact on individuals with CF, recommendations have not been validated specifically in CF. Evidence from CF-specific RCTs suggest probiotics are likely to improve GI inflammation and reduce the incidence of pulmonary exacerbations. Yet, the highly variable, low-quality data has been insufficient to determine ideal strains, dosage, and treatment duration, constraining the implementation of probiotics into routine CF care.

The evidence for prebiotics lags behind that for probiotics. There was only one clinical trial for the use of prebiotics in CF, and the results were mixed. Some individuals demonstrated successful microbial modulation and an increased production of butyrate and propionate. However, a subset of CF reactions exhibited enterococcal overgrowth, resulting in lactate accumulation and reduced SCFA biosynthesis [22]. The altered microbiome and predisposition to pathogenic overgrowth in CF highlights the need for standardised preparations of well-characterised prebiotics to be investigated specifically within the CF population to adequately evaluate safety and efficacy.

The increased energy needs of individuals with CF are currently being meet through "energy-dense, nutrient-poor" diets with excess saturated fats and inadequate fibre intakes [113]. As previously discussed, high-fibre, low-fat diets are associated with improved inflammation, immune modulation, and gut barrier function outcomes. While simply adding more dietary fibre to existing CF diets is tantalizing, interventional dietary trials have demonstrated that increasing dietary fibre intake does not necessarily translate to increased SCFA production or improvements in disease outcomes [220,270]. This is pertinent to members of the CF population, who are likely to have altered SCFA-generating pathways [7,56,60]. High-quality randomized trials with well-defined dietary components are essential to provide justification for modulating microbiome–host effects through diet. Considering the uptake of highly effective modulator therapies, a re-evaluation of dietary recommendations with a focus on diet quality and individual energy requirements is also recommended.

As outlined in previous reviews, there is a need for large, well-designed, longitudinal and multicentred clinical trials to effectively evaluate the safety and efficacy of probiotics in CF. While this is also true for prebiotics, there is a paucity of prebiotics research in CF and a plethora of substrates with prebiotic potential. The use of organoids, cell lines, and or animal models may be an economic option to demonstrate beneficial effects and mechanisms of action across a variety of compounds before proceeding to clinical trials. Likewise, there are a variety of dietary interventions and supplements that may have beneficial host–microbiome effects. High-quality, randomized studies with well-defined compounds are needed to evaluate safety and efficacy in CF before dietary interventions or supplements can be utilised to modulate the CF microbiome.

11. Conclusions

Although the exact mechanisms are not yet fully elucidated, the host–microbiome interactions in CF are critical to the incidence of GI inflammation and disease.

Targeted diet-based therapies provide an opportunity to modulate the altered CF microbiome to counter the early disruption to microbiome progression, and could transform GI and respiratory disease outcomes. The expansion of metagenomic, proteomic, and transcriptomic analyses continues to illuminate CF specific taxonomic and functional alterations. This knowledge will be critical to the development of next-generation precision probiotics and prebiotics. For now, there is insufficient evidence to support the safe and effective use of prebiotics in CF, but probiotics and a re-evaluation of the CF diet may be beneficial. Critically, there is a need for personalised approaches that understand an individual's baseline microbiota and can manage potential microbiome-modulating therapies alongside existing medications and complex nutritional needs.

Author Contributions: Conceptualization, J.M.v.D. and C.Y.O.; writing-original draft preparation, J.M.v.D. and R.Y.T.; writing-review and editing J.M.v.D. and C.Y.O.; visualization J.M.v.D.; supervision C.Y.O. All authors have read and agreed to the published version of the manuscript.

Funding: This research received no external funding.

Institutional Review Board Statement: Not applicable.

Data Availability Statement: Not applicable.

Conflicts of Interest: The authors declare no conflict of interest.

References

1. Riordan, J.R.; Rommens, J.M.; Kerem, B.-S.; Alon, N.; Rozmahel, R.; Grzelczak, Z.; Zielenski, J.; Lok, S.; Plavsic, N.; Chou, J.-L.; et al. Identification of the cystic fibrosis gene: Cloning and characterization of complementary DNA. *Science* **1989**, *245*, 1066–1073. [CrossRef] [PubMed]
2. Riordan, J.R. CFTR function and prospects for therapy. *Annu. Rev. Biochem.* **2008**, *77*, 701–726. [CrossRef] [PubMed]
3. Rowbotham, N.J.; Smith, S.; A Leighton, P.; Rayner, O.C.; Gathercole, K.; Elliott, Z.C.; Nash, E.F.; Daniels, T.; A Duff, A.J.; Collins, S.; et al. The top 10 research priorities in cystic fibrosis developed by a partnership between people with CF and healthcare providers. *Thorax* **2018**, *73*, 388–390. [CrossRef]
4. Tabori, H.; Arnold, C.; Jaudszus, A.; Mentzel, H.-J.; Renz, D.M.; Reinsch, S.; Lorenz, M.; Michl, R.; Gerber, A.; Lehmann, T.; et al. Abdominal symptoms in cystic fibrosis and their relation to genotype, history, clinical and laboratory findings. *PLoS ONE* **2017**, *12*, e0174463. [CrossRef]
5. Ooi, C.Y.; Durie, P.R. Cystic fibrosis from the gastroenterologist's perspective. *Nat. Rev. Gastroenterol. Hepatol.* **2016**, *13*, 175–185. [CrossRef]
6. Ooi, C.Y.; Pang, T.; Leach, S.T.; Katz, T.; Day, A.S.; Jaffe, A. Fecal Human β-Defensin 2 in Children with Cystic Fibrosis: Is There a Diminished Intestinal Innate Immune Response? *Dig. Dis. Sci.* **2015**, *60*, 2946–2952. [CrossRef]
7. Coffey, M.J.; Nielsen, S.; Wemheuer, B.; Kaakoush, N.O.; Garg, M.; Needham, B.; Pickford, R.; Jaffe, A.; Thomas, T.; Ooi, C.Y. Gut Microbiota in Children With Cystic Fibrosis: A Taxonomic and Functional Dysbiosis. *Sci. Rep.* **2019**, *9*, 18593. [CrossRef] [PubMed]
8. Garg, M.; Ooi, C.Y. The Enigmatic Gut in Cystic Fibrosis: Linking Inflammation, Dysbiosis, and the Increased Risk of Malignancy. *Curr. Gastroenterol. Rep.* **2017**, *19*, 6. [CrossRef] [PubMed]
9. Maisonneuve, P.; Marshall, B.C.; Knapp, E.A.; Lowenfels, A.B. Cancer risk in cystic fibrosis: A 20-year nationwide study from the United States. *J. Natl. Cancer Inst.* **2013**, *105*, 122–129. [CrossRef]
10. Maisonneuve, P.; Fitzsimmons, S.C.; Neglia, J.; Campbell, P.W.; Lowenfels, A.B. Cancer risk in nontransplanted and transplanted cystic fibrosis patients: A 10-year study. *J. Natl. Cancer Inst.* **2003**, *95*, 381–387. [CrossRef]
11. Meyer, K.C.; Francois, M.L.; Thomas, H.K.; Radford, K.L.; Hawes, D.S.; Mack, T.L.; Cornwell, R.D.; Maloney, J.D.; De Oliveira, N.C. Colon cancer in lung transplant recipients with CF: Increased risk and results of screening. *J. Cyst. Fibros.* **2011**, *10*, 366–369. [CrossRef]
12. Burke, D.; Fouhy, F.; Harrison, M.J.; Rea, M.C.; Cotter, P.D.; O'Sullivan, O.; Stanton, C.; Hill, C.; Shanahan, F.; Plant, B.J.; et al. The altered gut microbiota in adults with cystic fibrosis. *BMC Microbiol.* **2017**, *17*, 58. [CrossRef]
13. Dhaliwal, J.; Leach, S.; Katz, T.; Nahidi, L.; Pang, T.; Lee, J.; Strachan, R.; Day, A.S.; Jaffe, A.; Ooi, C.Y. Intestinal inflammation and impact on growth in children with cystic fibrosis. *J. Pediatr. Gastroenterol. Nutr.* **2015**, *60*, 521–526. [CrossRef]
14. Pang, T.; Leach, S.T.; Katz, T.; Jaffe, A.; Day, A.S.; Ooi, C.Y. Elevated fecal M2-pyruvate kinase in children with cystic fibrosis: A clue to the increased risk of intestinal malignancy in adulthood? *J. Gastroenterol. Hepatol.* **2015**, *30*, 866–871. [CrossRef]
15. Coffey, M.J.; Garg, M.; Homaira, N.; Jaffe, A.; Ooi, C.Y. A systematic cochrane review of probiotics for people with cystic fibrosis. *Paediatr. Respir. Rev.* **2021**, *39*, 61–64. [CrossRef]
16. Bruzzese, E.; Callegari, M.L.; Raia, V.; Viscovo, S.; Scotto, R.; Ferrari, S.; Morelli, L.; Buccigrossi, V.; Vecchio, A.L.; Ruberto, E.; et al. Disrupted intestinal microbiota and intestinal inflammation in children with cystic fibrosis and its restoration with Lactobacillus GG: A randomised clinical trial. *PLoS ONE* **2014**, *9*, e87796. [CrossRef] [PubMed]
17. Fallahi, G.; Motamed, F.; Yousefi, A.; Shafieyoun, A.; Najafi, M.; Khodadad, A.; Farhmand, F.; Ahmadvand, A.; Rezaei, N. The effect of probiotics on fecal calprotectin in patients with cystic fibrosis. *Turk. J. Pediatr.* **2013**, *55*, 475–478.

18. Del Campo, R.; Garriga, M.; Agrimbau, J.; Lamas, A.; Maiz, L.; Canton, R.; Suarez, L. Improvement of intestinal comfort in cystic fibrosis patients after probiotics consumption. *J. Cyst. Fibros.* **2009**, *8*, S89. [CrossRef]
19. Bruzzese, E.; Raia, V.; Spagnuolo, M.I.; Volpicelli, M.; De Marco, G.; Maiuri, L.; Guarino, A. Effect of Lactobacillus GG supplementation on pulmonary exacerbations in patients with cystic fibrosis: A pilot study. *Clin. Nutr.* **2007**, *26*, 322–328. [CrossRef]
20. Di Nardo, G.; Oliva, S.; Menichella, A.; Pistelli, R.; De Biase, R.V.; Patriarchi, F.; Cucchiara, S.; Stronati, L. Lactobacillus reuteri ATCC55730 in cystic fibrosis. *J. Pediatr. Gastroenterol. Nutr.* **2014**, *58*, 81–86. [CrossRef]
21. Jafari, S.-A.; Mehdizadeh-Hakkak, A.; Kianifar, H.-R.; Hebrani, P.; Ahanchian, H.; Abbasnejad, E. Effects of probiotics on quality of life in children with cystic fibrosis: A randomized controlled trial. *Iran. J. Pediatrics* **2013**, *23*, 669.
22. Wang, Y.; Leong, L.E.; Keating, R.L.; Kanno, T.; Abell, G.C.; Mobegi, F.M.; Choo, J.M.; Wesselingh, S.L.; Mason, A.J.; Burr, L.D.; et al. Opportunistic bacteria confer the ability to ferment prebiotic starch in the adult cystic fibrosis gut. *Gut Microbes* **2019**, *10*, 367–381. [CrossRef] [PubMed]
23. Venkatasubramanian, J.; Ao, M.; Rao, M.C. Ion transport in the small intestine. *Curr. Opin. Gastroenterol.* **2010**, *26*, 123–128. [CrossRef] [PubMed]
24. De Lisle, R.C.; Borowitz, D. The cystic fibrosis intestine. *Cold Spring Harb. Perspect. Med.* **2013**, *3*, a009753. [CrossRef] [PubMed]
25. Liou, T.G. The Clinical Biology of Cystic Fibrosis Transmembrane Regulator Protein: Its Role and Function in Extrapulmonary Disease. *Chest* **2019**, *155*, 605–616. [CrossRef] [PubMed]
26. Scott, P.; Anderson, K.; Singhania, M.; Cormier, R. Cystic Fibrosis, CFTR, and Colorectal Cancer. *Int. J. Mol. Sci.* **2020**, *21*, 2891. [CrossRef]
27. Xue, R.; Gu, H.; Qiu, Y.; Guo, Y.; Korteweg, C.; Huang, J.; Gu, J. Expression of Cystic Fibrosis Transmembrane Conductance Regulator in Ganglia of Human Gastrointestinal Tract. *Sci. Rep.* **2016**, *6*, 30926. [CrossRef] [PubMed]
28. Yeh, K.M.; Johansson, O.; Le, H.; Rao, K.; Markus, I.; Perera, D.S.; Lubowski, D.Z.; King, D.W.; Zhang, L.; Chen, H.; et al. Cystic fibrosis transmembrane conductance regulator modulates enteric cholinergic activities and is abnormally expressed in the enteric ganglia of patients with slow transit constipation. *J. Gastroenterol.* **2019**, *54*, 994–1006. [CrossRef]
29. Saxby, N.; Painter, C.; Kench, A.; King, S.; Crowder, T.; van der Haak, N.; The Australian and New Zealand Cystic Fibrosis Nutrition Guideline Authorship Group. *Nutrition Guidelines for Cystic Fibrosis in Australia and New Zealand*; Bell, S.C., Ed.; Thoracic Society of Australia and New Zealand: Sydney, Australia, 2017.
30. Li, L.; Somerset, S. Digestive system dysfunction in cystic fibrosis: Challenges for nutrition therapy. *Dig. Liver Dis.* **2014**, *46*, 865–874. [CrossRef]
31. Durie, P.R.; Forstner, G.G. Pathophysiology of the exocrine pancreas in cystic fibrosis. *J. R. Soc. Med.* **1989**, *82* (Suppl. 16), 2–10.
32. Ooi, C.Y.; Durie, P.R. Cystic fibrosis transmembrane conductance regulator (CFTR) gene mutations in pancreatitis. *J. Cyst. Fibros.* **2012**, *11*, 355–362. [CrossRef]
33. Ooi, C.Y.; Dorfman, R.; Cipolli, M.; Gonska, T.; Castellani, C.; Keenan, K.; Freedman, S.D.; Zielenski, J.; Berthiaume, Y.; Corey, M.; et al. Type of CFTR mutation determines risk of pancreatitis in patients with cystic fibrosis. *Gastroenterology* **2011**, *140*, 153–161. [CrossRef]
34. Wilschanski, M.; Durie, P.R. Patterns of GI disease in adulthood associated with mutations in the CFTR gene. *Gut* **2007**, *56*, 1153–1163. [CrossRef] [PubMed]
35. Jaclyn, R.B.; Kenneth, J.F.; Simon, C.L.; Rhonda, G.P.; Scott, C.B.; Billy, B.; Giuseppe, C.; Carlo, C.; Marco, C.; Carla, C.; et al. Genetic modifiers of liver disease in cystic fibrosis. *JAMA* **2009**, *302*, 1076–1083. [CrossRef]
36. Gelfond, D.; Borowitz, D. Gastrointestinal complications of cystic fibrosis. *Clin. Gastroenterol. Hepatol.* **2013**, *11*, 333–342, quiz e30-1. [CrossRef] [PubMed]
37. Ahmed, I.; Roy, B.C.; Khan, S.A.; Septer, S.; Umar, S. Microbiome, Metabolome and Inflammatory Bowel Disease. *Microorganisms* **2016**, *4*, 20. [CrossRef]
38. Clemente, J.C.; Manasson, J.; Scher, J.U. The role of the gut microbiome in systemic inflammatory disease. *BMJ* **2018**, *360*, j5145. [CrossRef] [PubMed]
39. Rothschild, D.; Weissbrod, O.; Barkan, E.; Kurilshikov, A.; Korem, T.; Zeevi, D.; Costea, P.I.; Godneva, A.; Kalka, I.N.; Bar, N.; et al. Environment dominates over host genetics in shaping human gut microbiota. *Nature* **2018**, *555*, 210–215. [CrossRef] [PubMed]
40. Bäckhed, F.; Roswall, J.; Peng, Y.; Feng, Q.; Jia, H.; Kovatcheva-Datchary, P.; Li, Y.; Xia, Y.; Xie, H.; Zhong, H.; et al. Dynamics and Stabilization of the Human Gut Microbiome during the First Year of Life. *Cell Host Microbe* **2015**, *17*, 690–703. [CrossRef]
41. Bokulich, N.A.; Chung, J.; Battaglia, T.; Henderson, N.; Jay, M.; Li, H.; Lieber, A.D.; Wu, F.; Perez-Perez, G.I.; Chen, Y.; et al. Antibiotics, birth mode, and diet shape microbiome maturation during early life. *Sci. Transl. Med.* **2016**, *8*, 343ra82. [CrossRef]
42. David, L.A.; Maurice, C.F.; Carmody, R.N.; Gootenberg, D.B.; Button, J.E.; Wolfe, B.E.; Ling, A.V.; Devlin, A.S.; Varma, Y.; Fischbach, M.A.; et al. Diet rapidly and reproducibly alters the human gut microbiome. *Nature* **2014**, *505*, 559–563. [CrossRef] [PubMed]
43. Wu, G.D.; Chen, J.; Hoffmann, C.; Bittinger, K.; Chen, Y.-Y.; Keilbaugh, S.A.; Bewtra, M.; Knights, D.; Walters, W.A.; Knight, R.; et al. Linking long-term dietary patterns with gut microbial enterotypes. *Science* **2011**, *334*, 105–108. [CrossRef] [PubMed]
44. Jakobsson, H.E.; Jernberg, C.; Andersson, A.F.; Sjölund-Karlsson, M.; Jansson, J.K.; Engstrand, L. Short-term antibiotic treatment has differing long-term impacts on the human throat and gut microbiome. *PLoS ONE* **2010**, *5*, e9836. [CrossRef]

45. Falony, G.; Joossens, M.; Vieira-Silva, S.; Wang, J.; Darzi, Y.; Faust, K.; Kurilshikov, A.; Bonder, M.J.; Valles-Colomer, M.; Vandeputte, D.; et al. Population-level analysis of gut microbiome variation. *Science* **2016**, *352*, 560–564. [CrossRef] [PubMed]
46. Jackson, M.A.; Goodrich, J.K.; Maxan, M.-E.; Freedberg, D.E.; Abrams, J.A.; Poole, A.; Sutter, J.L.; Welter, D.; Ley, R.; Bell, J.; et al. Proton pump inhibitors alter the composition of the gut microbiota. *Gut* **2016**, *65*, 749–756. [CrossRef]
47. Sommer, F.; Anderson, J.M.; Bharti, R.; Raes, J.; Rosenstiel, P. The resilience of the intestinal microbiota influences health and disease. *Nat. Rev. Microbiol.* **2017**, *15*, 630–638. [CrossRef]
48. Valdes, A.; Walter, J.; Segal, E.; Spector, T.D. Role of the gut microbiota in nutrition and health. *BMJ* **2018**, *361*, k2179. [CrossRef]
49. Kho, Z.Y.; Lal, S.K. The Human Gut Microbiome—A Potential Controller of Wellness and Disease. *Front. Microbiol.* **2018**, *9*, 1835. [CrossRef]
50. Thursby, E.; Juge, N. Introduction to the human gut microbiota. *Biochem. J.* **2017**, *474*, 1823–1836. [CrossRef] [PubMed]
51. Cho, I.; Blaser, M.J. The human microbiome: At the interface of health and disease. *Nat. Rev. Genet.* **2012**, *13*, 260–270. [CrossRef]
52. Miragoli, F.; Federici, S.; Ferrari, S.; Minuti, A.; Rebecchi, A.; Bruzzese, E.; Buccigrossi, V.; Guarino, A.; Callegari, M.L. Impact of cystic fibrosis disease on archaea and bacteria composition of gut microbiota. *FEMS Microbiol. Ecol.* **2017**, *93*, fiw230. [CrossRef] [PubMed]
53. Vernocchi, P.; Del Chierico, F.; Russo, A.; Majo, F.; Rossitto, M.; Valerio, M.; Casadei, L.; La Storia, A.; De Filippis, F.; Rizzo, C.; et al. Gut microbiota signatures in cystic fibrosis: Loss of host CFTR function drives the microbiota enterophenotype. *PLoS ONE* **2018**, *13*, e0208171. [CrossRef] [PubMed]
54. Nielsen, S.; Needham, B.; Leach, S.T.; Day, A.S.; Jaffe, A.; Thomas, T.; Ooi, C. Disrupted progression of the intestinal microbiota with age in children with cystic fibrosis. *Sci. Rep.* **2016**, *6*, 24857. [CrossRef] [PubMed]
55. Hayden, H.S.; Eng, A.; Pope, C.E.; Brittnacher, M.J.; Vo, A.T.; Weiss, E.J.; Hager, K.R.; Martin, B.D.; Leung, D.H.; Heltshe, S.L.; et al. Fecal dysbiosis in infants with cystic fibrosis is associated with early linear growth failure. *Nat. Med.* **2020**, *26*, 215–221. [CrossRef] [PubMed]
56. Eng, A.; Hayden, H.S.; Pope, C.E.; Brittnacher, M.J.; Vo, A.T.; Weiss, E.J.; Hager, K.R.; Leung, D.H.; Heltshe, S.L.; Raftery, D.; et al. Infants with cystic fibrosis have altered fecal functional capacities with potential clinical and metabolic consequences. *BMC Microbiol.* **2021**, *21*, 247. [CrossRef]
57. Antosca, K.M.; Chernikova, D.A.; Price, C.E.; Ruoff, K.L.; Li, K.; Guill, M.F.; Sontag, N.R.; Morrison, H.G.; Hao, S.; Drumm, M.L.; et al. Altered Stool Microbiota of Infants with Cystic Fibrosis Shows a Reduction in Genera Associated with Immune Programming from Birth. *J. Bacteriol.* **2019**, *201*, e00274-19. [CrossRef] [PubMed]
58. Ooi, C.Y.; Syed, S.A.; Rossi, L.; Garg, M.; Needham, B.; Avolio, J.; Young, K.; Surette, M.G.; Gonska, T. Impact of CFTR modulation with Ivacaftor on Gut Microbiota and Intestinal Inflammation. *Sci. Rep.* **2018**, *8*, 17834. [CrossRef] [PubMed]
59. Manor, O.; Levy, R.; Pope, C.E.; Hayden, H.S.; Brittnacher, M.J.; Carr, R.; Radey, M.C.; Hager, K.R.; Heltshe, S.L.; Ramsey, B.W.; et al. Metagenomic evidence for taxonomic dysbiosis and functional imbalance in the gastrointestinal tracts of children with cystic fibrosis. *Sci. Rep.* **2016**, *6*, 22493. [CrossRef]
60. Thavamani, A.; Salem, I.; Sferra, T.; Sankararaman, S. Impact of Altered Gut Microbiota and Its Metabolites in Cystic Fibrosis. *Metabolites* **2021**, *11*, 123. [CrossRef] [PubMed]
61. Hoffman, L.R.; Pope, C.E.; Hayden, H.S.; Heltshe, S.; Levy, R.; McNamara, S.; Jacobs, M.A.; Rohmer, L.; Radey, M.; Ramsey, B.W.; et al. Escherichia coli dysbiosis correlates with gastrointestinal dysfunction in children with cystic fibrosis. *Clin. Infect. Dis.* **2014**, *58*, 396–399. [CrossRef]
62. Meeker, S.M.; Mears, K.S.; Sangwan, N.; Brittnacher, M.J.; Weiss, E.J.; Treuting, P.M.; Tolley, N.; Pope, C.E.; Hager, K.R.; Vo, A.T.; et al. CFTR dysregulation drives active selection of the gut microbiome. *PLoS Pathog.* **2020**, *16*, e1008251. [CrossRef]
63. Matamouros, S.; Hayden, H.S.; Hager, K.R.; Brittnacher, M.J.; Lachance, K.; Weiss, E.J.; Pope, C.E.; Imhaus, A.-F.; McNally, C.P.; Borenstein, E.; et al. Adaptation of commensal proliferating Escherichia coli to the intestinal tract of young children with cystic fibrosis. *Proc. Natl. Acad. Sci. USA* **2018**, *115*, 1605–1610. [CrossRef] [PubMed]
64. Kristensen, M.; Prevaes, S.M.; Kalkman, G.; Tramper-Stranders, G.A.; Hasrat, R.; de Winter-de Groot, K.M.; Janssens, H.M.; Tiddens, H.A.; van Westreenen, M.; Sanders, E.A.; et al. Development of the gut microbiota in early life: The impact of cystic fibrosis and antibiotic treatment. *J. Cyst. Fibros.* **2020**, *19*, 553–561. [CrossRef] [PubMed]
65. Loman, B.R.; Shrestha, C.L.; Thompson, R.; Groner, J.A.; Mejias, A.; Ruoff, K.L.; O'Toole, G.A.; Bailey, M.T.; Kopp, B.T. Age and environmental exposures influence the fecal bacteriome of young children with cystic fibrosis. *Pediatr. Pulmonol.* **2020**, *55*, 1661–1670. [CrossRef] [PubMed]
66. De Freitas, M.B.; Moreira, E.A.M.; Tomio, C.; Moreno, Y.M.F.; Daltoé, F.; Barbosa, E.; Neto, N.L.; Buccigrossi, V.; Guarino, A. Altered intestinal microbiota composition, antibiotic therapy and intestinal inflammation in children and adolescents with cystic fibrosis. *PLoS ONE* **2018**, *13*, e0198457. [CrossRef] [PubMed]
67. Duytschaever, G.; Huys, G.; Bekaert, M.; Boulanger, L.; De Boeck, K.; Vandamme, P. Dysbiosis of bifidobacteria and Clostridium cluster XIVa in the cystic fibrosis fecal microbiota. *J. Cyst. Fibros.* **2013**, *12*, 206–215. [CrossRef]
68. Adriaanse, M.P.M.; Van Der Sande, L.J.T.M.; Neucker, A.M.V.D.; Menheere, P.P.C.A.; Dompeling, E.; Buurman, W.A.; Vreugdenhil, A.C.E. Evidence for a Cystic Fibrosis Enteropathy. *PLoS ONE* **2015**, *10*, e0138062. [CrossRef]
69. Ellemunter, H.; Engelhardt, A.; Schüller, K.; Steinkamp, G. Fecal Calprotectin in Cystic Fibrosis and Its Relation to Disease Parameters: A Longitudinal Analysis for 12 Years. *J. Pediatr. Gastroenterol. Nutr.* **2017**, *65*, 438–442. [CrossRef]

70. Enaud, R.; Hooks, K.B.; Barre, A.; Barnetche, T.; Hubert, C.; Massot, M.; Bazin, T.; Clouzeau, H.; Bui, S.; Fayon, M.; et al. Intestinal Inflammation in Children with Cystic Fibrosis Is Associated with Crohn's-Like Microbiota Disturbances. *J. Clin. Med.* **2019**, *8*, 645. [CrossRef]
71. Flass, T.; Tong, S.; Frank, D.N.; Wagner, B.; Robertson, C.; Kotter, C.V.; Sokol, R.J.; Zemanick, E.; Accurso, F.; Hoffenberg, E.; et al. Intestinal lesions are associated with altered intestinal microbiome and are more frequent in children and young adults with cystic fibrosis and cirrhosis. *PLoS ONE* **2015**, *10*, e0116967. [CrossRef]
72. Parisi, G.F.; Papale, M.; Rotolo, N.; Aloisio, D.; Tardino, L.; Scuderi, M.G.; Di Benedetto, V.; Nenna, R.; Midulla, F.; Leonardi, S. Severe disease in Cystic Fibrosis and fecal calprotectin levels. *Immunobiology* **2017**, *222*, 582–586. [CrossRef]
73. Rumman, N.; Sultan, M.; El-Chammas, K.; Goh, V.; Salzman, N.; Quintero, D.; Werlin, S. Calprotectin in cystic fibrosis. *BMC Pediatr.* **2014**, *14*, 133. [CrossRef]
74. Smyth, R.L.; Croft, N.M.; O'Hea, U.; Marshall, T.G.; Ferguson, A. Intestinal inflammation in cystic fibrosis. *Arch. Dis. Child.* **2000**, *82*, 394–399. [CrossRef]
75. Werlin, S.L.; Benuri-Silbiger, I.; Kerem, E.; Adler, S.N.; Goldin, E.; Zimmerman, J.; Malka, N.; Cohen, L.; Armoni, S.; Yatzkan-Israelit, Y.; et al. Evidence of intestinal inflammation in patients with cystic fibrosis. *J. Pediatr. Gastroenterol. Nutr.* **2010**, *51*, 304–308. [CrossRef] [PubMed]
76. Brecelj, J.; Zidar, N.; Jeruc, J.; Orel, R. Morphological and Functional Assessment of Oesophageal Mucosa Integrity in Children With Cystic Fibrosis. *J. Pediatr. Gastroenterol. Nutr.* **2016**, *62*, 757–764. [CrossRef] [PubMed]
77. Munck, A. Cystic fibrosis: Evidence for gut inflammation. *Int. J. Biochem. Cell Biol.* **2014**, *52*, 180–183. [CrossRef]
78. Lisle, R.C.D.; Roach, E.; Jansson, K. Effects of laxative and N-acetylcysteine on mucus accumulation, bacterial load, transit, and inflammation in the cystic fibrosis mouse small intestine. *Am. J. Physiol. -Gastrointest. Liver Physiol.* **2007**, *293*, G577–G584. [CrossRef]
79. Vij, N.; Mazur, S.; Zeitlin, P.L. CFTR is a negative regulator of NFkappaB mediated innate immune response. *PLoS ONE* **2009**, *4*, e4664. [CrossRef]
80. Henen, S.; Denton, C.; Teckman, J.; Borowitz, D.; Patel, D. Review of Gastrointestinal Motility in Cystic Fibrosis. *J. Cyst. Fibros.* **2021**, *20*, 578–585. [CrossRef] [PubMed]
81. Debray, M.; El Mourabit, H.; Merabtene, F.; Brot, L.; Ulveling, D.; Chrétien, Y.; Rainteau, M.; Moszer, I.; Wendum, M.; Sokol, H.; et al. Diet-Induced Dysbiosis and Genetic Background Synergize With Cystic Fibrosis Transmembrane Conductance Regulator Deficiency to Promote Cholangiopathy in Mice. *Hepatol. Commun.* **2018**, *2*, 1533–1549. [CrossRef]
82. Kim, K.-A.; Gu, W.; Lee, I.-A.; Joh, E.-H.; Kim, D.-H. High fat diet-induced gut microbiota exacerbates inflammation and obesity in mice via the TLR4 signaling pathway. *PLoS ONE* **2012**, *7*, e47713. [CrossRef]
83. Crawford, M.; Whisner, C.; Al-Nakkash, L.; Sweazea, K.L. Six-Week High-Fat Diet Alters the Gut Microbiome and Promotes Cecal Inflammation, Endotoxin Production, and Simple Steatosis without Obesity in Male Rats. *Lipids* **2019**, *54*, 119–131. [CrossRef]
84. Knoop, K.A.; McDonald, K.G.; Kulkarni, D.H.; Newberry, R.D. Antibiotics promote inflammation through the translocation of native commensal colonic bacteria. *Gut* **2016**, *65*, 1100–1109. [CrossRef]
85. Debyser, G.; Mesuere, B.; Clement, L.; Van de Weygaert, J.; Van Hecke, P.; Duytschaever, G.; Aerts, M.; Dawyndt, P.; De Boeck, K.; Vandamme, P.; et al. Faecal proteomics: A tool to investigate dysbiosis and inflammation in patients with cystic fibrosis. *J. Cyst. Fibros.* **2016**, *15*, 242–250. [CrossRef] [PubMed]
86. Parada Venegas, D.; De La Fuente, M.K.; Landskron, G.; González, M.J.; Quera, R.; Dijkstra, G.; Harmsen, H.J.M.; Faber, K.N.; Hermoso, M.A. Short Chain Fatty Acids (SCFAs)-Mediated Gut Epithelial and Immune Regulation and Its Relevance for Inflammatory Bowel Diseases. *Front. Immunol.* **2019**, *10*, 277. [CrossRef]
87. Silva, Y.P.; Bernardi, A.; Frozza, R.L. Role Short-Chain Fat. Acids Gut Microbiota Gut-Brain Communication. *Front. Endocrinol. (Lausanne)* **2020**, *11*, 25. [CrossRef] [PubMed]
88. Dengler, F.; Kraetzig, A.; Gabel, G. Butyrate Protects Porcine Colon Epithelium from Hypoxia-Induced Damage on a Functional Level. *Nutrients* **2021**, *13*, 305. [CrossRef] [PubMed]
89. Kelly, C.J.; Zheng, L.; Campbell, E.L.; Saeedi, B.; Scholz, C.C.; Bayless, A.J.; Wilson, K.E.; Glover, L.E.; Kominsky, D.J.; Magnuson, A.; et al. Crosstalk between Microbiota-Derived Short-Chain Fatty Acids and Intestinal Epithelial HIF Augments Tissue Barrier Function. *Cell Host Microbe* **2015**, *17*, 662–671. [CrossRef] [PubMed]
90. Lee, C.; Kim, B.G.; Kim, J.H.; Chun, J.; Im, J.P.; Kim, J.S. Sodium butyrate inhibits the NF-kappa B signaling pathway and histone deacetylation, and attenuates experimental colitis in an IL-10 independent manner. *Int. Immunopharmacol.* **2017**, *51*, 47–56. [CrossRef]
91. Yap, Y.A.; McLeod, K.H.; I McKenzie, C.; Gavin, P.G.; Davalos-Salas, M.; Richards, J.L.; Moore, R.J.; Lockett, T.J.; Clarke, J.M.; Eng, V.V.; et al. An acetate-yielding diet imprints an immune and anti-microbial programme against enteric infection. *Clin. Transl. Immunol.* **2021**, *10*, e1233. [CrossRef] [PubMed]
92. Diao, H.; Jiao, A.R.; Yu, B.; Mao, X.B.; Chen, D.W. Gastric infusion of short-chain fatty acids can improve intestinal barrier function in weaned piglets. *Genes Nutr.* **2019**, *14*, 4. [CrossRef]
93. Liu, L.; Sun, D.; Mao, S.; Zhu, W.; Liu, J. Infusion of sodium butyrate promotes rumen papillae growth and enhances expression of genes related to rumen epithelial VFA uptake and metabolism in neonatal twin lambs. *J. Anim. Sci.* **2019**, *97*, 909–921. [CrossRef]
94. Scales, B.S.; Dickson, R.P.; Huffnagle, G.B. A tale of two sites: How inflammation can reshape the microbiomes of the gut and lungs. *J. Leukoc. Biol.* **2016**, *100*, 943–950. [CrossRef]

95. Winter, S.E.; Baumler, A.J. Dysbiosis in the inflamed intestine: Chance favors the prepared microbe. *Gut Microbes* **2014**, *5*, 71–73. [CrossRef] [PubMed]
96. Lupp, C.; Robertson, M.L.; Wickham, M.; Sekirov, I.; Champion, O.L.; Gaynor, E.C.; Finlay, B.B. Host-mediated inflammation disrupts the intestinal microbiota and promotes the overgrowth of Enterobacteriaceae. *Cell Host Microbe* **2007**, *2*, 204. [CrossRef]
97. Corey, M.; McLaughlin, F.; Williams, M.; Levison, H. A comparison of survival, growth, and pulmonary function in patients with cystic fibrosis in Boston and Toronto. *J. Clin. Epidemiol.* **1988**, *41*, 583–591. [CrossRef]
98. Culhane, S.; George, C.; Pearo, B.; Spoede, E. Malnutrition in cystic fibrosis: A review. *Nutr. Clin. Pract.* **2013**, *28*, 676–683. [CrossRef] [PubMed]
99. Gaskin, K.J. Nutritional care in children with cystic fibrosis: Are our patients becoming better? *Eur. J. Clin. Nutr.* **2013**, *67*, 558–564. [CrossRef]
100. Turck, D.; Braegger, C.P.; Colombo, C.; Declercq, D.; Morton, A.; Pancheva, R.; Robberecht, E.; Stern, M.; Strandvik, B.; Wolfe, S.; et al. ESPEN-ESPGHAN-ECFS guidelines on nutrition care for infants, children, and adults with cystic fibrosis. *Clin. Nutr.* **2016**, *35*, 557–577. [CrossRef] [PubMed]
101. Borowitz, D.; Baker, R.D.; Stallings, V. Consensus report on nutrition for pediatric patients with cystic fibrosis. *J. Pediatr. Gastroenterol. Nutr.* **2002**, *35*, 246–259. [CrossRef]
102. Sanders, D.B.; Zhang, Z.; Farrell, P.M.; Lai, H.J. Early life growth patterns persist for 12 years and impact pulmonary outcomes in cystic fibrosis. *J. Cyst. Fibros.* **2018**, *17*, 528–535. [CrossRef]
103. Steinkamp, G.; Wiedemann, B. Relationship between nutritional status and lung function in cystic fibrosis: Cross sectional and longitudinal analyses from the German CF quality assurance (CFQA) project. *Thorax* **2002**, *57*, 596–601. [CrossRef]
104. Peterson, M.L.; Jacobs, D.R., Jr.; Milla, C.E. Longitudinal changes in growth parameters are correlated with changes in pulmonary function in children with cystic fibrosis. *Pediatrics* **2003**, *112 Pt 1*, 588–592. [CrossRef]
105. Konstan, M.W.; Butler, S.M.; Wohl, M.E.B.; Stoddard, M.; Matousek, R.; Wagener, J.S.; Johnson, C.A.; Morgan, W.J. Growth and nutritional indexes in early life predict pulmonary function in cystic fibrosis. *J. Pediatr.* **2003**, *142*, 624–630. [CrossRef] [PubMed]
106. Yen, E.H.; Quinton, H.; Borowitz, D. Better nutritional status in early childhood is associated with improved clinical outcomes and survival in patients with cystic fibrosis. *J. Pediatr.* **2013**, *162*, 530–535.e1. [CrossRef]
107. Hanna, R.M.; Weiner, D.J. Overweight and obesity in patients with cystic fibrosis: A center-based analysis. *Pediatr. Pulmonol.* **2015**, *50*, 35–41. [CrossRef]
108. Forrester, D.L.; Knox, A.J.; Smyth, A.R.; Fogarty, A.W. Measures of body habitus are associated with lung function in adults with cystic fibrosis: A population-based study. *J. Cyst. Fibros.* **2013**, *12*, 284–289. [CrossRef]
109. Forte, G.C.; Pereira, J.S.; Drehmer, M.; Simon, M.I.S.D.S. Anthropometric and dietary intake indicators as predictors of pulmonary function in cystic fibrosis patients. *J. Bras. Pneumol.* **2012**, *38*, 470–476. [CrossRef] [PubMed]
110. Pedreira, C.; Robert, R.; Dalton, V.; Oliver, M.; Carlin, J.; Robinson, P.; Cameron, F. Association of body composition and lung function in children with cystic fibrosis. *Pediatr. Pulmonol.* **2005**, *39*, 276–280. [CrossRef] [PubMed]
111. Ashkenazi, M.; Nathan, N.; Sarouk, I.; Bar Aluma, B.E.; Dagan, A.; Bezalel, Y.; Keler, S.; Vilozni, D.; Efrati, O. Nutritional Status in Childhood as a Prognostic Factor in Patients with Cystic Fibrosis. *Lung* **2019**, *197*, 371–376. [CrossRef]
112. Stallings, V.A.; Stark, L.J.; Robinson, K.A.; Feranchak, A.P.; Quinton, H. Evidence-based practice recommendations for nutrition-related management of children and adults with cystic fibrosis and pancreatic insufficiency: Results of a systematic review. *J. Am. Diet. Assoc.* **2008**, *108*, 832–839. [CrossRef]
113. Sutherland, R.; Katz, T.; Liu, V.; Quintano, J.; Brunner, R.; Tong, C.W.; Collins, C.E.; Ooi, C.Y. Dietary intake of energy-dense, nutrient-poor and nutrient-dense food sources in children with cystic fibrosis. *J. Cyst. Fibros.* **2018**, *17*, 804–810. [CrossRef] [PubMed]
114. Poulimeneas, D.; Grammatikopoulou, M.G.; Devetzi, P.; Petrocheilou, A.; Kaditis, A.G.; Papamitsou, T.; Doudounakis, S.E.; Vassilakou, T. Adherence to Dietary Recommendations, Nutrient Intake Adequacy and Diet Quality among Pediatric Cystic Fibrosis Patients: Results from the GreeCF Study. *Nutrients* **2020**, *12*, 3126. [CrossRef]
115. Calvo-Lerma, J.; Hulst, J.; Boon, M.; Martins, T.; Ruperto, M.; Colombo, C.; Fornés-Ferrer, V.; Woodcock, S.; Claes, I.; Asseiceira, I.; et al. The Relative Contribution of Food Groups to Macronutrient Intake in Children with Cystic Fibrosis: A European Multicenter Assessment. *J. Acad. Nutr. Diet.* **2019**, *119*, 1305–1319. [CrossRef] [PubMed]
116. Woestenenk, J.; Castelijns, S.; Van Der Ent, C.; Houwen, R. Dietary intake in children and adolescents with cystic fibrosis. *Clin. Nutr.* **2014**, *33*, 528–532. [CrossRef]
117. Smith, C.; Winn, A.; Seddon, P.; Ranganathan, S. A fat lot of good: Balance and trends in fat intake in children with cystic fibrosis. *J. Cyst. Fibros.* **2012**, *11*, 154–157. [CrossRef] [PubMed]
118. Panagopoulou, P.; Fotoulaki, M.; Nikolaou, A.; Nousia-Arvanitakis, S.; Maria, F. Prevalence of malnutrition and obesity among cystic fibrosis patients. *Pediatr. Int.* **2014**, *56*, 89–94. [CrossRef]
119. Stephenson, A.L.; A Mannik, L.; Walsh, S.; Brotherwood, M.; Robert, R.; Darling, P.B.; Nisenbaum, R.; Moerman, J.; Stanojevic, S. Longitudinal trends in nutritional status and the relation between lung function and BMI in cystic fibrosis: A population-based cohort study. *Am. J. Clin. Nutr.* **2013**, *97*, 872–877. [CrossRef] [PubMed]
120. Gramegna, A.; Aliberti, S.; Contarini, M.; Savi, D.; Sotgiu, G.; Majo, F.; Saderi, L.; Lucidi, V.; Amati, F.; Pappalettera, M.; et al. Overweight and obesity in adults with cystic fibrosis: An Italian multicenter cohort study. *J. Cyst. Fibros.* **2021**; *in press*. [CrossRef]

121. Bonhoure, A.; Boudreau, V.; Litvin, M.; Colomba, J.; Bergeron, C.; Mailhot, M.; Tremblay, F.; Lavoie, A.; Rabasa-Lhoret, R. Overweight, obesity and significant weight gain in adult patients with cystic fibrosis association with lung function and cardiometabolic risk factors. *Clin. Nutr.* **2020**, *39*, 2910–2916. [CrossRef]
122. Harindhanavudhi, T.; Wang, Q.; Dunitz, J.; Moran, A.; Moheet, A. Prevalence and factors associated with overweight and obesity in adults with cystic fibrosis: A single-center analysis. *J. Cyst. Fibros.* **2020**, *19*, 139–145. [CrossRef]
123. Bass, R.; Brownell, J.N.; Stallings, V.A. The Impact of Highly Effective CFTR Modulators on Growth and Nutrition Status. *Nutrients* **2021**, *13*, 2907. [CrossRef]
124. Stallings, V.A.; Sainath, N.; Oberle, M.; Bertolaso, C.; Schall, J.I. Energy Balance and Mechanisms of Weight Gain with Ivacaftor Treatment of Cystic Fibrosis Gating Mutations. *J. Pediatr.* **2018**, *201*, 229–237.e4. [CrossRef]
125. Wainwright, C.E.; Elborn, J.S.; Ramsey, B.W. Lumacaftor-Ivacaftor in Patients with Cystic Fibrosis Homozygous for Phe508del CFTR. *N. Engl. J. Med.* **2015**, *373*, 1783–1784. [CrossRef]
126. Rowe, S.M.; Heltshe, S.L.; Gonska, T.; Donaldson, S.H.; Borowitz, D.; Gelfond, D.; Sagel, S.D.; Khan, U.; Mayer-Hamblett, N.; Van Dalfsen, J.M.; et al. Clinical mechanism of the cystic fibrosis transmembrane conductance regulator potentiator ivacaftor in G551D-mediated cystic fibrosis. *Am. J. Respir. Crit. Care Med.* **2014**, *190*, 175–184. [CrossRef] [PubMed]
127. Ding, S.; Chi, M.M.; Scull, B.P.; Rigby, R.; Schwerbrock, N.M.J.; Magness, S.; Jobin, C.; Lund, P.K. High-fat diet: Bacteria interactions promote intestinal inflammation which precedes and correlates with obesity and insulin resistance in mouse. *PLoS ONE* **2010**, *5*, e12191. [CrossRef]
128. Gulhane, M.; Murray, L.; Lourie, R.; Tong, H.; Sheng, Y.H.; Wang, R.; Kang, A.; Schreiber, V.; Wong, K.Y.; Magor, G.; et al. High Fat Diets Induce Colonic Epithelial Cell Stress and Inflammation that is Reversed by IL-22. *Sci. Rep.* **2016**, *6*, 28990. [CrossRef] [PubMed]
129. Enaud, R.; Prevel, R.; Ciarlo, E.; Beaufils, F.; Wieërs, G.; Guery, B.; Delhaes, L. The Gut-Lung Axis in Health and Respiratory Diseases: A Place for Inter-Organ and Inter-Kingdom Crosstalks. *Front. Cell. Infect. Microbiol.* **2020**, *10*, 9. [CrossRef] [PubMed]
130. Dang, A.T.; Marsland, B.J. Microbes, metabolites, and the gut-lung axis. *Mucosal. Immunol.* **2019**, *12*, 843–850. [CrossRef] [PubMed]
131. Corrêa-Oliveira, R.; Fachi, J.L.; Vieira, A.; Sato, F.T.; Vinolo, M.A.R. Regulation of immune cell function by short-chain fatty acids. *Clin. Transl. Immunol.* **2016**, *5*, e73. [CrossRef] [PubMed]
132. Sun, M.; Wu, W.; Liu, Z.; Cong, Y. Microbiota metabolite short chain fatty acids, GPCR, and inflammatory bowel diseases. *J. Gastroenterol.* **2017**, *52*, 1–8. [CrossRef]
133. Hoen, A.G.; Li, J.; Moulton, L.A.; O'Toole, G.A.; Housman, M.L.; Koestler, D.C.; Guill, M.F.; Moore, J.; Hibberd, P.L.; Morrison, H.; et al. Associations between gut microbial colonization in early life and respiratory outcomes in cystic fibrosis. *J. Pediatrics* **2015**, *167*, 138–147.e3. [CrossRef]
134. Lai, H.-C.; Lin, T.-L.; Chen, T.-W.; Kuo, Y.-L.; Chang, C.-J.; Wu, T.-R.; Shu, C.-C.; Tsai, Y.-H.; Swift, S.; Lu, C.-C. Gut microbiota modulates COPD pathogenesis: Role of anti-inflammatory *Parabacteroides goldsteinii* lipopolysaccharide. *Gut* **2022**, *71*, 309–321. [CrossRef]
135. Bolia, R.; Ooi, C.Y.; Lewindon, P.; Bishop, J.; Ranganathan, S.; Harrison, J.; Ford, K.; Van Der Haak, N.; Oliver, M.R. Practical approach to the gastrointestinal manifestations of cystic fibrosis. *J. Paediatr. Child. Health* **2018**, *54*, 609–619. [CrossRef] [PubMed]
136. Beaufils, F.; Mas, E.; Mittaine, M.; Addra, M.; Fayon, M.; Delhaes, L.; Clouzeau, H.; Galode, F.; Lamireau, T.; Bui, S.; et al. Increased Fecal Calprotectin Is Associated with Worse Gastrointestinal Symptoms and Quality of Life Scores in Children with Cystic Fibrosis. *J. Clin. Med.* **2020**, *9*, 4080. [CrossRef] [PubMed]
137. Sathe, M.; Huang, R.; Heltshe, S.; Eng, A.; Borenstein, E.; Miller, S.I.; Hoffman, L.; Gelfond, D.; Leung, D.H.; Borowitz, D.; et al. Gastrointestinal Factors Associated With Hospitalization in Infants With Cystic Fibrosis: Results from the BONUS Study. *J. Pediatr. Gastroenterol. Nutr.* **2021**, *73*, 395–402. [CrossRef] [PubMed]
138. Yamada, A.; Komaki, Y.; Komaki, F.; Micic, D.; Zullow, S.; Sakuraba, A. Risk of gastrointestinal cancers in patients with cystic fibrosis: A systematic review and meta-analysis. *Lancet Oncol.* **2018**, *19*, 758–767. [CrossRef]
139. Bernstein, C.N.; Blanchard, J.F.; Kliewer, E.; Wajda, A. Cancer risk in patients with inflammatory bowel disease: A population-based study. *Cancer* **2001**, *91*, 854–862. [CrossRef]
140. Rutter, M.; Saunders, B.; Wilkinson, K.; Rumbles, S.; Schofield, G.; Kamm, M.; Williams, C.; Price, A.; Talbot, I.; Forbes, A. Severity of inflammation is a risk factor for colorectal neoplasia in ulcerative colitis. *Gastroenterology* **2004**, *126*, 451–459. [CrossRef]
141. Ullman, T.A.; Itzkowitz, S.H. Intestinal inflammation and cancer. *Gastroenterology* **2011**, *140*, 1807–1816. [CrossRef]
142. Terzić, J.; Grivennikov, S.; Karin, E.; Karin, M. Inflammation and colon cancer. *Gastroenterology* **2010**, *138*, 2101–2114.e5. [CrossRef]
143. Arthur, J.C.; Perez-Chanona, E.; Mühlbauer, M.; Tomkovich, S.; Uronis, J.M.; Fan, T.-J.; Campbell, B.J.; Abujamel, T.; Dogan, B.; Rogers, A.B.; et al. Intestinal inflammation targets cancer-inducing activity of the microbiota. *Science* **2012**, *338*, 120–123. [CrossRef] [PubMed]
144. Pleguezuelos-Manzano, C.; Puschhof, J.; Rosendahl Huber, A.; Van Hoeck, A.; Wood, H.M.; Nomburg, J.; Gurjao, C.; Manders, F.; Dalmasso, G.; Stege, P.B.; et al. Mutational signature in colorectal cancer caused by genotoxic pks(+) *E coli.*. *Nature* **2020**, *580*, 269–273. [CrossRef] [PubMed]
145. Martin, H.M.; Campbell, B.J.; Hart, C.; Mpofu, C.; Nayar, M.; Singh, R.; Englyst, H.; Williams, H.F.; Rhodes, J.M. Enhanced Escherichia coli adherence and invasion in Crohn's disease and colon cancer. *Gastroenterology* **2004**, *127*, 80–93. [CrossRef] [PubMed]

146. Tang, Y.; Chen, Y.; Jiang, H.; Robbins, G.T.; Nie, D. G-protein-coupled receptor for short-chain fatty acids suppresses colon cancer. *Int. J. Cancer* **2011**, *128*, 847–856. [CrossRef] [PubMed]
147. Donohoe, D.R.; Holley, D.; Collins, L.B.; Montgomery, S.A.; Whitmore, A.C.; Hillhouse, A.; Curry, K.P.; Renner, S.W.; Greenwalt, A.; Ryan, E.P.; et al. A gnotobiotic mouse model demonstrates that dietary fiber protects against colorectal tumorigenesis in a microbiota- and butyrate-dependent manner. *Cancer Discov.* **2014**, *4*, 1387–1397. [CrossRef]
148. Tian, Y.; Xu, Q.; Sun, L.; Ye, Y.; Ji, G. Short-chain fatty acids administration is protective in colitis-associated colorectal cancer development. *J. Nutr. Biochem.* **2018**, *57*, 103–109. [CrossRef]
149. Milner, E.; Stevens, B.; An, M.; Lam, V.; Ainsworth, M.; Dihle, P.; Stearns, J.; Dombrowski, A.; Rego, D.; Segars, K. Utilizing Probiotics for the Prevention and Treatment of Gastrointestinal Diseases. *Front. Microbiol.* **2021**, *12*, 689958. [CrossRef]
150. Santo, C.E.; Caseiro, C.; Martins, M.; Monteiro, R.; Brandão, I. Gut microbiota, in the halfway between nutrition and lung function. *Nutrients* **2021**, *13*, 1716. [CrossRef]
151. Araya, M.; Morelli, L.; Reid, G.; Sanders, M.E.; Stanton, C.; Pineiro, M. *Joint FAO/WHO Working Group Report on Drafting Guidelines for the Evaluation of Probiotics in Food*; World Health Organization: London, Ontario; Food and Agriculture Organization of the United Nations: Quebec City, QC, Canada, 2002.
152. Sanders, M.E.; Merenstein, D.J.; Reid, G.; Gibson, G.R.; Rastall, R.A. Probiotics and prebiotics in intestinal health and disease: From biology to the clinic. *Nat. Rev. Gastroenterol. Hepatol.* **2019**, *16*, 605–616. [CrossRef]
153. Sanders, M.E.; Merenstein, D.J.; Reid, G.; Gibson, G.R.; Rastall, R.A. Author Correction: Probiotics and prebiotics in intestinal health and disease: From biology to the clinic. *Nat. Rev. Gastroenterol. Hepatol.* **2019**, *16*, 642. [CrossRef]
154. Coffey, M.J.; Garg, M.; Homaira, N.; Jaffé, A.; Ooi, C.Y. Probiotics for people with cystic fibrosis. *Cochrane Database Syst. Rev.* **2018**. [CrossRef]
155. Cani, P.D.; de Vos, W.M. Next-Generation Beneficial Microbes: The Case of Akkermansia muciniphila. *Front. Microbiol.* **2017**, *8*, 1765. [CrossRef]
156. Arrieta, M.-C.; Stiemsma, L.T.; Amenyogbe, N.; Brown, E.M.; Finlay, B. The Intestinal Microbiome in Early Life: Health and Disease. *Front. Immunol.* **2014**, *5*, 427. [CrossRef]
157. Bergmann, K.R.; Liu, S.X.L.; Tian, R.; Kushnir, A.; Turner, J.R.; Li, H.L.; Chou, P.M.; Weber, C.R.; De Plaen, I.G. Bifidobacteria Stabilize Claudins at Tight Junctions and Prevent Intestinal Barrier Dysfunction in Mouse Necrotizing Enterocolitis. *Am. J. Pathol.* **2013**, *182*, 1595–1606. [CrossRef]
158. Goldenberg, J.Z.; Yap, C.; Lytvyn, L.; Lo, C.K.-F.; Beardsley, J.; Mertz, D.; Johnston, B.C. Probiotics for the prevention of Clostridium difficile-associated diarrhea in adults and children. *Cochrane Database Syst. Rev.* **2017**, *12*, Cd006095. [CrossRef]
159. Deng, K.; Chen, T.; Wu, Q.; Xin, H.; Wei, Q.; Hu, P.; Wang, X.; Wang, X.; Wei, H.; Shah, N.P. In vitro and in vivo examination of anticolonization of pathogens by Lactobacillus paracasei FJ861111.1. *J. Dairy Sci.* **2015**, *98*, 6759–6766. [CrossRef]
160. Su, G.; Ko, C.W.; Bercik, P.; Falck-Ytter, Y.; Sultan, S.; Weizman, A.V.; Morgan, R.L. AGA Clinical Practice Guidelines on the Role of Probiotics in the Management of Gastrointestinal Disorders. *Gastroenterology* **2020**, *159*, 697–705. [CrossRef]
161. Rivière, A.; Selak, M.; Lantin, D.; Leroy, F.; De Vuyst, L. Bifidobacteria and Butyrate-Producing Colon Bacteria: Importance and Strategies for Their Stimulation in the Human Gut. *Front. Microbiol.* **2016**, *7*, 979. [CrossRef]
162. Maldonado-Gómez, M.X.; Martínez, I.; Bottacini, F.; O'Callaghan, A.; Ventura, M.; van Sinderen, D.; Hillmann, B.; Vangay, P.; Knights, D.; Hutkins, R.W.; et al. Stable Engraftment of Bifidobacterium longum AH1206 in the Human Gut Depends on Individualized Features of the Resident Microbiome. *Cell Host Microbe* **2016**, *20*, 515–526. [CrossRef]
163. Van Baarlen, P.; Wells, J.M.; Kleerebezem, M. Regulation of intestinal homeostasis and immunity with probiotic lactobacilli. *Trends Immunol.* **2013**, *34*, 208–215. [CrossRef]
164. Rivière, A.; Gagnon, M.; Weckx, S.; Roy, D.; De Vuyst, L. Mutual Cross-Feeding Interactions between Bifidobacterium longum subsp. longum NCC2705 and Eubacterium rectale ATCC 33656 Explain the Bifidogenic and Butyrogenic Effects of Arabinoxylan Oligosaccharides. *Appl. Environ. Microbiol.* **2015**, *81*, 7767–7781. [CrossRef]
165. Hegarty, J.W.; Guinane, C.; Ross, R.P.; Hill, C.; Cotter, P.D. Bacteriocin production: A relatively unharnessed probiotic trait? *F1000Res* **2016**, *5*, 2587. [CrossRef]
166. Bali, V.; Panesar, P.S.; Bera, M.B.; Kennedy, J.F. Bacteriocins: Recent Trends and Potential Applications. *Crit. Rev. Food Sci. Nutr.* **2016**, *56*, 817–834. [CrossRef]
167. Mokoena, M.P. Lactic Acid Bacteria and Their Bacteriocins: Classification, Biosynthesis and Applications against Uropathogens: A Mini-Review. *Molecules* **2017**, *22*, 1255. [CrossRef]
168. Collado, M.C.; Grześkowiak, Ł.; Salminen, S. Probiotic strains and their combination inhibit in vitro adhesion of pathogens to pig intestinal mucosa. *Curr. Microbiol.* **2007**, *55*, 260–265. [CrossRef]
169. Fijan, S.; Šulc, D.; Steyer, A. Study of the In Vitro Antagonistic Activity of Various Single-Strain and Multi-Strain Probiotics against Escherichia coli. *Int. J. Environ. Res. Public Health* **2018**, *15*, 1539. [CrossRef]
170. Zheng, D.; Liwinski, T.; Elinav, E. Interaction between microbiota and immunity in health and disease. *Cell Res.* **2020**, *30*, 492–506. [CrossRef]
171. Rowland, I.; Gibson, G.; Heinken, A.; Scott, K.; Swann, J.; Thiele, I.; Tuohy, K. Gut microbiota functions: Metabolism of nutrients and other food components. *Eur. J. Nutr.* **2018**, *57*, 1–24. [CrossRef] [PubMed]
172. Klaenhammer, T.R.; Kleerebezem, M.; Kopp, M.V.; Rescigno, M. The impact of probiotics and prebiotics on the immune system. *Nat. Rev. Immunol.* **2012**, *12*, 728–734. [CrossRef]

173. Sichetti, M.; De Marco, S.; Pagiotti, R.; Traina, G.; Pietrella, D. Anti-inflammatory effect of multistrain probiotic formulation (*L. rhamnosus*, *B. lactis*, and *B. longum*). *Nutrition* **2018**, *53*, 95–102. [CrossRef] [PubMed]
174. Pothoulakis, C. Review article: Anti-inflammatory mechanisms of action of Saccharomyces boulardii. *Aliment Pharmacol. Ther.* **2009**, *30*, 826–833. [CrossRef]
175. Costeloe, K.; Hardy, P.; Juszczak, E.; Wilks, M.; Millar, M.R. Bifidobacterium breve BBG-001 in very preterm infants: A randomised controlled phase 3 trial. *Lancet* **2016**, *387*, 649–660. [CrossRef]
176. Przemska-Kosicka, A.; Childs, C.E.; Enani, S.; Maidens, C.; Dong, H.; Bin Dayel, I.; Tuohy, K.; Todd, S.; Gosney, M.A.; Yaqoob, P. Effect of a synbiotic on the response to seasonal influenza vaccination is strongly influenced by degree of immunosenescence. *Immun. Ageing* **2016**, *13*, 6. [CrossRef]
177. Vitetta, L.; Vitetta, G.; Hall, S. Immunological Tolerance and Function: Associations Between Intestinal Bacteria, Probiotics, Prebiotics, and Phages. *Front. Immunol.* **2018**, *9*, 2240. [CrossRef]
178. Cristofori, F.; Dargenio, V.N.; Dargenio, C.; Miniello, V.L.; Barone, M.; Francavilla, R. Anti-Inflammatory and Immunomodulatory Effects of Probiotics in Gut Inflammation: A Door to the Body. *Front. Immunol.* **2021**, *12*, 178. [CrossRef]
179. Sanders, M.E.; Benson, A.; Lebeer, S.; Merenstein, D.J.; Klaenhammer, T.R. Shared mechanisms among probiotic taxa: Implications for general probiotic claims. *Curr. Opin. Biotechnol.* **2018**, *49*, 207–216. [CrossRef]
180. Fredua-Agyeman, M.; Stapleton, P.; Basit, A.W.; Beezer, A.E.; Gaisford, S. In vitro inhibition of Clostridium difficile by commercial probiotics: A microcalorimetric study. *International J. Pharm.* **2017**, *517*, 96–103. [CrossRef]
181. Vieira, A.; Galvão, I.; Macia, L.; Sernaglia, M.; Vinolo, M.A.; Garcia, C.; Tavares, L.P.; Amaral, F.A.; Sousa, L.; Martins, F.; et al. Dietary fiber and the short-chain fatty acid acetate promote resolution of neutrophilic inflammation in a model of gout in mice. *J. Leukoc. Biol.* **2017**, *101*, 275–284. [CrossRef]
182. Smith, P.M.; Howitt, M.R.; Panikov, N.; Michaud, M.; Gallini, C.A.; Bohlooly-Y, M.; Glickman, J.N.; Garrett, W.S. The microbial metabolites, short-chain fatty acids, regulate colonic Treg cell homeostasis. *Sci. (N. Y.)* **2013**, *341*, 569–573. [CrossRef]
183. Salinas, E.; Reyes-Pavón, D.; Cortes-Perez, N.G.; Torres-Maravilla, E.; Bitzer-Quintero, O.K.; Langella, P.; Bermúdez-Humarán, L.G. Bioactive Compounds in Food as a Current Therapeutic Approach to Maintain a Healthy Intestinal Epithelium. *Microorganisms* **2021**, *9*, 1634. [CrossRef]
184. Anderson, R.C.; Cookson, A.L.; McNabb, W.C.; Park, Z.; McCann, M.J.; Kelly, W.J.; Roy, N.C. Lactobacillus plantarum MB452 enhances the function of the intestinal barrier by increasing the expression levels of genes involved in tight junction formation. *BMC Microbiol.* **2010**, *10*, 316. [CrossRef]
185. La Fata, G.; Weber, P.; Mohajeri, M.H. Probiotics and the Gut Immune System: Indirect Regulation. *Probiotics Antimicrob. Proteins* **2018**, *10*, 11–21. [CrossRef]
186. Kluijfhout, S.; Trieu, T.V.; Vandenplas, Y. Efficacy of the Probiotic Probiotical Confirmed in Acute Gastroenteritis. *Pediatr. Gastroenterol. Hepatol. Nutr.* **2020**, *23*, 464–471. [CrossRef]
187. Mack, D.R.; Michail, S.; Wei, S.; McDougall, L.; Hollingsworth, M.A. Probiotics inhibit enteropathogenic E. coli adherence in vitro by inducing intestinal mucin gene expression. *Am. J. Physiol.* **1999**, *276*, G941–G950. [CrossRef]
188. Wang, X.; Zhang, P.; Zhang, X. Probiotics Regulate Gut Microbiota: An Effective Method to Improve Immunity. *Molecules* **2021**, *26*, 6076. [CrossRef]
189. Mujagic, Z.; De Vos, P.; Boekschoten, M.V.; Govers, C.; Pieters, H.-J.H.M.; De Wit, N.J.W.; Bron, P.A.; Masclee, A.A.M.; Troost, F.J. The effects of Lactobacillus plantarum on small intestinal barrier function and mucosal gene transcription; a randomized double-blind placebo controlled trial. *Sci. Rep.* **2017**, *7*, 40128. [CrossRef]
190. Kim, Y.K.; Shin, C. The Microbiota-Gut-Brain Axis in Neuropsychiatric Disorders: Pathophysiological Mechanisms and Novel Treatments. *Curr. Neuropharmacol.* **2018**, *16*, 559–573. [CrossRef]
191. Reid, G. Disentangling What We Know About Microbes and Mental Health. *Front. Endocrinol.* **2019**, *10*, 81. [CrossRef]
192. EFSA Panel on Dietetic Products, Nutrition and Allergies (NDA). Scientific Opinion on the substantiation of health claims related to live yoghurt cultures and improved lactose digestion (ID 1143, 2976) pursuant to Article 13(1) of Regulation (EC) No 1924/2006. *EFSA J.* **2010**, *8*, 1763. [CrossRef]
193. Bruzzese, E.; Raia, V.; Ruberto, E.; Scotto, R.; Giannattasio, A.; Bruzzese, D.; Cavicchi, M.C.; Francalanci, M.; Colombo, C.; Faelli, N.; et al. Lack of efficacy of Lactobacillus GG in reducing pulmonary exacerbations and hospital admissions in children with cystic fibrosis: A randomised placebo controlled trial. *J. Cyst. Fibros.* **2018**, *17*, 375–382. [CrossRef]
194. Di Benedetto, L.; Raia, V.; Pastore, A.; Albano, F.; Spagnuolo, M.I.; De Vizia, B.; Guarino, A. Lactobacillus casei strain gg as adjunctive treatment to children with cystic fibrosis. *J. Pediatric Gastroenterol. Nutr.* **1998**, *26*, 542. [CrossRef]
195. Del Campo, R.; Garriga, M.; Pérez-Aragón, A.; Guallarte, P.; Lamas, A.; Máiz, L.; Bayón, C.; Roy, G.; Cantón, R.; Zamora, J.; et al. Improvement of digestive health and reduction in proteobacterial populations in the gut microbiota of cystic fibrosis patients using a Lactobacillus reuteri probiotic preparation: A double blind prospective study. *J. Cyst. Fibros.* **2014**, *13*, 716–722. [CrossRef]
196. Van Biervliet, S.; Hauser, B.; Verhulst, S.; Stepman, H.; Delanghe, J.; Warzee, J.-P.; Pot, B.; Vandewiele, T.; Wilschanski, M. Probiotics in cystic fibrosis patients: A double blind crossover placebo controlled study: Pilot study from the ESPGHAN Working Group on Pancreas/CF. *Clin. Nutr. ESPEN* **2018**, *27*, 59–65. [CrossRef]
197. Anderson, J.L.; Miles, C.; Tierney, A.C. Effect of probiotics on respiratory, gastrointestinal and nutritional outcomes in patients with cystic fibrosis: A systematic review. *J. Cyst. Fibros.* **2017**, *16*, 186–197. [CrossRef]

198. Neri, L.C.L.; Taminato, M.; Filho, L.S. Systematic Review of Probiotics for Cystic Fibrosis Patients: Moving Forward. *J. Pediatr. Gastroenterol. Nutr.* **2019**, *68*, 394–399. [CrossRef]
199. Nikniaz, Z.; Nikniaz, L.; Bilan, N.; Somi, M.H.; Faramarzi, E. Does probiotic supplementation affect pulmonary exacerbation and intestinal inflammation in cystic fibrosis: A systematic review of randomized clinical trials. *World J. Pediatrics* **2017**, *13*, 307–313. [CrossRef]
200. Van Biervliet, S.; Declercq, D.; Somerset, S. Clinical effects of probiotics in cystic fibrosis patients: A systematic review. *Clin. Nutr. ESPEN* **2017**, *18*, 37–43. [CrossRef]
201. Ananthan, A.; Balasubramanian, H.; Rao, S.; Patole, S. Probiotic supplementation in children with cystic fibrosis-a systematic review. *Eur. J. Pediatr.* **2016**, *175*, 1255–1266. [CrossRef]
202. Peredo-Lovillo, A.; Romero-Luna, H.E.; Jiménez-Fernández, M. Health promoting microbial metabolites produced by gut microbiota after prebiotics metabolism. *Food Res. Int.* **2020**, *136*, 109473. [CrossRef]
203. Gibson, G.R.; Hutkins, R.; Sanders, M.E.; Prescott, S.L.; Reimer, R.A.; Salminen, S.J.; Scott, K.; Stanton, C.; Swanson, K.S.; Cani, P.D.; et al. Expert consensus document: The International Scientific Association for Probiotics and Prebiotics (ISAPP) consensus statement on the definition and scope of prebiotics. *Nat. Rev. Gastroenterol. Hepatol.* **2017**, *14*, 491–502. [CrossRef]
204. Charalampopoulos, D.; Rastall, R.A. Prebiotics in foods. *Curr. Opin. Biotechnol.* **2012**, *23*, 187–191. [CrossRef] [PubMed]
205. Markowiak, P.; Śliżewska, K. Effects of Probiotics, Prebiotics, and Synbiotics on Human Health. *Nutrients* **2017**, *9*, 1021. [CrossRef]
206. Carlson, J.L.; Erickson, J.M.; Hess, J.M.; Gould, T.J.; Slavin, J.L. Prebiotic Dietary Fiber and Gut Health: Comparing the in Vitro Fermentations of Beta-Glucan, Inulin and Xylooligosaccharide. *Nutrients* **2017**, *9*, 1361. [CrossRef]
207. Slavin, J. Fiber and prebiotics: Mechanisms and health benefits. *Nutrients* **2013**, *5*, 1417–1435. [CrossRef]
208. Krumbeck, J.A.; Rasmussen, H.E.; Hutkins, R.W.; Clarke, J.; Shawron, K.; Keshavarzian, A.; Walter, J. Probiotic Bifidobacterium strains and galactooligosaccharides improve intestinal barrier function in obese adults but show no synergism when used together as synbiotics. *Microbiome* **2018**, *6*, 121. [CrossRef]
209. Shoaf, K.; Mulvey, G.L.; Armstrong, G.D.; Hutkins, R.W. Prebiotic galactooligosaccharides reduce adherence of enteropathogenic Escherichia coli to tissue culture cells. *Infect. Immun.* **2006**, *74*, 6920–6928. [CrossRef] [PubMed]
210. Beserra, B.T.S.; Fernandes, R.; do Rosario, V.A.; Mocellin, M.C.; Kuntz, M.G.F.; Trindade, E.B.S.M. A systematic review and meta-analysis of the prebiotics and synbiotics effects on glycaemia, insulin concentrations and lipid parameters in adult patients with overweight or obesity. *Clin. Nutr.* **2015**, *34*, 845–858. [CrossRef] [PubMed]
211. Zheng, H.J.; Guo, J.; Wang, Q.; Wang, L.; Wang, Y.; Zhang, F.; Huang, W.-J.; Zhang, W.; Liu, W.J.; Wang, Y. Probiotics, prebiotics, and synbiotics for the improvement of metabolic profiles in patients with chronic kidney disease: A systematic review and meta-analysis of randomized controlled trials. *Crit. Rev. Food Sci. Nutr.* **2021**, *61*, 577–598. [CrossRef]
212. Fernandes, R.; Rosario, V.A.D.; Mocellin, M.C.; Kuntz, M.G.; Trindade, E.B. Effects of inulin-type fructans, galacto-oligosaccharides and related synbiotics on inflammatory markers in adult patients with overweight or obesity: A systematic review. *Clin. Nutr.* **2017**, *36*, 1197–1206. [CrossRef] [PubMed]
213. Chethan, G.E.; Garkhal, J.; Sircar, S.; Malik, Y.P.S.; Mukherjee, R.; Sahoo, N.R.; Agarwal, R.K.; De, U.K. Immunomodulatory potential of β-glucan as supportive treatment in porcine rotavirus enteritis. *Vet. Immunol. Immunopathol.* **2017**, *191*, 36–43. [CrossRef]
214. Moro, G.; Arslanoglu, S.; Stahl, B.; Jelinek, J.; Wahn, U.; Boehm, G. A mixture of prebiotic oligosaccharides reduces the incidence of atopic dermatitis during the first six months of age. *Arch. Dis. Child.* **2006**, *91*, 814–819. [CrossRef]
215. Sierra, C.; Bernal, M.-J.; Blasco-Alonso, J.; Martínez, R.; Dalmau, J.; Ortuño, I.; Espín, B.; Vasallo, M.-I.; GIL Ortega, D.; Vidal, M.-L.; et al. Prebiotic effect during the first year of life in healthy infants fed formula containing GOS as the only prebiotic: A multicentre, randomised, double-blind and placebo-controlled trial. *Eur. J. Nutr.* **2015**, *54*, 89–99. [CrossRef]
216. Wilms, E.; An, R.; Smolinska, A.; Stevens, Y.; Weseler, A.R.; Elizalde, M.; Drittij, M.-J.; Ioannou, A.; van Schooten, F.J.; Smidt, H.; et al. Galacto-oligosaccharides supplementation in prefrail older and healthy adults increased faecal bifidobacteria, but did not impact immune function and oxidative stress. *Clin. Nutr.* **2021**, *40*, 3019–3031. [CrossRef]
217. Vulevic, J.; Juric, A.; Tzortzis, G.; Gibson, G.R. A mixture of trans-galactooligosaccharides reduces markers of metabolic syndrome and modulates the fecal microbiota and immune function of overweight adults. *J. Nutr.* **2013**, *143*, 324–331. [CrossRef] [PubMed]
218. Vulevic, J.; Juric, A.; Walton, G.E.; Claus, S.; Tzortzis, G.; Toward, R.E.; Gibson, G.R. Influence of galacto-oligosaccharide mixture (B-GOS) on gut microbiota, immune parameters and metabonomics in elderly persons. *Br. J. Nutr.* **2015**, *114*, 586–595. [CrossRef] [PubMed]
219. Maretti, C.; Cavallini, G. The association of a probiotic with a prebiotic (Flortec, Bracco) to improve the quality/quantity of spermatozoa in infertile patients with idiopathic oligoasthenoteratospermia: A pilot study. *Andrology* **2017**, *5*, 439–444. [CrossRef]
220. Wolter, M.; Grant, E.T.; Boudaud, M.; Steimle, A.; Pereira, G.V.; Martens, E.C.; Desai, M.S. Leveraging diet to engineer the gut microbiome. *Nat. Rev. Gastroenterol. Hepatol.* **2021**, *18*, 885–902. [CrossRef] [PubMed]
221. Grabinger, T.; Garzon, J.F.G.; Hausmann, M.; Geirnaert, A.; Lacroix, C.; Hennet, T. Alleviation of Intestinal Inflammation by Oral Supplementation With 2-Fucosyllactose in Mice. *Front. Microbiol.* **2019**, *10*, 1385. [CrossRef]
222. Tian, M.; Li, D.; Ma, C.; Feng, Y.; Hu, X.; Chen, F. Barley Leaf Insoluble Dietary Fiber Alleviated Dextran Sulfate Sodium-Induced Mice Colitis by Modulating Gut Microbiota. *Nutrients* **2021**, *13*, 846. [CrossRef] [PubMed]

223. Llewellyn, S.R.; Britton, G.J.; Contijoch, E.J.; Vennaro, O.H.; Mortha, A.; Colombel, J.-F.; Grinspan, A.; Clemente, J.C.; Merad, M.; Faith, J.J. Interactions Between Diet and the Intestinal Microbiota Alter Intestinal Permeability and Colitis Severity in Mice. *Gastroenterology* **2018**, *154*, 1037–1046.e2. [CrossRef]
224. Chen, H.; Chen, D.; Qin, W.; Liu, Y.; Che, L.; Huang, Z.; Luo, Y.; Zhang, Q.; Lin, D.; Liu, Y.; et al. Wheat bran components modulate intestinal bacteria and gene expression of barrier function relevant proteins in a piglet model. *Int. J. Food Sci. Nutr.* **2017**, *68*, 65–72. [CrossRef]
225. Zhang, T.; Ding, C.; Zhao, M.; Dai, X.; Yang, J.; Li, Y.; Gu, L.; Wei, Y.; Gong, J.; Zhu, W.; et al. Sodium Butyrate Reduces Colitogenic Immunoglobulin A-Coated Bacteria and Modifies the Composition of Microbiota in IL-10 Deficient Mice. *Nutrients* **2016**, *8*, 728. [CrossRef]
226. Valcheva, R.; Koleva, P.; Martínez, I.; Walter, J.; Gänzle, M.G.; Dieleman, L.A. Inulin-type fructans improve active ulcerative colitis associated with microbiota changes and increased short-chain fatty acids levels. *Gut Microbes* **2019**, *10*, 334–357. [CrossRef]
227. Xie, X.; He, Y.; Li, H.; Yu, D.; Na, L.; Sun, T.; Zhang, D.; Shi, X.; Xia, Y.; Jiang, T.; et al. Effects of prebiotics on immunologic indicators and intestinal microbiota structure in perioperative colorectal cancer patients. *Nutrition* **2019**, *61*, 132–142. [CrossRef]
228. Fritsch, J.; Garces, L.; Quintero, M.A.; Pignac-Kobinger, J.; Santander, A.M.; Fernández, I.; Ban, Y.J.; Kwon, D.; Phillips, M.C.; Knight, K.; et al. Low-Fat, High-Fiber Diet Reduces Markers of Inflammation and Dysbiosis and Improves Quality of Life in Patients With Ulcerative Colitis. *Clin. Gastroenterol. Hepatol.* **2021**, *19*, 1189–1199.e30. [CrossRef] [PubMed]
229. Laffin, M.; Perry, T.; Park, H.; Hotte, N.; Fedorak, R.N.; Thiesen, A.; Dicken, B.; Madsen, K.L. Prebiotic Supplementation Following Ileocecal Resection in a Murine Model is Associated With a Loss of Microbial Diversity and Increased Inflammation. *Inflamm. Bowel. Dis.* **2017**, *24*, 101–110. [CrossRef] [PubMed]
230. Rodriguez-Palacios, A.; Harding, A.; Menghini, P.; Himmelman, C.; Retuerto, M.; Nickerson, K.P.; Lam, M.; Croniger, C.M.; McLean, M.; Durum, S.K.; et al. The Artificial Sweetener Splenda Promotes Gut Proteobacteria, Dysbiosis, and Myeloperoxidase Reactivity in Crohn's Disease-Like Ileitis. *Inflamm. Bowel. Dis.* **2018**, *24*, 1005–1020. [CrossRef]
231. Liu, F.; Li, P.; Chen, M.; Luo, Y.; Prabhakar, M.; Zheng, H.; He, Y.; Qi, Q.; Long, H.; Zhang, Y.; et al. Fructooligosaccharide (FOS) and Galactooligosaccharide (GOS) Increase Bifidobacterium but Reduce Butyrate Producing Bacteria with Adverse Glycemic Metabolism in healthy young population. *Sci. Rep.* **2017**, *7*, 11789. [CrossRef] [PubMed]
232. Bernard, H.; Desseyn, J.-L.; Bartke, N.; Kleinjans, L.; Stahl, B.; Belzer, C.; Knol, J.; Gottrand, F.; Husson, M.-O. Dietary pectin-derived acidic oligosaccharides improve the pulmonary bacterial clearance of Pseudomonas aeruginosa lung infection in mice by modulating intestinal microbiota and immunity. *J. Infect. Dis.* **2015**, *211*, 156–165. [CrossRef] [PubMed]
233. Furusawa, Y.; Obata, Y.; Fukuda, S.; Endo, T.A.; Nakato, G.; Takahashi, D.; Nakanishi, Y.; Uetake, C.; Kato, K.; Kato, T.; et al. Commensal microbe-derived butyrate induces the differentiation of colonic regulatory T cells. *Nature* **2013**, *504*, 446–450. [CrossRef] [PubMed]
234. So, D.; Whelan, K.; Rossi, M.; Morrison, M.; Holtmann, G.; Kelly, J.T.; Shanahan, E.R.; Staudacher, H.M.; Campbell, K.L. Dietary fiber intervention on gut microbiota composition in healthy adults: A systematic review and meta-analysis. *Am. J. Clin. Nutr.* **2018**, *107*, 965–983. [CrossRef]
235. Jang, Y.O.; Kim, O.-H.; Kim, S.J.; Lee, S.H.; Yun, S.; Lim, S.E.; Yoo, H.J.; Shin, Y.; Lee, S.W. High-fiber diets attenuate emphysema development via modulation of gut microbiota and metabolism. *Sci. Rep.* **2021**, *11*, 7008. [CrossRef]
236. Devkota, S.; Wang, Y.; Musch, M.W.; Leone, V.; Fehlner-Peach, H.; Nadimpalli, A.; Antonopoulos, D.A.; Jabri, B.; Chang, E.B. Dietary-fat-induced taurocholic acid promotes pathobiont expansion and colitis in Il10−/− mice. *Nature* **2012**, *487*, 104–108. [CrossRef] [PubMed]
237. Williams, N.C.; Johnson, M.A.; Shaw, D.; Spendlove, I.; Vulevic, J.; Sharpe, G.R.; Hunter, K.A. A prebiotic galactooligosaccharide mixture reduces severity of hyperpnoea-induced bronchoconstriction and markers of airway inflammation. *Br. J. Nutr.* **2016**, *116*, 798–804. [CrossRef]
238. Kolodziejczyk, A.A.; Zheng, D.; Elinav, E. Diet–microbiota interactions and personalized nutrition. *Nat. Rev. Microbiol.* **2019**, *17*, 742–753. [CrossRef] [PubMed]
239. Asnicar, F.; Berry, S.E.; Valdes, A.M.; Nguyen, L.H.; Piccinno, G.; Drew, D.A.; Leeming, E.; Gibson, R.; Le Roy, C.; Al Khatib, H.; et al. Microbiome connections with host metabolism and habitual diet from 1,098 deeply phenotyped individuals. *Nat. Med.* **2021**, *27*, 321–332. [CrossRef]
240. Lawson, M.A.E.; O'Neill, I.J.; Kujawska, M.; Javvadi, S.G.; Wijeyesekera, A.; Flegg, Z.; Chalklen, L.; Hall, L.J. Breast milk-derived human milk oligosaccharides promote Bifidobacterium interactions within a single ecosystem. *ISME J.* **2020**, *14*, 635–648. [CrossRef]
241. Fanning, S.; Hall, L.J.; Cronin, M.; Zomer, A.; Mac Sharry, J.; Goulding, D.; O'Connell Motherway, M.; Shanahan, F.; Nally, K.; Dougan, G.; et al. Bifidobacterial surface-exopolysaccharide facilitates commensal-host interaction through immune modulation and pathogen protection. *Proc. Natl. Acad. Sci. USA* **2012**, *109*, 2108–2113. [CrossRef]
242. Beaumont, M.; Paës, C.; Mussard, E.; Knudsen, C.; Cauquil, L.; Aymard, P.; Barilly, C.; Gabinaud, B.; Zemb, O.; Fourre, S.; et al. Gut microbiota derived metabolites contribute to intestinal barrier maturation at the suckling-to-weaning transition. *Gut Microbes* **2020**, *11*, 1268–1286. [CrossRef] [PubMed]
243. Zegarra-Ruiz, D.F.; El Beidaq, A.; Iñiguez, A.J.; Di Ricco, M.; Manfredo Vieira, S.; Ruff, W.E.; Mubiru, D.; Fine, R.L.; Sterpka, J.; Greiling, T.M.; et al. A Diet-Sensitive Commensal Lactobacillus Strain Mediates TLR7-Dependent Systemic Autoimmunity. *Cell Host Microbe* **2019**, *25*, 113–127.e6. [CrossRef] [PubMed]

244. Mahmoud, T.I.; Wang, J.; Karnell, J.L.; Wang, Q.; Wang, S.; Naiman, B.; Gross, P.; Brohawn, P.Z.; Morehouse, C.; Aoyama, J.; et al. Autoimmune manifestations in aged mice arise from early-life immune dysregulation. *Sci. Transl. Med.* **2016**, *8*, 361ra137. [CrossRef]
245. Cox, L.M.; Yamanishi, S.; Sohn, J.; Alekseyenko, A.V.; Leung, J.M.; Cho, I.; Kim, S.G.; Li, H.; Gao, Z.; Mahana, D.; et al. Altering the Intestinal Microbiota during a Critical Developmental Window Has Lasting Metabolic Consequences. *Cell* **2014**, *158*, 705–721. [CrossRef]
246. AL Nabhani, Z.; Dulauroy, S.; Marques, R.; Cousu, C.; Al Bounny, S.; Déjardin, F.; Sparwasser, T.; Bérard, M.; Cerf-Bensussan, N.; Eberl, G. A Weaning Reaction to Microbiota Is Required for Resistance to Immunopathologies in the Adult. *Immunity* **2019**, *50*, 1276–1288.e5. [CrossRef] [PubMed]
247. Kalbermatter, C.; Trigo, N.F.; Christensen, S.; Ganal-Vonarburg, S.C. Maternal Microbiota, Early Life Colonization and Breast Milk Drive Immune Development in the Newborn. *Front. Immunol.* **2021**, *12*, 1768. [CrossRef] [PubMed]
248. Vatanen, T.; Franzosa, E.A.; Schwager, R.; Tripathi, S.; Arthur, T.D.; Vehik, K.; Lernmark, Å.; Hagopian, W.A.; Rewers, M.J.; She, J.-X.; et al. The human gut microbiome in early-onset type 1 diabetes from the TEDDY study. *Nature* **2018**, *562*, 589–594. [CrossRef] [PubMed]
249. Rodríguez, J.M.; Murphy, K.; Stanton, C.; Ross, R.P.; Kober, O.I.; Juge, N.; Avershina, E.; Rudi, K.; Narbad, A.; Jenmalm, M.C.; et al. The composition of the gut microbiota throughout life, with an emphasis on early life. *Microb. Ecol. Health Dis.* **2015**, *26*, 26050. [CrossRef] [PubMed]
250. Murphy, E.A.; Velazquez, K.T.; Herbert, K.M. Influence of high-fat diet on gut microbiota: A driving force for chronic disease risk. *Curr. Opin. Clin. Nutr. Metab. Care* **2015**, *18*, 515–520. [CrossRef]
251. Lee, S.M.; Kim, N.; Yoon, H.; Nam, R.H.; Lee, D.H. Microbial Changes and Host Response in F344 Rat Colon Depending on Sex and Age Following a High-Fat Diet. *Front. Microbiol.* **2018**, *9*, 2236. [CrossRef]
252. Tomas, J.; Mulet, C.; Saffarian, A.; Cavin, J.-B.; Ducroc, R.; Regnault, B.; Tan, C.K.; Duszka, K.; Burcelin, R.; Wahli, W.; et al. High-fat diet modifies the PPAR-gamma pathway leading to disruption of microbial and physiological ecosystem in murine small intestine. *Proc. Natl. Acad. Sci. USA* **2016**, *113*, E5934–E5943. [CrossRef]
253. Anitha, M.; Reichardt, F.; Tabatabavakili, S.; Nezami, B.G.; Chassaing, B.; Mwangi, S.; Vijay-Kumar, M.; Gewirtz, A.; Srinivasan, S. Intestinal dysbiosis contributes to the delayed gastrointestinal transit in high-fat diet fed mice. *Cell. Mol. Gastroenterol. Hepatol.* **2016**, *2*, 328–339. [CrossRef]
254. Thorburn, A.N.; Macia, L.; Mackay, C.R. Diet, metabolites, and "western-lifestyle" inflammatory diseases. *Immunity* **2014**, *40*, 833–842. [CrossRef]
255. Macia, L.; Tan, J.; Vieira, A.T.; Leach, K.; Stanley, D.; Luong, S.; Maruya, M.; McKenzie, C.I.; Hijikata, A.; Wong, C.; et al. Metabolite-sensing receptors GPR43 and GPR109A facilitate dietary fibre-induced gut homeostasis through regulation of the inflammasome. *Nat. Commun.* **2015**, *6*, 6734. [CrossRef]
256. Ruff, W.E.; Greiling, T.M.; Kriegel, M.A. Host–microbiota interactions in immune-mediated diseases. *Nat. Rev. Microbiol.* **2020**, *18*, 521–538. [CrossRef] [PubMed]
257. Desai, M.S.; Seekatz, A.M.; Koropatkin, N.M.; Kamada, N.; Hickey, C.A.; Wolter, M.; Pudlo, N.A.; Kitamoto, S.; Terrapon, N.; Muller, A.; et al. A Dietary Fiber-Deprived Gut Microbiota Degrades the Colonic Mucus Barrier and Enhances Pathogen Susceptibility. *Cell* **2016**, *167*, 1339–1353.e21. [CrossRef] [PubMed]
258. Louis, P.; Flint, H.J. Formation of propionate and butyrate by the human colonic microbiota. *Environ. Microbiol.* **2017**, *19*, 29–41. [CrossRef] [PubMed]
259. O'Keefe, S.J.D.; Li, J.V.; Lahti, L.; Ou, J.; Carbonero, F.; Mohammed, K.; Posma, J.M.; Kinross, J.; Wahl, E.; Ruder, E.; et al. Fat, fibre and cancer risk in African Americans and rural Africans. *Nat. Commun.* **2015**, *6*, 6342. [CrossRef]
260. Berer, K.; Martínez, I.; Walker, A.; Kunkel, B.; Schmitt-Kopplin, P.; Walter, J.; Krishnamoorthy, G. Dietary non-fermentable fiber prevents autoimmune neurological disease by changing gut metabolic and immune status. *Sci. Rep.* **2018**, *8*, 10431. [CrossRef] [PubMed]
261. Chen, K.; Chen, H.; Faas, M.M.; De Haan, B.J.; Li, J.; Xiao, P.; Zhang, H.; Diana, J.; De Vos, P.; Sun, J. Specific inulin-type fructan fibers protect against autoimmune diabetes by modulating gut immunity, barrier function, and microbiota homeostasis. *Mol. Nutr. Food Res.* **2017**, *61*, 1601006. [CrossRef] [PubMed]
262. Madan, J.C.; Koestler, D.C.; Stanton, B.A.; Davidson, L.; Moulton, L.A.; Housman, M.L.; Moore, J.H.; Guill, M.F.; Morrison, H.G.; Sogin, M.L.; et al. Serial analysis of the gut and respiratory microbiome in cystic fibrosis in infancy: Interaction between intestinal and respiratory tracts and impact of nutritional exposures. *MBio* **2012**, *3*, e00251-e12. [CrossRef]
263. Gill, S.K.; Rossi, M.; Bajka, B.; Whelan, K. Dietary fibre in gastrointestinal health and disease. *Nat. Rev. Gastroenterol. Hepatol.* **2021**, *18*, 101–116. [CrossRef] [PubMed]
264. Halnes, I.; Baines, K.J.; Berthon, B.S.; MacDonald-Wicks, L.K.; Gibson, P.G.; Wood, L.G. Soluble Fibre Meal Challenge Reduces Airway Inflammation and Expression of GPR43 and GPR41 in Asthma. *Nutrients* **2017**, *9*, 57. [CrossRef] [PubMed]
265. Wood, L.G.; Garg, M.L.; Gibson, P.G. A high-fat challenge increases airway inflammation and impairs bronchodilator recovery in asthma. *J. Allergy Clin. Immunol.* **2011**, *127*, 1133–1140. [CrossRef]
266. Simon, M.I.S.D.S.; Molle, R.D.; Silva, F.M.; Rodrigues, T.W.; Feldmann, M.; Forte, G.C.; Marostica, P.J.C. Antioxidant Micronutrients and Essential Fatty Acids Supplementation on Cystic Fibrosis Outcomes: A Systematic Review. *J. Acad. Nutr. Diet* **2020**, *120*, 1016–1033.e1. [CrossRef]

267. Kanhere, M.; He, J.; Chassaing, B.; Ziegler, T.R.; Alvarez, J.A.; Ivie, E.A.; Hao, L.; Hanfelt, J.; Gewirtz, A.T.; Tangpricha, V. Bolus weekly Vitamin D3 supplementation impacts gut and airway microbiota in adults with cystic fibrosis: A double-blind, randomized, placebo-controlled clinical trial. *J. Clin. Endocrinol. Metab.* **2018**, *103*, 564–574. [CrossRef] [PubMed]
268. Sullivan, J. Probiotic use and antibiotic associated gastrointestinal symptoms in pediatric patients with cystic fibrosis. *Pediatric Pulmonol.* **2015**, *50* (Suppl. 41), 407.
269. De Simone, C. The Unregulated Probiotic Market. *Clin. Gastroenterol. Hepatol.* **2019**, *17*, 809–817. [CrossRef] [PubMed]
270. Wastyk, H.C.; Fragiadakis, G.K.; Perelman, D.; Dahan, D.; Merrill, B.D.; Yu, F.B.; Topf, M.; Gonzalez, C.G.; Van Treuren, W.; Han, S.; et al. Gut-microbiota-targeted diets modulate human immune status. *Cell* **2021**, *184*, 4137–4153.e14. [CrossRef]

Article

Utilization of the Healthy Eating Index in Cystic Fibrosis

Rosara Milstein Bass [1,2], Alyssa Tindall [1] and Saba Sheikh [2,3,*]

[1] Division of Gastroenterology, Hepatology and Nutrition, The Children's Hospital of Philadelphia, Philadelphia, PA 19104, USA; bassr@chop.edu (R.M.B.); tindalla@chop.edu (A.T.)
[2] Department of Pediatrics, University of Pennsylvania, Philadelphia, PA 19104, USA
[3] Division of Pulmonary and Sleep Medicine, The Children's Hospital of Philadelphia, Philadelphia, PA 19104, USA
* Correspondence: sheikhs@chop.edu; Tel.: +1-518-423-1730

Abstract: (1) Background: Malnutrition has been a hallmark of cystic fibrosis (CF) for some time, and improved nutritional status is associated with improved outcomes. While individuals with CF historically required higher caloric intake than the general population, new CF therapies and improved health in this population suggest decreased metabolic demand and prevalence of overweight and obesity have increased. This study aimed to (a) examine diet quality in a population of young adults with CF using the Healthy Eating Index, a measure of diet quality in accordance with the U.S. Dietary Guidelines for Americans and (b) evaluate and describe how subcomponents of the HEI might apply to individuals with CF (2) Methods: 3-day dietary recalls from healthy adolescents and young adults with CF were obtained and scored based on the Healthy Eating Index (3) Results: Dietary recalls from 26 (14M/12F) adolescents and young adults with CF (ages 16–23), were obtained. Individuals with CF had significantly lower HEI scores than the general population and lower individual component scores for total vegetables, greens and beans, total fruits, whole fruits, total protein, seafood and plant protein and sodium (p values < 0.01 for all). (4) Conclusion: Dietary quality was poor in these healthy adolescents and young adults with CF. Given the increased prevalence of overweight and obesity in CF, updated dietary guidance is urgently needed for this population. The Healthy Eating Index may be a valuable tool for evaluating dietary quality in CF.

Keywords: cystic fibrosis; nutrition; healthy eating index; dietary guidelines

1. Introduction

Malnutrition has been a hallmark of cystic fibrosis since the disease was first described in 1938 [1]. While exocrine pancreatic insufficiency is the primary driver of malnutrition in individuals with CF, abnormalities of energy metabolism and gastrointestinal and hepatobiliary function also contribute [2,3]. The importance of improving nutritional status was first identified by Corey et al. in 1988, who identified that taller and heavier individuals with CF on unrestricted diet and appropriate pancreatic enzyme supplementation had improved survival compared to their peers, despite similar pulmonary function [4]. Subsequent analyses confirmed the relationship between z-scores for weight and BMI and longitudinal changes in forced expiratory volume in one second within 60% to 140% of predicted values (FEV_1%) [5]. The clearly defined relationship between nutritional status and overall health has driven the most recent consensus recommendations for nutrition in people with CF, which endorse maintenance of normal ranges of weight and stature for age (BMI $\geq$ 50th percentile in children, $\geq$22 in adult women and $\geq$23 in adult men) through routine energy intake of 110–200% of standards for healthy population [6]. Meeting nutritional goals has historically been a challenge for individuals with CF. However, more recent studies of healthy youth and young adults with CF, have demonstrated consistent achievement of the recommended caloric recommendations often through a diet that is high in saturated fat and added sugar and low in fiber [7–11].

In fact, it is likely that some individuals with CF are consuming calories in excess of their metabolic needs, as overweight and obesity are now being reported in individuals with CF [12,13]. Overweight and obesity is expected to increase in prevalence in the CF population, as highly effective cystic fibrosis transmembrane regulator (CFTR) modulator use becomes widespread. Highly effective CFTR modulators have been associated with increased weight and BMI, and are now available for approximately 90% of individuals with CF [14–17]. The changing body composition of the CF population warrants additional investigation within the CF research community regarding (1) the impact of overweight and obesity on health in individuals with CF and (2) the need for better understanding as to how dietary quality can impact nutritional status and health outcomes in this population.

Additional guidance for nutrition in CF, designed to build upon the existing 2008 guidelines was provided by McDonald et al. and recommend an "age-appropriate, healthy diet, that emphasizes culturally appropriate foods associated with positive health outcomes in the general population, including vegetables, fruits, whole grains, seafood, eggs, beans, peas, nuts and seeds, dairy products, and meat and poultry" [18]. These specific food groups were not mentioned in the prior guidelines, and represent a shift in focus from diet quantity to diet quality. The Healthy Eating Index (HEI) incorporates all of these features and provides a metric for the assessment of dietary quality in the general population. The HEI is a measure of diet quality in accordance with U.S. Dietary Guidelines for Americans, independent of quantity [19]. While there is inconclusive and limited evidence regarding specific nutrients/food groups on outcomes in CF, it is likely that in aggregate, an overall healthy diet is beneficial for individuals with CF, and the HEI tool may provide a framework for dietary assessment and counseling in this population. This study aimed to (a) leverage the HEI for examination of diet quality in a population of young adults with CF and (b) evaluate and describe how subcomponents of the HEI scoring system might apply to individuals with CF.

2. Materials and Methods

2.1. Study Population

Individuals with CF from the CF Center at the Children's Hospital of Philadelphia and the Hospital of the University of Pennsylvania were recruited for an ongoing longitudinal study examining body composition and muscle function in CF (ClinicalTrials.gov identifier: NCT02776098). Inclusion criteria for the trial were age 16–23 and confirmed diagnosis of CF. Exclusion criteria included chronic glucocorticoid use, organ transplantation, severe CF pulmonary disease (forced expiratory volume (FEV) 1%-predicted < 40%) and established diagnosis of CF related diabetes. While not part of recruitment criteria, the CF cohort was found to be was relatively healthy overall, as defined by median FEV_1 98% of predicted and median BMI 23.7 for males and 24.9 for females.

The dietary data for the general population were obtained from nationally representative survey data—the What We Eat In America (WWEIA) portion of the 2015–2016 National Health and Nutrition Examination Survey (NHANES) [20]. The WWEIA is the dietary intake interview component of the NHANES. In 2015–2016, 7918 Americans ages 2 and above (3877M/4041F) reported the types and quantities of foods and beverages they consume in a 24-h period through this interview. The Center for Nutrition Policy and Promotion calculated average HEI-2015 scores for the American population using this information [19].

2.2. Diet Recall

Three 24-h dietary recall interviews (2 weekdays and 1 weekend day) were collected by a Registered Dietitian Nutritionist via phone from each study participant. Interviews were conducted within the 3 week period following the study visit. The dates and times of the recalls were unannounced (unscheduled) so that participants did not change their normal eating pattern.

The 24-h dietary and supplement recall interviews elicited a detailed summary of all foods, beverages and dietary supplements consumed by participants during a complete 24-h period (from midnight-to-midnight) for the day preceding the interview. Information was obtained on the time of each eating occasion, the type of meal (breakfast, lunch, supper, snack), the location of the meal (home, school, other) and what and how much was consumed.

Data from these interviews were entered into Nutrition Data System for Research (NDSR) (Version NDS2021, MN, USA) for analysis.

2.3. Calculation of an HEI Score

HEI scores for the CF population were calculated using Stata 17.0. Dietary recall data were evaluated using percentage estimated energy requirement (%EER) and recall days were included if %EER was >30%. Subjects were included if at least two recall days met this criteria. NDSR output files 04 and 09 were used to calculate total HEI score and individual component scores according to the Healthy Eating Index-2015 scoring standards, which were also used in the NHANES population [21]. Mean scores were calculated for participants by taking the mean score of their HEI and component scores across their recall days.

2.4. Statistical Analysis

Stata 17.0 was used to analyze study data. Subjects were grouped by age according to the 2020–2025 Dietary Guidelines for Americans (14–18 and 19–30 years). Normality of the data was assessed using q–q plots [22]. The Kolmogorov–Smirnov test for nonparametric data was used to compare the mean total HEI score and mean component scores between the two age groups. One-sample t-tests were used to compare mean HEI scores and components with nationally representative mean HEI scores and component scores.

3. Results

3.1. Subject Characteristics

26 adolescents and young adults with CF (median age 19.8 (range: 16.5–23.2) and 14M/12F) were enrolled in the study. Of these subjects, 20 were pancreatic insufficient (PI) and 6 were pancreatic sufficient (PS). Subjects were generally healthy, as evidenced by median FEV_1% predicted of 98% (71–129), and average BMI 23.7 (19.9–33.8) for males and 24.9 (20.2–28.4) for females (BMIs of >23 for males and >22 for females are associated with improved outcomes in CF). Five Subjects ages 16–18 had significantly lower weight ($p = 0.009$) and height ($p = 0.04$) than subjects ages 19–26, but better pulmonary function ($p = 0.004$), all likely associated with younger age (Table 1). Nine subjects (6M/3F) had BMI below CFF targets of 23 for males and 22 for females. Twelve subjects (6M/6F) had BMI > 25, which the WHO defines as overweight, and 1 subject (M) had BMI > 30, which the WHO defines as obese.

3.2. HEI Scores for CF vs. General Population

Individuals with CF had significantly lower scores than the general populations for total HEI score and within the following categories of the HEI (Table 2): total vegetables; greens and beans; total fruits; whole fruits; total protein; seafood and plant protein; and sodium. No difference in HEI metrics was seen between younger (age 16–18) and older (age 19–23) individuals with CF (Table 2) or when either of these age groups were compared to the US population (all $p > 0.05$, data not shown).

Table 1. Subject Characteristics by Age: Younger individuals with Cystic Fibrosis (CF) had lower weight ($p = 0.009$) and height ($p = 0.04$), and higher forced expiratory volume in one second (FEV$_1$)% predicted ($p = 0.004$) than older individuals with CF. Gender distribution, pancreatic status and BMI were not different between the two age groups (all $p > 0.05$).

	Total (n = 26)	Ages 16–18 (n = 9)	Ages 19–23 (n = 17)	p-Value
Sex	14M/12F	3M/6F	11M/6F	Pr = 0.127
Weight	64.9 [50.6; 107.5]	60.5 [50.6; 71.0]	69.8 [57.3; 107.5]	0.009
Height	166.7 [148.8; 185.1]	163.1 [151.4; 173.6]	171.8 [148.8; 185.1]	0.04
BMI	24.1 [19.9; 33.8]	22.1 [20.2; 26.8]	25.5 [19.9; 33.8]	0.07
Pancreatic Insufficiency	20Y/6N	6Y/3N	14Y/3N	Pr = 0.366
FEV$_1$% Predicted	98 [71; 129]	109 [94; 120]	87 [71; 129]	0.004

Table 2. HEI scores CF vs. reference: Individuals with CF had significantly lower total HEI scores than the US population ($p < 0.0001$) and lower individual component scores for total vegetables, greens and beans, total fruits, whole fruits, total protein, seafood and plant protein and sodium (all $p < 0.01$). While younger (ages 16–18) individuals with CF tended to have lower total and component HEI scores than older (ages 19–26) individuals with CF, these differences were not statistically significant (all $p > 0.05$). * indicates clinical significance, $p < 0.05$.

Variable	Possible Score	U.S. Population	Mean ± SD Score			p-Values	
			CF Subjects			Within CF: 16–18 Y vs. 19–23 Y	U.S. vs. All CF Subjects
			All Subjects	16–18 Y (n = 7)	19–23 Y (n = 19)		
HEI total score	100	59	46.0 ± 12.1	45.0 ± 9.5	46.0 ± 13.2	0.61	<0.0001 *
Total vegetables	5	3.3	2.2 ± 1.1	1.8 ± 1.2	2.4 ± 1.1	0.34	0.0001 *
Greens and beans	5	3.1	0.8 ± 1.3	1.0 ± 1.0	0.7 ± 1.4	0.62	<0.0001 *
Total fruits	5	4.2	1.3 ± 1.4	1.6 ± 1.7	1.2 ± 1.3	1	<0.0001 *
Whole fruits	5	2.9	1.0 ± 1.3	0.9 ± 1.0	1.0 ± 1.4	1	<0.0001 *
Dairy	10	6	6.3 ± 3.1	6.7 ± 3.5	6.2 ± 3.0	0.62	0.58
Total protein	5	5	3.7 ± 1.0	3.6 ± 1.4	3.7 ± 0.8	0.68	<0.0001 *
Seafood and plant protein	5	5	1.5 ± 1.8	2.1 ± 2.1	1.3 ± 1.7	0.43	<0.0001 *
Whole grains	10	3	3.2 ± 2.7	2.9 ± 3.1	3.4 ± 2.6	0.43	0.65
Fatty acids	10	4.1	3.7 ± 2.6	3.4 ± 1.2	3.8 ± 3.0	0.72	0.4
Refined grains	10	6.4	5.5 ± 2.8	5.3 ± 3.0	5.6 ± 2.8	0.85	0.11
Sodium	10	3.7	5.0 ± 2.4	5.1 ± 2.5	5.0 ± 2.5	0.62	0.01 *
Added sugars	10	6.8	6.2 ± 3.0	6.5 ± 2.4	6.0 ± 3.2	0.66	0.28
Saturated fats	10	5.1	5.7 ± 2.8	4.6 ± 2.3	6.1 ± 2.9	0.18	0.28

4. Discussion

This study examined dietary quality in a cohort of young, healthy adults with CF. Our data showed that younger individuals had poor dietary quality as measured by the Healthy Eating Index, and scores are lower than what have been reported for a healthy reference population. We found that although not statistically significant, 16–18 year olds tended to have worse diet quality than 19–23 year olds, which is consistent with trends seen in the general population.

While multiple studies have examined dietary quality in CF, this study is unique as we leveraged the HEI, which specifically examines how dietary intake aligns with the recommendations for the Dietary Guidelines for Americans [19]. Alvarez et al. have also utilized this score to measure dietary quality in an older population (ages 18–50) of individuals with CF, and found individuals with CF had lower HEI scores as compared to age matched controls. The present study differs as our data were collected using a 3 day diet history obtained by a Registered Dietitian Nutritionist as opposed to a patient dietary records, we leveraged the NHANES database as a reference sample, and our CF population was healthier, as evidenced by their higher average FEV$_1$ (98% vs. 74%) and higher average

BMI (24.1 vs. 21.6). Although Alvarez et al. did not find a significant difference between CF and controls when examining specific elements of the HEI, we found that individuals with CF had lower scores than the general population in total HEI score and almost all HEI components. Specific dietary components have not been specifically linked with outcomes in CF; however, it is important to consider the deficits in the current CF diet, and how the benefits of each of the HEI components may specifically apply to individuals with CF.

4.1. Fruits, Vegetables and Legumes and Cystic Fibrosis

Our study found individuals with CF scored lower than average for consumption of vegetables, fruits and greens and beans. This is consistent with prior studies of intake in CF. In a cross-sectional study of 24 clinically stable adults with CF (mean FEV_1% 60, BMI 22 kg/m^2), none of the subjects met the recommended five servings of vegetables, and only 8.3% met daily fruit intake recommendations [23]. In an observational study by Calvo et al., which examined dietary patterns in 207 children with CF (ages 2–17) from six European CF centers, median intake of vegetable products was 1.0–2.8 times per day, fruits from 1.0–1.5 times per day and legumes from 0–0.2 times per day, all well below recommended daily intake [11].

Similar findings were seen in a Greek study of dietary quality in CF, only 16.7% of females (n = 44) and 17.1% of males (n = 32) consumed fruit more than once daily and only 14.6% of females and 20% of males consumed vegetables more than once daily. Additionally, this same study found that only 41.7% and 40% of females and males, respectively, consumed pulses on more than one occasion per week [7].

As fruits, vegetables and legumes are rich in fiber, not unexpectedly, overall daily fiber intake has been shown to be lower than recommended in individuals with CF, with median intakes ranging from 10.5–17.0 g/day in a study by Calvo et al. [11]. When Gavin et al. compared fiber intake in youth with CF to age-matched controls, mean daily fiber intake was lower in CF [24].

In addition to fruit, vegetable and legume fiber sources, whole grains are also an important source of dietary fiber. While individuals with CF still scored below target in this category, intake was not significantly lower than the general population, nor were intakes of refined grains or added sugar higher than the general population. No studies have specifically examined intake of whole grains as compared to refined grains in individuals with CF, but Sutherland et al. found significantly higher intake of refined carbohydrates including confectionary, packaged snacks, baked products and sweetened drinks in youth with CF as compared to age- and gender-matched controls.8 Additionally, Tierney et al. found in people with CF, 29% of daily caloric intake came from discretionary foods (packaged sweets, sugar sweetened beverages etc.), compared with 25% of non-CF adults.

The role of fiber in the CF diet has not been fully established. In a study of 68 children with CF, who were grouped based on presence of no, mild/moderate or severe abdominal pain, grams of fiber per kilogram body weight was significantly lower in the severe abdominal pain group than the others [24]. Contradictory results were seen by Proesman et al. who did not find a relationship between fiber intake and gastrointestinal problems in patients with CF [25].

Individuals with CF have been shown to have dyslipidemia, with elevated triglycerides and low HDL [26]. Dietary fiber has been shown to be inversely related to triglyceride levels in non-CF adults with overweight and obesity [27]. A cross-sectional study conducted in China found a dose-response relationship between increased dietary fiber intake and increase of HDL cholesterol in males [28]. Future studies should examine whether dietary fiber intake, similarly impacts lipid profiles of people with CF.

The quality of carbohydrate intake by individuals with CF, may also have implications for glycemic controls, as carbohydrates, total sugars, added sugars and dietary glucose load were found to be significantly positively associated with measures of glycemic variability in a cohort of adults with CF who were not on insulin or other glucose-lowering therapies [29].

4.2. Dairy and Cystic Fibrosis

Individuals with CF had suboptimal scores for dairy intake, but interestingly, scores were slightly higher than the general population. Currently, there are no specific guidelines for calcium intake in CF, and in the United States, it is recommended that people with CF adhere to the 1997 Institute of Medicine (IOM) guidelines for calcium intake for the general population [16,30]. Only 48.5% of Australian adults with CF met daily requirements for dairy intake [23]. In a study by Calvo et al., average calcium intake ranged from 614–1022 mg daily, below 1997 IOM recommendations for 1300–1500 mg daily [30]. Interestingly on food frequency questionnaires, dairy products were cited as the most frequently consumed food group among people with CF [11]. Vitamin D, which also plays an important role in bone health, is monitored annually, and as most individuals with CF are unable to meet needs through dietary intake alone, regardless of dairy consumption, dosing is based on serum levels, with adjustments as needed to achieve a target level of >30 ng/mL; however, there is a lack of international consensus on this goal [31,32].

With improved longevity in CF, CF bone disease (CFBD) is an emerging problem and can impact both pulmonary function and overall health [32]. As calcium and vitamin D play critical roles in bone formation and maintenance, it is important to better understand (a) the role that these specific nutrients play in the pathogenesis of CFBD and (b) the optimal intake that is necessary for individuals with CF, as this may be different than the general population.

4.3. Protein and Cystic Fibrosis

In our study, both total and seafood/plant protein were suboptimal in CF and lower than the reference population. Current recommendations are for 20% of calories in the CF diet come from protein [33]. Additionally, it has been shown that individuals with CF have lower lean muscle mass and increased systemic inflammation, which both are associated with increased dietary protein needs [34]. In a study by Calvo et al., protein intake was adequate in people with CF according to these recommendations, but in an earlier longitudinal study by Smith et al., intake was consistently less than the recommended 20% of total caloric intake [11,35]. Additionally, in patients with CF and exocrine pancreatic insufficiency (EPI), protein digestibility is severely impaired. While PERT can improve digestibility, the process remains severely delayed [36]. Thus, further studies are needed to examine (a) whether increased protein intake can improve lean body mass in people with CF and whether (b) protein digestion can be optimized in people with EPI.

Finally protein source deserves additional consideration. Even when protein intake is adequate in people with CF, our study echoed prior findings that these needs are primarily met through intake of meat and dairy products as opposed to plant based protein [11]. In a systematic review and meta-analysis of protein intake and all-cause mortality in the general population, intake of plant protein was associated with a lower risk of all cause and cardiovascular disease mortality. This inverse relationship remained significant in studies that controlled for energy, BMI and macronutrient intake [37]. It will be important to understand if these findings also apply in the CF population.

4.4. Saturated Fat and Cystic Fibrosis

A high fat diet has been a hallmark of nutritional therapy for CF, as fat malabsorption can occur even with optimal enzyme supplementation [33]. Historically, the recommendations for fat intake were not typically achieved in people with CF; however, in more recent years, these recommendations are consistently being met and exceeded. Our study found that individuals with CF did not consume significantly more total or saturated fat than the general population.

While a low fat diet has been associated with increased mortality in CF, a high fat diet without attention to type of dietary fat may also be problematic.4 Multiple studies have shown high saturated fat intake in individuals with CF [11,35]. Despite high total fat intake, individuals with CF are at risk of essential fatty acid deficiency. The composition of dietary fat intake has been shown to play a role in development of this condition, with intake of mono and polyunsaturated fat intake related to higher levels of essential fatty acids in CF [38].

Saturated fat intake has also been shown to play a role in the pathogenesis of cardiovascular disease. While individuals with CF were previously not to be thought to be at increased risk of cardiovascular disease, more recently, higher augmentation index, a composite vascular parameter of both arterial stiffness and global peripheral wave reflection, was shown to be higher in adults with CF [39]. These studies suggest that while a high fat diet likely remains important in CF, additional attention should be paid to the composition of dietary fat, with increased intake of mono and polyunsaturated fats, and decreased consumption of saturated fat.

4.5. Strengths, Limitations and Future Directions

This study is the first, to our knowledge, to examine the dietary quality of children and young adults using specific sub-scores of components of the HEI, which is pertinent to understanding dietary patterns during periods of growth and development in CF. Dietary data for this study were collected by Registered Dietitian Nutritionists using a 24-h recall method, not participant diet records. Additionally, all participants' dietary data met the minimum requirements for inclusion based on %EER. Limitations include a cross-sectional study design; a small, geographically localized sample size; and no matched control group. Examining dietary data in a larger, more representative sample of people with CF would allow for greater generalizability. Additionally, while interviews regarding dietary intake are vulnerable to inaccurate reporting by subjects, the use of a trained diet technician in data collection allows for increased detail that is often missed when subjects report dietary intake independently.

5. Conclusions

In summary, a cohort of young, relatively healthy individuals with CF generally had poor dietary quality as measured by the Healthy Eating Index. In the context of increased prevalence of overweight and obesity, and improved longevity, additional research is warranted to determine the most appropriate dietary recommendations to support overall health in people with CF.

Author Contributions: R.M.B.: conceptualization, methodology, investigation, data curation, writing—original draft preparation. A.T.: conceptualization, methodology, investigation, formal analysis, data curation, writing—review and editing. S.S.: conceptualization, methodology, investigation, data collection, data curation, writing—review and editing. All authors have read and agreed to the published version of the manuscript.

Funding: This research was funded by the NIDDK, grant number K23 DK107937 (to S.S.); Penn and CHOP Centers for Human Phenomic Science, grant number UL1 TR001878; and the Cystic Fibrosis Foundation, grant number BASS20D0 (to R.B.).

Institutional Review Board Statement: The study was conducted in accordance with the Declaration of Helsinki, and approved by the Institutional Review Board (or Ethics Committee) of Children's Hospital of Philadelphia (IRB 15-012279, approval date: 26 October 2015).

Informed Consent Statement: Informed consent was obtained from all subjects involved in the study.

Data Availability Statement: Data for CF participants are not publicly available. Dietary data for the general population were obtained from What We Eat In America (WWEIA) portion of the 2015–2016 National Health and Nutrition Examination Survey (NHANES).

Conflicts of Interest: The authors declare no conflict of interest.

References

1. Andersen, D.H. Cystic Fibrosis of the Pancreas and Its Relation to Celiac Disease: A Clinical and Pathologic Study. *Am. J. Dis. Child.* **1938**, *56*, 344–399. [CrossRef]
2. Wilschanski, M.; Novak, I. The Cystic Fibrosis of Exocrine Pancreas. *Cold Spring Harb. Perspect. Med.* **2013**, *3*, a009746. [CrossRef] [PubMed]
3. Bass, R.; Brownell, J.; Stallings, V. The Impact of Highly Effective CFTR Modulators on Growth and Nutrition Status. *Nutrients* **2021**, *13*, 2907. [CrossRef] [PubMed]
4. Corey, M.; McLaughlin, F.J.; Williams, M.; Levison, H. A Comparison of Survival, Growth, and Pulmonary Function in Patients with Cystic Fibrosis in Boston and Toronto. *J. Clin. Epidemiol.* **1988**, *41*, 583–591. [CrossRef]
5. Zemel, B.S.; Jawad, A.F.; FitzSimmons, S.; Stallings, V.A. Longitudinal Relationship among Growth, Nutritional Status, and Pulmonary Function in Children with Cystic Fibrosis: Analysis of the Cystic Fibrosis Foundation National CF Patient Registry. *J. Pediatr.* **2000**, *137*, 374–380. [CrossRef]
6. Stallings, V.A.; Stark, L.J.; Robinson, K.A.; Feranchak, A.P.; Quinton, H. Evidence-Based Practice Recommendations for Nutrition-Related Management of Children and Adults with Cystic Fibrosis and Pancreatic Insufficiency: Results of a Systematic Review. *J. Am. Diet. Assoc.* **2008**, *108*, 832–839. [CrossRef]
7. Poulimeneas, D.; Grammatikopoulou, M.G.; Devetzi, P.; Petrocheilou, A.; Kaditis, A.G.; Papamitsou, T.; Doudounakis, S.E.; Vassilakou, T. Adherence to Dietary Recommendations, Nutrient Intake Adequacy and Diet Quality among Pediatric Cystic Fibrosis Patients: Results from the GreeCF Study. *Nutrients* **2020**, *12*, 3126. [CrossRef]
8. Sutherland, R.; Katz, T.; Liu, V.; Quintano, J.; Brunner, R.; Tong, C.W.; Collins, C.E.; Ooi, C.Y. Dietary Intake of Energy-Dense, Nutrient-Poor and Nutrient-Dense Food Sources in Children with Cystic Fibrosis. *J. Cyst. Fibros.* **2018**, *17*, 804–810. [CrossRef]
9. Woestenenk, J.W.; Castelijns, S.J.A.M.; van der Ent, C.K.; Houwen, R.H.J. Dietary Intake in Children and Adolescents with Cystic Fibrosis. *Clin. Nutr.* **2014**, *33*, 528–532. [CrossRef]
10. Kawchak, D.A.; Zhao, H.; Scanlin, T.F.; Tomezsko, J.L.; Cnaan, A.; Stallings, V.A. Longitudinal, Prospective Analysis of Dietary Intake in Children with Cystic Fibrosis. *J. Pediatr.* **1996**, *129*, 119–129. [CrossRef]
11. Calvo-Lerma, J.; Hulst, J.; Boon, M.; Martins, T.; Ruperto, M.; Colombo, C.; Fornés-Ferrer, V.; Woodcock, S.; Claes, I.; Asseiceira, I.; et al. The Relative Contribution of Food Groups to Macronutrient Intake in Children with Cystic Fibrosis: A European Multicenter Assessment. *J. Acad. Nutr. Diet.* **2019**, *119*, 1305–1319. [CrossRef]
12. Hanna, R.M.; Weiner, D.J. Overweight and Obesity in Patients with Cystic Fibrosis: A Center-Based Analysis. *Pediatr. Pulmonol.* **2015**, *50*, 35–41. [CrossRef] [PubMed]
13. Harindhanavudhi, T.; Wang, Q.; Dunitz, J.; Moran, A.; Moheet, A. Prevalence and Factors Associated with Overweight and Obesity in Adults with Cystic Fibrosis: A Single-Center Analysis. *J. Cyst. Fibros.* **2020**, *19*, 139–145. [CrossRef] [PubMed]
14. Heijerman, H.G.M.; McKone, E.F.; Downey, D.G.; Van Braeckel, E.; Rowe, S.M.; Tullis, E.; Mall, M.A.; Welter, J.J.; Ramsey, B.W.; McKee, C.M.; et al. Efficacy and Safety of the Elexacaftor plus Tezacaftor plus Ivacaftor Combination Regimen in People with Cystic Fibrosis Homozygous for the F508del Mutation: A Double-Blind, Randomised, Phase 3 Trial. *Lancet* **2019**, *394*, 1940–1948. [CrossRef]
15. Middleton, P.G.; Mall, M.A.; Dřevínek, P.; Lands, L.C.; McKone, E.F.; Polineni, D.; Ramsey, B.W.; Taylor-Cousar, J.L.; Tullis, E.; Vermeulen, F.; et al. Elexacaftor-Tezacaftor-Ivacaftor for Cystic Fibrosis with a Single Phe508del Allele. *N. Engl. J. Med.* **2019**, *381*, 1809–1819. [CrossRef] [PubMed]
16. Borowitz, D.; Lubarsky, B.; Wilschanski, M.; Munck, A.; Gelfond, D.; Bodewes, F.; Schwarzenberg, S.J. Nutritional Status Improved in Cystic Fibrosis Patients with the G551D Mutation After Treatment with Ivacaftor. *Dig. Dis. Sci.* **2016**, *61*, 198–207. [CrossRef] [PubMed]
17. Davies, J.C.; Cunningham, S.; Harris, W.T.; Lapey, A.; Regelmann, W.E.; Sawicki, G.S.; Southern, K.W.; Robertson, S.; Green, Y.; Cooke, J.; et al. Safety, Pharmacokinetics, and Pharmacodynamics of Ivacaftor in Patients Aged 2-5 Years with Cystic Fibrosis and a CFTR Gating Mutation (KIWI): An Open-Label, Single-Arm Study. *Lancet Respir. Med.* **2016**, *4*, 107–115. [CrossRef]
18. McDonald, C.M.; Bowser, E.K.; Farnham, K.; Alvarez, J.A.; Padula, L.; Rozga, M. Dietary Macronutrient Distribution and Nutrition Outcomes in Persons with Cystic Fibrosis: An Evidence Analysis Center Systematic Review. *J. Acad. Nutr. Diet.* **2021**, *121*, 1574–1590.e3. [CrossRef]
19. Krebs-Smith, S.M.; Pannucci, T.E.; Subar, A.F.; Kirkpatrick, S.I.; Lerman, J.L.; Tooze, J.A.; Wilson, M.M.; Reedy, J. Update of the Healthy Eating Index: HEI-2015. *J. Acad. Nutr. Diet.* **2018**, *118*, 1591–1602. [CrossRef]
20. Center for Disease Control and Prevention. *National Health and Nutrition Examination Survey Data, 2015–2016*; U.S. Department of Health and Human Services: Hyattsville, MD, USA.
21. Available online: https://www.fns.usda.gov/How-Hei-Scored (accessed on 17 December 2021).
22. U.S. Department of Agriculture; U.S. Department of Health and Human Services. *Dietary Guidelines for Americans, 2020–2025*, 9th ed.; U.S. Government Publishing Office: Washington, DC, USA. Available online: DietaryGuidelines.Gov (accessed on 19 October 2021).
23. Tierney, A.C.; Fong-To, S.A.; Clode, M.; Casamento, J.; Yuen, A.T.W.; King, S. 234 Vegetable, Fruit, Dairy and Discretionary Food Intake of Cystic Fibrosis (CF) Patients and Comparison with the General Population: A Cross-Sectional Study. *J. Cyst. Fibros.* **2015**, *14*, S118. [CrossRef]

24. Gavin, J.; Ellis, J.; Dewar, A.L.; Rolles, C.J.; Connett, G.J. Dietary Fibre and the Occurrence of Gut Symptoms in Cystic Fibrosis. *Arch. Dis. Child.* **1997**, *76*, 35–37. [CrossRef]
25. Proesmans, M.; De Boeck, K. Evaluation of Dietary Fiber Intake in Belgian Children with Cystic Fibrosis: Is There a Link with Gastrointestinal Complaints? *J. Pediatr. Gastroenterol. Nutr.* **2002**, *35*, 610–614. [CrossRef] [PubMed]
26. Nowak, J.K.; Szczepanik, M.; Wojsyk-Banaszak, I.; Mądry, E.; Wykrętowicz, A.; Krzyżanowska-Jankowska, P.; Drzymała-Czyż, S.; Nowicka, A.; Pogorzelski, A.; Sapiejka, E.; et al. Cystic Fibrosis Dyslipidaemia: A Cross-Sectional Study. *J. Cyst. Fibros.* **2019**, *18*, 566–571. [CrossRef] [PubMed]
27. Hannon, B.A.; Thompson, S.V.; Edwards, C.G.; Skinner, S.K.; Niemiro, G.M.; Burd, N.A.; Holscher, H.D.; Teran-Garcia, M.; Khan, N.A. Dietary Fiber Is Independently Related to Blood Triglycerides Among Adults with Overweight and Obesity. *Curr. Dev. Nutr.* **2019**, *3*, nzy094. [CrossRef] [PubMed]
28. Zhou, Q.; Wu, J.; Jie, T.; Wang, J.-J.; Lu, C.-H.; Wang, P.-X. Beneficial Effect of Higher Dietary Fiber Intake on Plasma HDL-C and TC/HDL-C Ratio among Chinese Rural-to-Urban Migrant Workers. *Int. J. Environ. Res. Public Health* **2015**, *12*, 4726–4738. [CrossRef] [PubMed]
29. Armaghanian, N.; Atkinson, F.; Taylor, N.; Kench, A.; Brand-Miller, J.; Markovic, T.; Steinbeck, K. Dietary Intake in Cystic Fibrosis and Its Role in Glucose Metabolism. *Clin. Nutr.* **2020**, *39*, 2495–2500. [CrossRef]
30. Institute of Medicine (US) Standing Committee on the Scientific Evaluation of Dietary Reference Intakes. *Dietary Reference Intakes for Calcium, Phosphorus, Magnesium, Vitamin D, and Fluoride*; The National Academies Collection: Reports funded by National Institutes of Health; National Academies Press (US): Washington, DC, USA, 1997.
31. Chesdachai, S.; Tangpricha, V. Treatment of Vitamin D Deficiency in Cystic Fibrosis. *J. Steroid Biochem. Mol. Biol.* **2016**, *164*, 36–39. [CrossRef]
32. Putman, M.S.; Anabtawi, A.; Le, T.; Tangpricha, V.; Sermet-Gaudelus, I. Cystic Fibrosis Bone Disease Treatment: Current Knowledge and Future Directions. *J. Cyst. Fibros.* **2019**, *18*, S56–S65. [CrossRef]
33. Matel, J.L.; Kerner, J.A. Nutritional Management of Cystic Fibrosis. *J. Parenter. Enter. Nutr.* **2012**, *36*, 60S–67S. [CrossRef]
34. Deutz, N.E.P.; Bauer, J.M.; Barazzoni, R.; Biolo, G.; Boirie, Y.; Bosy-Westphal, A.; Cederholm, T.; Cruz-Jentoft, A.; Krznaric, Z.; Nair, K.S.; et al. Protein Intake and Exercise for Optimal Muscle Function with Aging: Recommendations from the ESPEN Expert Group. *Clin. Nutr.* **2014**, *33*, 929–936. [CrossRef]
35. Smith, C.; Winn, A.; Seddon, P.; Ranganathan, S. A Fat Lot of Good: Balance and Trends in Fat Intake in Children with Cystic Fibrosis. *J. Cyst. Fibros.* **2012**, *11*, 154–157. [CrossRef] [PubMed]
36. Engelen, M.P.K.J.; Com, G.; Anderson, P.J.; Deutz, N.E.P. New Stable Isotope Method to Measure Protein Digestibility and Response to Pancreatic Enzyme Intake in Cystic Fibrosis. *Clin. Nutr.* **2014**, *33*, 1024–1032. [CrossRef] [PubMed]
37. Naghshi, S.; Sadeghi, O.; Willett, W.C.; Esmaillzadeh, A. Dietary Intake of Total, Animal, and Plant Proteins and Risk of All Cause, Cardiovascular, and Cancer Mortality: Systematic Review and Dose-Response Meta-Analysis of Prospective Cohort Studies. *BMJ* **2020**, *370*, m2412. [CrossRef] [PubMed]
38. Maqbool, A.; Schall, J.I.; Gallagher, P.R.; Zemel, B.S.; Strandvik, B.; Stallings, V.A. Relation Between Dietary Fat Intake Type and Serum Fatty Acid Status in Children With Cystic Fibrosis. *J. Pediatric Gastroenterol. Nutr.* **2012**, *55*, 605–611. [CrossRef]
39. Hull, J.H.; Garrod, R.; Ho, T.B.; Knight, R.K.; Cockcroft, J.R.; Shale, D.J.; Bolton, C.E. Increased Augmentation Index in Patients with Cystic Fibrosis. *Eur. Respir. J.* **2009**, *34*, 1322–1328. [CrossRef]

Systematic Review

Effects of Exercise on Nutritional Status in People with Cystic Fibrosis: A Systematic Review

William B. Nicolson [1], Julianna Bailey [2,3], Najlaa Z. Alotaibi [4], Stefanie Krick [2,3,*] and John D. Lowman [3,4,*]

1. Graduate Medical Education, Heersink School of Medicine, University of Alabama at Birmingham, Birmingham, AL 35294, USA; wbnicolson@uabmc.edu
2. Division of Pulmonary, Allergy and Critical Care Medicine, University of Alabama at Birmingham, Birmingham, AL 35294, USA; juliannabailey@uabmc.edu
3. Gregory Fleming Cystic Fibrosis Research Center, University of Alabama at Birmingham, Birmingham, AL 35294, USA
4. Department of Physical Therapy, University of Alabama at Birmingham, Birmingham, AL 35294, USA; nzalotai@uab.edu
* Correspondence: skrick@uabmc.edu (S.K.); jlowman@uab.edu (J.D.L.); Tel.: +1-205-975-3043 (S.K.); +1-205-934-5892 (J.D.L.)

Abstract: Background: Physical exercise is an important part of regular care for people with cystic fibrosis (CF). It is unknown whether such exercise has beneficial or detrimental effects on nutritional status (body composition). Thus, the objective of this review was to evaluate the effect of exercise on measures of nutritional status in children and adults with CF. Methods: Standardized reporting guidelines for systematic reviews were followed and the protocol was prospectively registered. Multiple databases were utilized (e.g., PubMed, Scopus, and CINHAL). Two reviewers independently reviewed titles/abstracts and then the full text for selected studies. Results: In total, 924 articles were originally identified; data were extracted from 4 eligible studies. These four studies included only children; pulmonary function ranged from severe to normal, and the majority of participants were at or below their recommended weight. Exercise training did not worsen nutritional status in any study; two studies that included resistance exercise reported an increase in fat-free mass. Three of the four studies also reported increased aerobic capacity and/or muscle strength. Conclusions: Exercise training can produce positive physiologic changes in children with CF without impairing their nutritional status. In fact, resistance exercise can help improve body mass. Much less is known about how exercise may affect adults or those who are overweight.

Keywords: cystic fibrosis; exercise; nutritional status; body mass index; body mass; anthropometric

Citation: Nicolson, W.B.; Bailey, J.; Alotaibi, N.Z.; Krick, S.; Lowman, J.D. Effects of Exercise on Nutritional Status in People with Cystic Fibrosis: A Systematic Review. *Nutrients* **2022**, *14*, 933. https://doi.org/10.3390/nu14050933

Academic Editors: Maria R. Mascarenhas and Jessica Alvarez

Received: 5 January 2022
Accepted: 16 February 2022
Published: 22 February 2022

Publisher's Note: MDPI stays neutral with regard to jurisdictional claims in published maps and institutional affiliations.

Copyright: © 2022 by the authors. Licensee MDPI, Basel, Switzerland. This article is an open access article distributed under the terms and conditions of the Creative Commons Attribution (CC BY) license (https://creativecommons.org/licenses/by/4.0/).

1. Introduction

Cystic fibrosis (CF) is a relatively rare genetic disease affecting over 30,000 people in the United States and more than 70,000 people worldwide [1], with a prevalence varying from country to country but being as high as 1 in 900 in parts of Canada to as low as 1 in 25,000 in Finland [2]. CF is caused by a mutation in the gene responsible for the cystic fibrosis transmembrane conductance regulator (CFTR). This protein is expressed in epithelial cells and serves to directly transport chloride and indirectly affects sodium and water transport. CFTR dysfunction leads to sticky mucus, causing mucus obstruction in various organs including the lungs, pancreas, liver, and intestines. Therefore, cystic fibrosis is a multisystem disease, leading to a decreased life expectancy and significantly impaired quality of life.

CF care requires a multidisciplinary team. It not only focuses on preserving pulmonary function, but also the organ-specific and systemic manifestations of the disease as mentioned above. Malnutrition is a common problem among CF patients, and it is a consequence of multiple factors. Poor bicarbonate secretion from the pancreas, mucosal abnormalities leading to poor intestinal wall function, and poor gut transit time are all thought to contribute to decreased fat absorption [3]. Patients with more pulmonary disease manifestations have a higher concentration of circulating inflammatory markers, which has been linked to decreased fat-free mass (FFM) and bone mineral density (BMD) [4]. People with CF also have an increased resting expenditure rate at baseline [5].

Given that people with CF struggle with malnutrition, their nutritional status, assessed via anthropometric measures, most commonly body mass index (BMI, for adults) or BMI percentile (for children), is also a primary focus of CF care. BMI has been identified as an independent predictor of mortality in cystic fibrosis, with one study demonstrating a hazard ratio of 5.5 (CI 1.8–16.8) for adolescents 12 to 14 years old with a BMI of 15.8 or less [6]. BMI also has implications for morbidity in patients with CF; a cross-sectional study demonstrated decreased FEV_1 in patients whose weight was less than 90% predicted [7]. Current CF guidelines recommend BMI goals for individuals with CF; children aged 2–20 are recommended to maintain a BMI $\geq$ 50th percentile, while adult women are recommended to maintain a BMI of 22–25 and adult men a BMI of 23–25 [8].

The morbidity and mortality of people with CF can also be predicted by their exercise capacity. It has been demonstrated that maximal $\dot{V}_{O_2}$ from cardiopulmonary exercise testing (CPET) can also serve as a predictor of mortality. Both Nixon et al. (1992) [9] and, more recently (2019), Hebestreit et al. [10] found a stepwise increase in survival for people with CF based on increased quantiles of percent predicted peak $\dot{V}_{O_2}$. Another study examined the longitudinal relationship between habitual physical activity and FEV_1, finding that those who were more physically active had a slower decline in FEV_1 [11].

Thus, there appears to be a potential conflict between nutritional status and exercise. Patients have an increased resting energy expenditure [5], and exercise would further increase total energy expenditure, perhaps worsening their nutritional status by causing additional weight loss. However, since patients with CF can improve their aerobic capacity through exercise, it remains unclear how exercise may affect their nutritional status/body composition. Most people think of exercise as a way to maintain or lose weight; thus, some people with CF that are underweight or at their goal weight may be reluctant to begin an exercise program. On the contrary, with the advent of highly effective modulator therapies, some patients are now concerned about gaining too much weight [12,13].

Thus, the goal of this systematic review is to help answer the question: do exercise and physical activity affect nutritional status in children and adults with cystic fibrosis? This question has clinical relevance due to the morbidity and mortality implications of malnutrition in this patient population and the perceived risk of weight loss in a population that has historically been underweight. In this new era of highly effective modulator therapies, there is now the potential risk for both normal weight obesity (increased fat mass with an otherwise normal body mass) as well as outright overweight and obese, especially as more people begin these drugs at a younger age and are on them for longer periods of time [12,13].

2. Materials and Methods

2.1. Systematic Review Design and Registration and Design

This systematic review was planned and conducted according to the Preferred Reporting Items for Systematic Reviews and Meta-Analyses guideline [14]. The protocol was registered in PROSPERO (CRD42021273303) [15].

2.2. Data Sources and Searches

After development of our population, intervention, comparator, outcome, and study design (PICOS) question, a medical librarian (MMB) developed a specific search strategy for multiple databases (PubMed, Scopus, Embase, CINHAL, SPORTDiscus and CENTRAL). The latest search was conducted on 20 August 2021 and all relevant records were imported into Covidence, an online software platform for conducting systematic reviews [16]. The search strategies used are in Supplementary Table S1.

2.3. Eligibility Criteria

Table 1 highlights the inclusion and exclusion criteria used in our review. The following nutritional status outcomes were considered: BMI, BMI percentile, BMI z-score, body mass, and fat-free mass.

Table 1. Summary of inclusion and exclusion criteria based on population/patient, intervention, comparator, outcome, and study design (PICOS).

PICOS Parameter	Inclusion Criteria	Exclusion Criteria
Population	Children and adults with cystic fibrosis (underweight, normal weight, or overweight)	Infants, toddlers and preschoolers (<5 years old)
Intervention	Exercise or physical activity	Passive exercise (e.g., stretching, range of motion)
Comparison	Non-exposed control group	
Outcome	Body mass index, body mass, body composition (e.g., fat-free mass)	
Study design	Randomized controlled trials	Language other than English, German, Spanish, or French

2.4. Study Selection

Titles and abstracts of all identified articles were independently assessed, in duplicate by 3 reviewers (WBN, NZA, and JDL) using Covidence [16]. Titles and abstracts that did not provide sufficient information on the inclusion and exclusion criteria were then selected for evaluation of the full text and were included according to the eligibility criteria. Disagreements between the 2 reviewers were resolved by consultation with a third reviewer.

2.5. Data Extraction

Data were extracted through a standardized spreadsheet (MS Excel, Microsoft Corporation, Seattle, WA, USA) created by the authors. Extracted data included publication details, study methodology, baseline participant characteristics, intervention description, and outcomes assessed. Disagreements were also resolved by consensus. The main outcomes were anthropometrics measures (e.g., BMI and body mass). Secondary outcomes included peak $\dot{V}_{O_2}$ and strength.

2.6. Quality Assessment

The Physiotherapy Evidence Database (PEDro) scale was used to evaluate methodological quality and risk of bias of the randomized controlled trials selected in this study. The quality assessment was performed by 3 independent reviewers (NZA, SK, and JB). Any items that were unclear were rated as a "no." Total scores were calculated based on 10 of the 11 items in the tool.

2.7. Data Synthesis and Analysis

Due to the heterogeneity of the outcomes collected, a meta-analysis could not be conducted. However, data extracted were quantitatively and qualitatively summarized in tabular format. BMI-for-age percentiles and weight-for-age percentiles were estimated by plotting values and extrapolating results on the National Center for Chronic Disease Prevention and Health Promotion's growth charts [17].

3. Results

3.1. Study Search Results and Selection

The search strategy resulted in 924 articles, of which 122 were considered relevant for a more detailed analysis; 4 of these studies met the eligibility criteria and were included in the systematic review [18–21]. Details of the selection process, including reasons for exclusion, are illustrated in Figure 1.

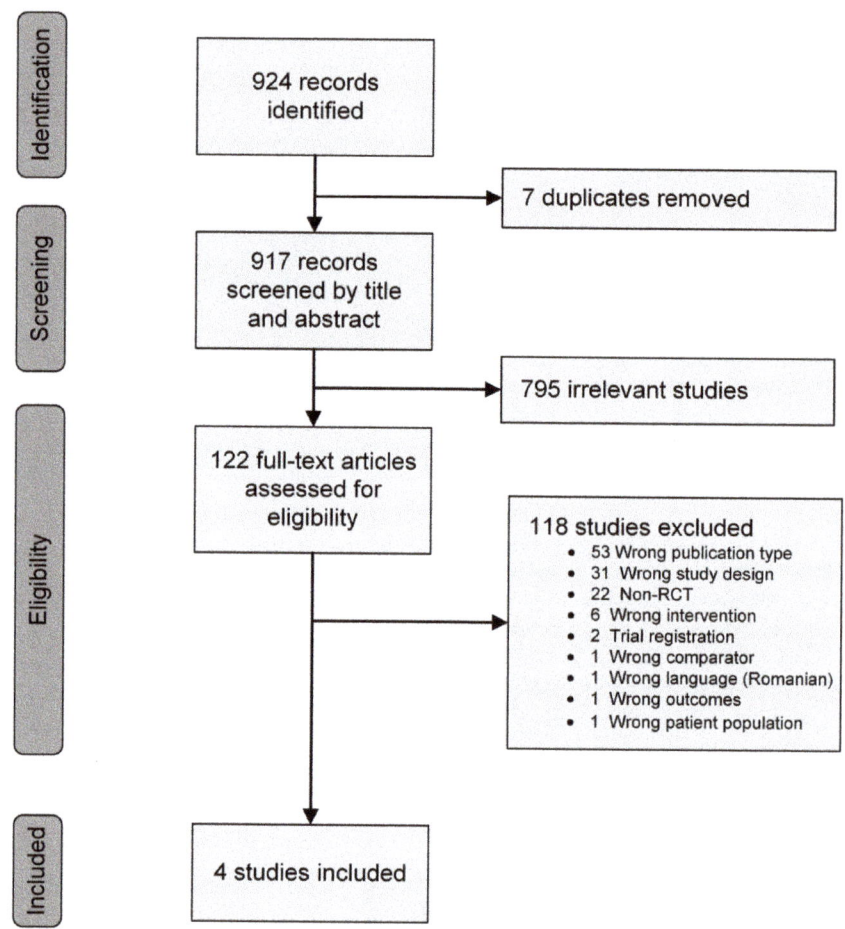

Figure 1. PRISMA flow diagram of the selection process and study search results.

3.2. Description of the Studies

Four studies evaluated participants' nutritional status using BMI or other anthropometric outcomes following active exercise and compared it with a control group [18–21]. Two of these reported either BMI or BMI z-scores [18,20]; three studies reported body mass and fat-free mass [19–21] and one reported upper-extremity skin fold and circumference [18]. No studies were conducted in North America. All four studies included children and no study included adults. Percent predicted FEV_1 for study participants ranged from severe to normal (Table 2), and one study reported using supplemental oxygen as needed [19]. Based on estimated mean age and body mass, 3 of the 4 studies were at or below the 25th percentile [19–21] and the other study was at the 50th percentile [18]; hence, almost all of the participants in the study were either normal or underweight. As recommended in a recent consensus statement [22], three used CPET results to guide the exercise prescription [19–21] (Table 3) and all four studies used CPET (i.e., peak $\dot{V}_{O_2}$) for outcome assessment (Table 4) [18–21].

Table 2. Study site and baseline population demographics (sex, age, pulmonary function and nutritional status) for all participants.

Ref.	First Author	Year	State/Country	Number of Participants (Female)	Age	FEV$_1$%	Body Mass (kg)	Weight for Age% [†]	BMI (kg/m^2)	BMI for Age% [§]
[19]	Selvadurai, H.C.	2002	New South Wales, Australia	66 (38)	13 (2)	57 (17)	38.0 (7.8)	16th	NA	NA
[20]	Santana-Sosa, E.	2012	Madrid, Spain	22 (9)	10.5 (2)	83 (11) *	37.0 (3.0)	65th	17.8	61st
[21]	Santana-Sosa, E.	2014	Madrid, Spain	20 (8)	10.5 (1)	73 (9) *	34.0 (3.8)	47th	16.1	34th
[18]	Hommerding, P.X.	2015	Rio Grande do Sul, Brazil	34 (14)	13 (3)	98 (20)	45.6 (15.3)	50th	NA	NA

Values reported as the mean (standard deviation) of combined intervention and control groups (NA: not available based on data provided). FEV$_1$%: percent predicted of forced expiratory volume in 1 s; BMI: body mass index. * Estimated values based on raw FEV (L/s) and other data reported in manuscript. [†] Estimated weight-for-age percentile based on reported sex proportion and mean age and body mass for all participants. [§] Estimated based on sex, age, and reported BMI using standardized growth charts.

Table 3. Characteristics of intervention setting and groups.

Ref.	Setting (Duration)	Exercise Group(s)		Control Group
[19]	Acute/Inpatient (~2–3 wks)	**Aerobic exercise training** *Mode:* treadmill or stationary cycling *Intensity:* 70% of HR$_{peak}$ *Duration:* 30 min *Frequency:* 5 d/wk *Other:* Supplemental oxygen was titrated to keep SpO$_2$ > 90% (if needed). Training was stopped if dyspnea ≥ 7 on Borg CR10 scale. Each session was individually supervised	**Resistance exercise training** *Mode:* Isotonic weight machines *Intensity:* 70% 1 RM *Duration:* 5 sets of 10 repetitions *Frequency:* 5 d/wk *Other:* Upper- and lower-extremity exercises (specific exercises and number not defined). Each session was individually supervised	No exercise

Table 3. Cont.

Ref.	Setting (Duration)	Exercise Group(s)	Control Group
[20]	Hospital-based, outpatient gym (8 weeks)	**Aerobic exercise training** *Mode:* Cycle ergometer *Intensity:* HR at ventilatory threshold (determined during exercise test) *Duration:* 20–40 min *Frequency:* 3 d/wk *Other:* HR monitor was worn during aerobic exercise. 10 min warmup on cycle. Each session was individually supervised **Resistance exercise training** *Mode:* Isotonic weight machines (bench press, shoulder press, leg extension, leg press, leg curl, abdominal crunch, low back extension, arm curl, elbow extension, seated row, and lateral pulldown) *Intensity:* Progressive, from 40 to 60% of 5 RM *Duration:* 3 circuits of 1 set of 12–15 repetitions of each exercise *Frequency:* 3 d/wk (following aerobic exercise session)	Chest physiotherapy twice daily and provided verbal instruction on the benefits of physical activity
[21]	Hospital-based, outpatient gym (8 weeks)	**Aerobic exercise** *Mode:* Cycle ergometer and "active playing" (i.e., running and soccer) *Intensity:* HR at ventilatory threshold (determined during exercise test) *Duration:* 20–40 min *Frequency:* 3 d/wk *Other:* HR monitor was worn during aerobic exercise. 10 min warmup on cycle. Each session was individually supervised **Resistance exercise** *Mode:* isotonic weight machine (leg press, pull down, leg extension, bench press, leg curl, seated row and abdominal crunch) *Intensity:* Progressive, beginning at 50% of 5 RM *Duration:* 3 circuits of 1 set of 12–15 repetitions of each exercise *Frequency:* 3 d/wk (following aerobic exercise) **Inspiratory muscle training** *Mode:* PowerBreathe threshold loading device *Intensity:* 40–50% of PI_{max}. *Duration:* 30 breaths *Frequency:* twice daily, 6–7 d/wk *Other:* One IMT session was performed during the 3 d/wk supervised sessions; the remaining IMT sessions were performed independently at home	Chest physiotherapy twice daily, IMT at 10% of PI_{max}, and provided instruction on the benefits of physical activity

Table 3. Cont.

Ref.	Setting (Duration)	Exercise Group(s)	Control Group
[18]	Home-based with tele-health follow-up every 2 weeks (3 months)	**Aerobic exercise** *Mode:* Self-selected per recommendations (e.g., walking, jogging, swimming, dancing skipping rope) *Intensity:* No recommendations given *Duration:* ≥20 min *Frequency:* at least 2 d/wk *Other:* Written manual of aerobic and stretching exercises provided	Verbal instructions regarding aerobic exercise which are part of routine outpatient care

Abbreviations: HR = heart rate; HRpeak = peak HR; SpO_2 = pulse oximetry saturation; Borg CR10 = Borg category ratio 10 scale; 1 RM = 1 repetition maximum; 5 RM = 5 repetition maximum; PImax = maximal inspiratory pressure; IMT = inspiratory muscle training.

Table 4. Nutritional status and physiologic outcomes of randomized controlled trials of exercise in people with CF.

Ref.	Nutritional Status Outcomes		Physiologic Outcomes	Conclusions
	BMI	Other		
[19]	Not reported/unable to calculate based on data reported	Δ body mass (kg): AET ↑ 0.80 (0.64) * RET ↑ 2.76 (0.70) * CTL ↑ 1.03 (0.58) * Δ fat-free mass (kg): AET ↑ 0.61 (0.37) * RET ↑ 2.40 (0.46) * CTL ↑ 0.60 (0.32) *	Δ $VO_{2\,peak}$ (mL/kg/min): AET ↑ 7.3 (6.3) * RET ↑ 0.7 (5.9) CTL ↓ 1.2 (6.2) Δ strength (Nm): AET ↑ 1.8 (6.2) RET ↑ 18.3 (7.0) * CTL ↓ 6.3 (6.1)	AET improved body composition (2%) and peak VO_2 (22%) RET improved body composition (7%) and was the only group to improve LE strength (18%) The CTL group had an improvement in body mass (2.7%), but an insignificant loss of strength and aerobic capacity
[20]	Δ BMI (kg/m^2): ET ↓ 0.1 CTL ↓ 0.1	Δ body mass (kg): ET ↑ 0.6 CTL ↑ 1.1 Δ fat-free mass (%): ET ↑ 1.3 CTL ↓ 0.2	Δ $VO_{2\,peak}$ (mL/kg/min): ET ↑ 3.9 (2–6) * CTL ↓ 2.2 (−5–0) Δ strength (kg): ET ↑ 10.5 (7–14) * CTL not reported	No significant changes in body composition variables. Peak VO_2 improved ~10% and strength ~25% in the ET group compared to a ~6% decrease in peak VO_2 and −2 to +5% change in strength of the CTL group

Table 4. Cont.

Ref.	Nutritional Status Outcomes		Physiologic Outcomes	Conclusions
	BMI	Other		
[21]	Not reported/unable to calculate based on data reported	Δ body mass (kg): ET ↑ 1.4 CTL ↑ 0.9 Δ fat-free mass (%): ET ↑ 1.0 CTL ↓ 0.1	Δ VO$_{2\text{peak}}$ (mL/kg/min): ET ↑ 6.9 * CTL ↓ 0.6 Δ strength (kg): ET ↑ 27 * CTL ↓ 1.3 Δ PI$_{max}$ (mm Hg): ET ↑ 39 * CTL ↑ 2.3	No significant changes in body mass, but fat-free mass increased in the ET group. Peak VO$_2$, LE strength, and inspiratory muscle strength increased 22%, 43%, and 58%, respectively, in the ET group. There were no significant changes in the CTL group
[18]	Δ BMI z-score: ET ↑ 0.2 (0.5) CTL ↑ 0.1 (0.2)	Δ Triceps skin fold ET ↑ 0.3 (1.3) CTL ↑ 0.1 (1.0) Δ Arm muscle circ. (cm) ET ↑ 0.1 (0.4) CTL ↓ 0.1 (0.2)	Δ VO$_{2\text{peak}}$ (mL/kg/min): ET ↑ 1.1 (4.6) CTL ↑ 2.3 (11.9)	In spite of self-reported increase in regular physical activity, there were no significant changes in any outcome measures in either group

Abbreviations: AET = aerobic exercise training group; RET = resistance exercise training group; ET = exercise training group; CTL = control group; Δ = change; ↑ = increase; ↓ = decrease; circ. = circumference. Data are presented as either the mean (SD) or the mean (95% CI); * indicates a significant change ($p < 0.05$).

Of these studies, one was performed in the hospital during an acute pulmonary exacerbation [19], two were performed in a hospital-based outpatient gym for children [20,21], and one was a home-based intervention with tele-rehab support [18] (Table 3). All four studies included aerobic exercise (AET), three at a moderate–vigorous intensity [19–21] and one with no specific intensity [18]. Two studies looked at AET as a separate intervention and two combined AET with resistance exercise training (RET) [20,21]. Selvadurai compared AET to RET and control [19], and Santana-Sosa's 2014 study combined inspiratory muscle training (IMT) [23] along with AET and RET compared to a control group [21]. The primary outcome was assessed at hospital discharge (2–3 weeks) in one study [19], 8 weeks in 2 studies [20,21] and 3 months in another [18].

Individual Study Descriptions

The earliest RCT to investigate the effects of exercise on nutritional status (body composition) in children was Selvadurai et al. (2002) [19]. They compared AET versus RET versus a control group during hospitalization for an acute pulmonary exacerbation. All groups received "intravenous antibiotics, chest physiotherapy, and nutritional supplementation." The respective training procedures for the exercise groups are described in Table 3. A maximal CPET using the modified Bruce protocol was used to guide the aerobic exercise prescription and as an outcome measure (peak $\dot{V}_{O_2}$). They also reported spirometry as an outcome. A daily 1 repetition maximum was used to guide the intensity of the resistance training group. Mean hospital length of stay for each group was ~19 days. Outcomes were assessed at discharge and 1 month post-discharge (1 month post-discharge results not shown).

Santana-Sosa et al.'s first study (2012) [20] examined a combination of AET and RET compared to a control group that only received verbal instructions regarding the benefits of exercise during an outpatient visit (Table 3). They also used results from a maximal CPET on a treadmill to guide intensity of AET (HR at the ventilatory threshold). The primary outcomes, assessed at 8 weeks, were peak $\dot{V}_{O_2}$ and muscle strength, but included body composition and pulmonary function as secondary outcomes.

A few years later (2014), Santana-Sosa et al. conducted a similar study [21] but included progressive IMT, using a threshold device (POWERbreathe), in addition to AET and RET (Table 3). Both of these studies also included a 4 week detraining phase (detraining results not shown).

The most recent study (2015), by Hommerding et al. [18] was a home-based intervention with tele-health support (Table 3). There was no direct supervision or reporting of adherence to the program, but, subjectively, the exercise group reported almost 4-fold more "regular physical activity"; however, only 35% of the exercise group reported exercising at least 3 days/wk, compared to 24% in the control group. Outcomes were assessed at three months. In addition to nutritional status outcomes (Table 4), they also reported spirometry and maximal CPET results (peak $\dot{V}_{O_2}$, exercise time, treadmill speed, and maximal HR).

3.3. Study Quality

All of the included studies had random allocation, baseline comparability, adequate follow up, between-group comparisons, and provided points estimates and variability (Table 5), with all scores ranging from 5 to 7. A PEDro score of 5 is considered "fair," and scores of 6–8 are considered "good" [24]. As is typical in most exercise studies, neither the participants nor the therapists were blinded to the intervention, but two did have blinded outcomes assessors and intention-to-treat analysis [20,21].

Table 5. Methodologic quality and statistical reporting assessment, using the PEDro scale, of randomized controlled trials evaluating the effect of exercise on anthropometric outcomes in people with CF.

PEDro Criteria	Selvadurai 2002 [19]	Santana-Sosa 2012 [20]	Santana-Sosa 2014 [21]	Hommerding 2015 [18]
1—Eligibility criteria	😊	😊	😊	☹
2—Random allocation	😊	😊	😊	😊
3—Concealed allocation	😊	☹	☹	☹
4—Baseline comparability	😊	😊	😊	😊
5—Blind subjects	☹	☹	☹	☹
6—Blind therapists	☹	☹	☹	☹
7—Blind assessors	☹	😊	😊	☹
8—Adequate follow-up	😊	😊	😊	😊
9—Intention-to-treat analysis	☹	😊	😊	☹
10—Between-group comparisons	😊	😊	😊	😊
11—Points estimates and variability	😊	😊	😊	😊
Total score	6	7	7	5

Note: Scores range from 0 to 10. Eligibility criteria (item 1) do not contribute to the total score.  indicates criteria was fulfilled; indicates criteria was not met.

3.4. Effects of Intervention

3.4.1. Nutritional Status Outcomes

One study reported raw BMI scores [20] while another reported BMI z-scores [18]; BMI z-scores are measures of relative weight adjusted for child age and sex [25]. One study saw a slight decrease in BMI in both groups [20] while the other reported a slight increase [18], but neither were clinically or statistically significant, either over time or between groups (Table 4). Body mass increased overtime in the intervention and control groups in the three studies in which it was assessed [19–21]; in two of these studies, the increase was insignificant [20,21] but the RET group in the Selvadurai study had both statistically significant and clinically meaningful increases in body mass (7.25%), which was all due to an increase in fat-free (i.e., muscle) mass [19]. Although statistically insignificant in Santana-Sosa's studies [20,21], fat-free mass increased in the exercise groups and decreased in the control groups.

Three of the four studies also included a follow-up period 4 weeks after the last supervised exercise session and post-training outcome assessment [19–21]. These results are mixed and likely the result of lack of standardization of this "detraining" period. Selvadurai reported that body mass and fat-free mass continued to rise in the aerobic training group, while body mass decreased in the resistance training group and remained stable in the control group; fat-free mass remained relatively unchanged in the resistance and control groups during the post-exercise period [19]. Santana-Sosa's initial study reported a stable weight, BMI and fat-free mass after 4 weeks of detraining [20], but their subsequent study reported a slight, but statistically insignificant, increase in body mass (0.5 kg) [21].

3.4.2. Physiologic Outcomes

Cardiorespiratory fitness, assessed as peak $\dot{V}_{O_2}$ from CPET, increased in the 3 studies that performed supervised AET at an appropriate intensity [19–21]; 2 of which reported an increase of over 20% (Table 4). It is also important to note that in spite of a potential learning effect, peak $\dot{V}_{O_2}$ decreased over time in the control groups of these three studies [19–21]. On the contrary, Hommerding did report an insignificant increase in peak $\dot{V}_{O_2}$ over 3 months in both the exercise and controls groups [18]. Whereas Selvadurai reported a stable or slightly higher peak $\dot{V}_{O_2}$ after detraining [19], both studies by Santana-Sosa reported a significant decline in peak $\dot{V}_{O_2}$ after cessation of regular training [20,21].

Strength was assessed in 3 studies using either an isokinetic dynomometer [19] or 5 RM on an isotonic weight machine [20,21]. Selvadurai reported an increase in lower-extremity strength of 18% in the RET but no significant change in AET or control groups [19] The studies by Santana-Sosa reported an increase in lower-extremity strength of 25 to 43% [20,21]. One study also trained and assessed inspiratory muscle performance; they reported an increase in PImax of 58% [21], which also seemed to be associated with an increase in peak $\dot{V}_{O_2}$ compared to their prior study (of similar design but without IMT) [20]. Surprisingly, the improvements in lower-extremity strength were preserved in all three studies that included a "detraining" phase [19–21]. Only Santana-Sosa's 2014 study reported a decrease in upper-extremity strength with "detraining" [21].

4. Discussion

To our knowledge, this is the first review to explicitly evaluate the effect of exercise on nutritional status in individuals with CF. Other systematic reviews have focused on exercise capacity, pulmonary function, and health-related quality of life [26]. Our review yielded only four relevant RCTs [18–21], none of which included adults, and there was an overall lack of uniformity in both the interventions provided as well as the outcomes assessed.

Despite having different exercise interventions and outcomes, none of these studies reported a statistically significant decrease in FFM, BMI, body mass, or triceps skin fold thickness [18–21]. This suggests that exercise, in the short term, in spite of a population that was mostly normal to underweight, does not negatively affect body composition in CF patients. In fact, Selvadurai, whose participants were the most malnourished (mean weight for age 16%) demonstrated that RET can improve body mass, body composition and muscle strength [19]; they were also able to demonstrate that AET led to larger increases in aerobic capacity and a slight, but statistically insignificant, increase in body mass compared to the control group; they ultimately suggested that a combined training program may be of most benefit to patients with CF. In their initial study (2012) [20], Santana Sosa did not notice any significant difference in BMI or FFM with a combination of AET and RET, but in their later study (2014) [21], they did find a significant increase in FFM in the exercise training group.

The mixed results of these studies may be due to multiple factors. They all used different exercise methods in their studies. In addition, they have a limited sample size in their study populations, with the largest study having 66 people [19]. Moreover, they only examined the results of their intervention over short periods of time, the longest being three months [18]. No RCTs were identified that examined changes in nutritional status related to exercise over long periods of time in a CF patient population. Since their exercise regimens were tightly controlled by either hospital admission or frequent phone calls, the lack of detrimental effects from exercise in these studies is likely a reliable outcome, despite variations in benefit. This lends credibility into studying the effects of exercise on body composition in CF patients over the long term. However, given the positive benefits of exercise on other parameters, assigning participants to a prolonged control group could be considered unethical.

Our results in children are similar to those found by Elice et al. in a matched cohort study of adults; they found that only 24% of those that exercised regularly had an altered BMI compared to 41% in those that did not exercise regularly [27]. More recently (2021), Van Biervliet reported on a prospective pre–post intervention study design for patients with CF (6 to 40 years old) to improve nutritional status and body composition; patients participated in a short-term (3 weeks), inpatient, physical exercise and nutritional intervention program [28]. Weight, BMI, and fat-free mass were improved in both children and adults; in addition, the number of adults classified as "malnourished" decreased from 41% to 24%, but was unchanged (24%) in children. To our knowledge, the largest exercise-related CF study was recently completed and published [29]. The ACTIVATE-CF study randomized 117 children ($\geq$12 years old) and adults to a 12 month partially supervised vigorous physical activity intervention [29]; although data on body composition were collected as a secondary outcome [30], these data are not yet reported.

Thus, based on the data we found in these four RCTs of exercise interventions in children with CF, as well as other non-RCT studies, there is no evidence, even in normal to underweight patients, that either AET or RET will worsen an individual's nutritional status; in fact, RET could help maintain or increase body mass and potentially lean body mass. Clinicians should counsel patients that are concerned about the speculative effects of exercise on their nutritional status and body composition that exercise is not detrimental and may even improve their nutritional status. The CF care team should continue to rely on the CF care team's registered dietitian to provide appropriate individualized nutrition care plans that compliment exercise regimens to help patients meet their personal goals related to weight and body composition (e.g., the team reported by Van Biervliet included a physician, dietician, psychologist, social worker and physical therapist [28]). In addition, both AET and RET have additional benefits for patients with CF (increased aerobic capacity and strength), benefits which are associated with a positive prognosis.

Hommerding demonstrated an increase in physical activity level in patients that had frequent follow-up for their exercise regimen [18]. For those working in multidisciplinary settings, referral to a physical therapist or an exercise specialist with experience with CF that can guide exercise regimens over time and as their health waxes and wanes would be of more benefit. There are standard guidelines on exercise testing [22], exercise prescription [31], and physical activity assessment [32] for clinicians working with individuals with CF.

We were surprised at the lack of high-quality RCTs that included nutritional status as an outcome of exercise interventions. Future research should not only include longer-term outcomes of exercise training on body composition in patients with CF, but should include adults as well as differentially look at the impact of exercise on patients that are underweight, normal weight and overweight, and should include alternatives to BMI and body mass in assessing nutritional status (e.g., dual-energy X-ray absorptiometry, skinfold thickness, bioelectrical impedance, and peripheral quantitative computed tomography) [12,33].

Supplementary Materials: The following supporting information can be downloaded at: https://www.mdpi.com/article/10.3390/nu14050933/s1, Table S1. Database search strategies.

Author Contributions: Conceptualization, W.B.N., J.D.L., S.K. and J.B.; methodology, J.D.L.; screening and eligibility assessment, W.B.N., J.D.L. and N.Z.A.; data extraction, W.B.N. and J.D.L.; quality assessment, N.Z.A., S.K. and J.B.; formal analysis, W.B.N. and J.D.L.; writing—original draft preparation, W.B.N. and J.D.L.; writing—review and editing, S.K. and J.B.; supervision, S.K. and J.D.L. All authors have read and agreed to the published version of the manuscript.

Funding: This research received no external funding.

Institutional Review Board Statement: Not applicable.

Informed Consent Statement: Not applicable.

Acknowledgments: Megan M. Bell, MSLS, Reference Librarian & Liaison to the School of Health Professions for The University of Alabama at Birmingham's Lister Hill Library of the Health Sciences, provided instrumental help in refining our PICO question, developing our search strategies and setting up our Covidence database. Laura Worley and Joshua Stuger assisted with an early pilot of this project and helped develop our PICOS question and review criteria.

Conflicts of Interest: The authors declare no conflict of interest.

References

1. Cystic Fibrosis Foundation. What Is Cystic Fibrosis? Available online: https://www.cff.org/intro-cf/about-cystic-fibrosis#what-is-cystic-fibrosis? (accessed on 4 January 2022).
2. O'Sullivan, B.P.; Freedman, S.D. Cystic fibrosis. *Lancet* **2009**, *373*, 1891–1904. [CrossRef]
3. Wouthuyzen-Bakker, M.; Bodewes, F.A.; Verkade, H.J. Persistent fat malabsorption in cystic fibrosis; lessons from patients and mice. *J. Cyst. Fibros.* **2011**, *10*, 150–158. [CrossRef]
4. Ionescu, A.A.; Nixon, L.S.; Luzio, S.; Lewis-Jenkins, V.; Evans, W.D.; Stone, M.D.; Owens, D.R.; Routledge, P.A.; Shale, D.J. Pulmonary function, body composition, and protein catabolism in adults with cystic fibrosis. *Am. J. Respir. Crit. Care Med.* **2002**, *165*, 495–500. [CrossRef]
5. Culhane, S.; George, C.; Pearo, B.; Spoede, E. Malnutrition in cystic fibrosis: A review. *Nutr. Clin. Pract.* **2013**, *28*, 676–683. [CrossRef]
6. Hulzebos, E.H.; Bomhof-Roordink, H.; van de Weert-van Leeuwen, P.B.; Twisk, J.W.; Arets, H.G.; van der Ent, C.K.; Takken, T. Prediction of mortality in adolescents with cystic fibrosis. *Med. Sci. Sports Exerc.* **2014**, *46*, 2047–2052. [CrossRef] [PubMed]
7. Steinkamp, G.; Wiedemann, B. Relationship between nutritional status and lung function in cystic fibrosis: Cross sectional and longitudinal analyses from the German CF quality assurance (CFQA) project. *Thorax* **2002**, *57*, 596–601. [CrossRef]
8. Stallings, V.A.; Stark, L.J.; Robinson, K.A.; Feranchak, A.P.; Quinton, H. Evidence-based practice recommendations for nutrition-related management of children and adults with cystic fibrosis and pancreatic insufficiency: Results of a systematic review. *J. Am. Diet. Assoc.* **2008**, *108*, 832–839. [CrossRef]
9. Nixon, P.A.; Orenstein, D.M.; Kelsey, S.F.; Doershuk, C.F. The Prognostic Value of Exercise Testing in Patients with Cystic Fibrosis. *N. Engl. J. Med.* **1992**, *327*, 1785–1788. [CrossRef]
10. Hebestreit, H.; Hulzebos, E.H.J.; Schneiderman, J.E.; Karila, C.; Boas, S.R.; Kriemler, S.; Dwyer, T.; Sahlberg, M.; Urquhart, D.S.; Lands, L.C.; et al. Cardiopulmonary Exercise Testing Provides Additional Prognostic Information in Cystic Fibrosis. *Am. J. Respir. Crit. Care Med.* **2019**, *199*, 987–995. [CrossRef]
11. Schneiderman, J.E.; Wilkes, D.L.; Atenafu, E.G.; Nguyen, T.; Wells, G.D.; Alarie, N.; Tullis, E.; Lands, L.C.; Coates, A.L.; Corey, M.; et al. Longitudinal relationship between physical activity and lung health in patients with cystic fibrosis. *Eur. Respir. J.* **2014**, *43*, 817–823. [CrossRef]
12. Gabel, M.E.; Fox, C.K.; Grimes, R.A.; Lowman, J.D.; McDonald, C.M.; Stallings, V.A.; Michel, S.H. Overweight and cystic fibrosis: An unexpected challenge. *Pediatr. Pulmonol.* **2022**, *57*, S40–S49. [CrossRef]
13. Bailey, J.; Rozga, M.; McDonald, C.M.; Bowser, E.K.; Farnham, K.; Mangus, M.; Padula, L.; Porco, K.; Alvarez, J.A. Effect of CFTR Modulators on Anthropometric Parameters in Individuals with Cystic Fibrosis: An Evidence Analysis Center Systematic Review. *J. Acad. Nutr. Diet.* **2021**, *121*, 1364–1378.e1362. [CrossRef]
14. Page, M.J.; McKenzie, J.E.; Bossuyt, P.M.; Boutron, I.; Hoffmann, T.C.; Mulrow, C.D.; Shamseer, L.; Tetzlaff, J.M.; Akl, E.A.; Brennan, S.E.; et al. The PRISMA 2020 statement: An updated guideline for reporting systematic reviews. *Syst. Rev.* **2021**, *10*, 89. [CrossRef] [PubMed]
15. Lowman, J.; Nicolson, W.; Alotaibi, N.; Krick, S.; Bailey, J. Exercise for Weight Management in People with Cystic Fibrosis: A Systematic Review. PROSPERO 2021 CRD42021273303. Available online: https://www.crd.york.ac.uk/prospero/display_record.php?ID=CRD42021273303 (accessed on 10 January 2022).
16. Covidence Systematic Review Software, Veritas Health Innovation, Melbourne, Australia. Available online: www.covidence.org (accessed on 10 January 2022).
17. National Center for Health Statistics in Collaboration with the National Center for Chronic Disease Prevention and Health Promotion. Clinical Growth Charts. Available online: http://www.cdc.gov/growthcharts (accessed on 4 January 2022).
18. Hommerding, P.X.; Baptista, R.R.; Makarewicz, G.T.; Schindel, C.S.; Donadio, M.V.; Pinto, L.A.; Marostica, P.J. Effects of an educational intervention of physical activity for children and adolescents with cystic fibrosis: A randomized controlled trial. *Respir. Care* **2015**, *60*, 81–87. [CrossRef] [PubMed]
19. Selvadurai, H.C.; Blimkie, C.J.; Meyers, N.; Mellis, C.M.; Cooper, P.J.; Van Asperen, P.P. Randomized controlled study of in-hospital exercise training programs in children with cystic fibrosis. *Pediatr. Pulmonol.* **2002**, *33*, 194–200. [CrossRef] [PubMed]

20. Santana-Sosa, E.; Groeneveld, I.F.; Gonzalez-Saiz, L.; López-Mojares, L.M.; Villa-Asensi, J.R.; Barrio Gonzalez, M.I.; Fleck, S.J.; Pérez, M.; Lucia, A. Intrahospital weight and aerobic training in children with cystic fibrosis: A randomized controlled trial. *Med. Sci. Sports Exerc.* **2012**, *44*, 2–11. [CrossRef] [PubMed]
21. Santana-Sosa, E.; Gonzalez-Saiz, L.; Groeneveld, I.F.; Villa-Asensi, J.R.; Barrio Gómez de Aguero, M.I.; Fleck, S.J.; López-Mojares, L.M.; Pérez, M.; Lucia, A. Benefits of combining inspiratory muscle with 'whole muscle' training in children with cystic fibrosis: A randomised controlled trial. *Br. J. Sports Med.* **2014**, *48*, 1513–1517. [CrossRef] [PubMed]
22. Hebestreit, H.; Arets, H.G.; Aurora, P.; Boas, S.; Cerny, F.; Hulzebos, E.H.; Karila, C.; Lands, L.C.; Lowman, J.D.; Swisher, A.; et al. Statement on Exercise Testing in Cystic Fibrosis. *Respiration* **2015**, *90*, 332–351. [CrossRef]
23. Shei, R.J.; Dekerlegand, R.L.; Mackintosh, K.A.; Lowman, J.D.; McNarry, M.A. Inspiration for the Future: The Role of Inspiratory Muscle Training in Cystic Fibrosis. *Sports Med. Open* **2019**, *5*, 36. [CrossRef]
24. Cashin, A.G.; McAuley, J.H. Clinimetrics: Physiotherapy Evidence Database (PEDro) Scale. *J. Physiother.* **2020**, *66*, 59. [CrossRef]
25. Must, A.; Anderson, S.E. Body mass index in children and adolescents: Considerations for population-based applications. *Int. J. Obes.* **2006**, *30*, 590–594. [CrossRef] [PubMed]
26. Radtke, T.; Nevitt, S.J.; Hebestreit, H.; Kriemler, S. Physical exercise training for cystic fibrosis. *Cochrane Database Syst. Rev.* **2017**, *11*, Cd002768. [CrossRef] [PubMed]
27. Elce, A.; Nigro, E.; Gelzo, M.; Iacotucci, P.; Carnovale, V.; Liguori, R.; Izzo, V.; Corso, G.; Castaldo, G.; Daniele, A.; et al. Supervised physical exercise improves clinical, anthropometric and biochemical parameters in adult cystic fibrosis patients: A 2-year evaluation. *Clin. Respir. J.* **2018**, *12*, 2228–2234. [CrossRef] [PubMed]
28. Van Biervliet, S.; Declercq, D.; Dereeper, S.; Vermeulen, D.; Würth, B.; De Guschtenaere, A. The effect of an intensive residential rehabilitation program on body composition in patients with cystic fibrosis. *Eur. J. Pediatr.* **2021**, *180*, 1981–1985. [CrossRef]
29. Hebestreit, H.; Kriemler, S.; Schindler, C.; Stein, L.; Karila, C.; Urquhart, D.S.; Orenstein, D.M.; Lands, L.C.; Schaeff, J.; Eber, E.; et al. Effects of a Partially Supervised Conditioning Program in Cystic Fibrosis: An International Multicenter, Randomized Controlled Trial (ACTIVATE-CF). *Am. J. Respir. Crit. Care Med.* **2022**, *205*, 330–339. [CrossRef]
30. Hebestreit, H.; Lands, L.C.; Alarie, N.; Schaeff, J.; Karila, C.; Orenstein, D.M.; Urquhart, D.S.; Hulzebos, E.H.J.; Stein, L.; Schindler, C.; et al. Effects of a partially supervised conditioning programme in cystic fibrosis: An international multi-centre randomised controlled trial (ACTIVATE-CF): Study protocol. *BMC Pulm. Med.* **2018**, *18*, 31. [CrossRef]
31. Swisher, A.K.; Hebestreit, H.; Mejia-Downs, A.; Lowman, J.D.; Gruber, W.; Nippins, M.; Alison, J.; Schneiderman, J. Exercise and Habitual Physical Activity for People With Cystic Fibrosis: Expert Consensus, Evidence-Based Guide for Advising Patients. *Cardiopulm. Phys. Ther. J.* **2015**, *26*, 85–98. [CrossRef]
32. Bradley, J.; O'Neill, B.; Kent, L.; Hulzebos, E.H.; Arets, B.; Hebestreit, H. Physical activity assessment in cystic fibrosis: A position statement. *J. Cyst. Fibros.* **2015**, *14*, e25–e32. [CrossRef]
33. McDonald, C.M.; Alvarez, J.A.; Bailey, J.; Bowser, E.K.; Farnham, K.; Mangus, M.; Padula, L.; Porco, K.; Rozga, M. Academy of Nutrition and Dietetics: 2020 Cystic Fibrosis Evidence Analysis Center Evidence-Based Nutrition Practice Guideline. *J. Acad. Nutr. Diet.* **2021**, *121*, 1591–1636.e1593. [CrossRef]

Review

Understanding Cystic Fibrosis Comorbidities and Their Impact on Nutritional Management

Dhiren Patel [1,*], Albert Shan [2], Stacy Mathews [2] and Meghana Sathe [3]

1. Division of Pediatric Gastroenterology, Hepatology and Nutrition, School of Medicine, Saint Louis University, St. Louis, MO 63104, USA
2. Department of Pediatric Gastroenterology, Hepatology and Nutrition, School of Medicine, Saint Louis University, St. Louis, MO 63104, USA; albert.shan@health.slu.edu (A.S.); stacy.mathews@health.slu.edu (S.M.)
3. Division of Pediatric Gastroenterology and Nutrition, University of Texas Southwestern, Children's Medical Center Dallas, Dallas, TX 75235, USA; meghana.sathe@utsouthwestern.edu
* Correspondence: dhiren.patel@health.slu.edu; Tel.: +1-314-577-5647

Abstract: Cystic fibrosis (CF) is a chronic, multisystem disease with multiple comorbidities that can significantly affect nutrition and quality of life. Maintaining nutritional adequacy can be challenging in people with cystic fibrosis and has been directly associated with suboptimal clinical outcomes. Comorbidities of CF can result in significantly decreased nutritional intake and intestinal absorption, as well as increased metabolic demands. It is crucial to utilize a multidisciplinary team with expertise in CF to optimize growth and nutrition, where patients with CF and their loved ones are placed in the center of the care model. Additionally, with the advent of highly effective modulators (HEMs), CF providers have begun to identify previously unrecognized nutritional issues, such as obesity. Here, we will review and summarize commonly encountered comorbidities and their nutritional impact on this unique population.

Keywords: cystic fibrosis; nutrition; comorbidities; multidisciplinary care; patient outcomes; growth and development; feeding difficulties

1. Introduction

Cystic fibrosis (CF) is a progressive, autosomal recessive genetic disorder with a mean prevalence of 0.74 in 10,000 persons in the US. Mutations in the cystic fibrosis transmembrane conductance regulator (CFTR) protein gene lead to disruptions in the transport of chloride out of cells. This, coupled with overactive epithelial sodium channels, leads to thickened mucus secretions throughout the body, including lungs, pancreas, liver, gallbladder, and intestines [1]. The above contributes to multisystemic comorbidities (Table 1) and makes maintaining adequate nutrition status challenging in CF patients by decreasing nutritional intake, increasing metabolic demands, and decreasing intestinal absorption [2]. It is crucial to utilize a multidisciplinary team with expertise in CF to optimize growth and nutrition, given that undernourished persons with CF are more likely to have poorer clinical outcomes, such as reduced pulmonary function [2,3]. Here, we will review commonly encountered comorbidities and their nutritional impact on this unique population.

Table 1. Comorbidities affecting nutrition in CF.

HEENT	Sinusitis
Respiratory	Chronic Lung Disease, Nocturnal Hypoxia
GI/Liver	Gastroesophageal reflux disease, gastroparesis, small intestine bacterial overgrowth, distal intestinal obstruction syndrome, constipation, pancreatitis, exocrine pancreas deficiency, enteropathies (e.g., celiac disease), cystic-fibrosis-related liver disease
Endocrine	Cystic-fibrosis-related diabetes
Psychosomatic	Disorders of eating, avoidant restrictive food intake disorder, disorders of gut–brain interaction
Micronutrients	Zinc deficiency, essential fatty acid deficiency, vitamin A, D, E, and iron deficiencies
HEM related	Overeating/obesity

2. Sinusitis

Cystic fibrosis chronic rhinosinusitis (CFCRS), a common finding in people with cystic fibrosis, is defined by at least 12 weeks of persistent sinus inflammation with signs and symptoms of sinusitis. In CF, there is a blunted cell membrane transport of chloride ions, leading to decreased amounts of water that crosses into mucosal secretions. This results in thick, inspissated mucus and poor mucociliary clearance, which can give rise to secondary bacterial colonization and recurrent sinopulmonary infections. Long-term mucosal infection and inflammation predispose patients with CF to chronic rhinosinusitis [4].

While some individuals may be asymptomatic, affected persons may present with headache, facial pain or pressure, anosmia or hyposmia, chronic nasal congestion, nasal discharge, polyps, or mucosal edema. In infants, as obligate nose breathers, nasal obstruction and congestion may affect the ability to feed [5]. Olfaction dysfunction may lead to poor appetite, food aversion, and subsequently suboptimal nutritional intake. In addition, the literature suggests that paranasal sinuses can harbor bacteria that lead to pulmonary exacerbations and further worsening of nutritional status, as discussed in the following section [6].

Treatments can include topical steroids, nasal saline irrigation, and mucolytic agents, such as dornase-alpha [7,8]. There are varied data supporting the use of antibiotics and endoscopic surgery [9]. There is also research supporting the role of CFTR protein modulation in treating CF-related sinus disease. Ivacaftor, a highly effective CFTR modulator therapy primarily utilized for gating mutations [10], was reported to improve CF-related chronic rhinosinusitis (CF CRC) symptoms in persons with medically and surgically intractable Cf CRC and has shown to improve food intake and appetite [11,12].

3. Lung Disease

There are significant data from the Cystic Fibrosis Foundation Patient Registry (CFFPR) regarding the correlation between pulmonary function, resting energy expenditure, and nutritional status [13]. Below, we will discuss literature highlighting the relationship between lung function and nutrition.

People with malnutrition have significantly lower mean vital capacity and forced expiratory volumes (FEV). One study found that adolescents who were malnourished presented with significant declines in pulmonary function, as measured by forced expiratory volume, compared to their peers. In one year, individuals who lost more than 5% of their weight had FEV values that were 16.5% lower than predicted. On the other hand, participants who gained weight had an increase in FEV that was 2.1% higher than predicted [14]. Another study found that at higher weights and with developmentally appropriate weight gain, children exhibit better improvements in average forced expiratory volume, which can be used as a surrogate marker of lung health [15]. In a study by Zemel et al., it was found that nutritional status was positively predictive of improvements in pulmonary function in children with mild pulmonary disease [13]. It is possible that deterioration of lung function leads to worsening nutritional status by way of decreasing intake and increasing nutritional requirements. An alternative is that deteriorating nutritional status may lead to poor pulmonary function by weakening respiratory musculature. Children with worsening pulmonary function test results should be evaluated for nutritional deficits, as it can be indicative of worsening disease.

Sino-pulmonary symptoms of CF can include persistent cough and nasal drainage. These can impact a child's ability to consume calories due to blunted olfaction and taste. It may also impact the process of eating when cough is persistent. Additionally, chronic bacterial infections, such as *Pseudomonas aeruginosa*, can lead to chronic inflammation. Chronic inflammation can be associated with adiposity, as well as reductions in linear growth [16]. Metabolic overload and adipocyte enlargement can trigger apoptosis and a subsequent inflammatory reaction. Additionally, research shows that excessive fat ingestion, particularly without concomitant antioxidant ingestion, may contribute to inflammation linked with obesity [17]. Welsh et al. suggests that increased adiposity leads to higher levels of C-reactive protein, a common marker of inflammation [18]. In a similar vein, weight loss in otherwise healthy obese patients is associated with decreases in C-reactive protein levels [19]. Further studies need to be conducted to better understand the relationship between obesity and inflammation in patients with CF.

4. Gastroesophageal Reflux Disease

Gastroesophageal reflux disease (GERD) is defined as troublesome symptoms and/or complications occurring when gastric contents pass into the esophagus [20]. It is significantly more common in children with CF [21], and data have suggested that infants with CF are four times more likely to have GERD than infants without CF [22]. GERD can be one of the most common and challenging-to-treat symptoms in patients with CF, with some estimates stating up to 80% of this population reporting GERD [23,24].

The development of GERD is multifactorial, including lower esophageal sphincter dysfunction and increased transient lower esophageal sphincter relaxations (TLESRs) [23,25,26]. In CF, other factors include increased negative thoracic inspiratory pressure from pulmonary disease, increased intra-abdominal pressure from coughing, poor clearance of esophageal digestive contents, gastric dysmotility [27], or CF treatment side effects, including a high-fat diet, gastrostomy feeds, medication side effects, and physiotherapy [25].

The diagnosis of GERD can be made clinically in a person who exhibits typical symptoms, including recurrent emesis or regurgitation, heartburn, abdominal or chest pain, respiratory symptoms, and poor weight gain, most marked in infants [28,29]. Barium swallowing study helps evaluate malrotation and other anatomic abnormalities. Upper gastrointestinal endoscopy may be valuable to exclude other causes or evaluate complications. While pH-metry itself has limited diagnostic utility, pH-Impedance (pH-MII) may help provide better symptom correlation with reflux events, especially when combined with high-resolution manometry (HRM). This could determine the role of reflux and aspiration in persons with diminished lung function [30,31].

The potential that GERD has in limiting caloric intake, including frequent emesis and challenges when introducing a high-fat diet, puts people with moderate-to-severe GERD at risk of malnutrition, impacting the quality of life and worsening disease severity. Currently, the association between GERD and poor pulmonary function is unclear [25]. Frequent proton pump inhibitors (PPI) use in patients with CF was associated with lower hemoglobin levels, possibly due to elevated gastric pH and decreased iron absorption [32].

There are no specific guidelines to manage GERD in patients with CF; treatment strategies, in general, include non-pharmacological, pharmacological treatment, and surgery [20]. In infants, thickened feedings, a trial of extensive hydrolyzed or amino-acid formula, and smaller volume feed at an increased frequency may help alleviate the symptom [20]. Head elevation or left lateral positioning may help treat older children. Avoidance of spicy, fried, and acidic food, as well as smaller but more frequent meals, may help symptom control. Weight control, especially in the era of HEM therapy, is helpful if the person is obese.

Commonly used pharmacological therapy, such as acid suppression, can be beneficial in reflux-related erosive esophagitis; however, these medications should be used cautiously, since utility in reducing the number of reflux episodes is unclear [33]. The facts on the effect of gastric acid suppression on pulmonary function are conflicting [34,35], with some data suggesting that acid suppression results in earlier and more frequent exacerbations [34], and other suggesting that acid suppression does not affect pulmonary function [35]. Baclofen and bethanechol may be used off label before consideration of surgical fundoplication, a common invasive anti-reflux surgical procedure [20]. The benefit of fundoplication is controversial but may be indicated in refractory cases. A retrospective study reported by Boesch et al. found no changes in pulmonary function or nutrition one year post-fundoplication [36]. In contrast, Shahid et al. found that in people with uncontrolled GERD and worsening lung function, Nissen fundoplication led to significant improvement in weight, fewer pulmonary exacerbations, and slowed the decline of lung function [37,38]. In lung transplant patients, fundoplication may reduce the risk of bronchiolitis obliterans [39]. Complications from fundoplication and GERD recurrence do happen, and consideration of the potential risks and benefits of the surgery need to be evaluated for each individual.

5. Gastroparesis

Gastroparesis (GP) is defined by constellation of various upper GI symptoms along with delayed gastric emptying in the absence of mechanical obstruction to the passage of content from the stomach to duodenum. Normal gastric emptying requires coordination between intrinsic neurons and extrinsic input from the central and autonomic nervous systems. Balanced excitatory and inhibitory signals and appropriated transmission lead to a series of motor events, including proximal stomach accommodation, antral contractions, pyloric sphincter relaxation, and antropyloric–duodenal coordination.

There are no data on the prevalence of the pediatric population so far. At-risk populations include malnutrition, eating disorders, functional GI disorders, gastrointestinal anatomical abnormalities, connective tissue disorders, postinfectious and chronic inflammatory processes [40]. The incidence of delayed or rapid gastric emptying within the CF population has not been consistent in the literature. In a retrospective cohort study, 9 out of 239 patients enrolled had CF, and 4 (44%) of the 9 CF patients had GP [41]. A systemic review conducted by Corral, J.E. et al. showed that patients with CF have a high frequency of GP of up to 38%, with a higher prevalence for patients older than 18 years, while younger patients with untreated pancreatic insufficiency had rapid gastric emptying [42].

While the pathophysiology of GP remains unclear, various mechanisms have been proposed, including altered gut hormones in response to meals, impaired ileal brake secondary to fat malabsorption, and abnormalities in the enteric nervous system [43–45]. Frequent use of opiates and anticholinergics can also decrease gastric emptying and intestinal transit time. Diabetes does not seem to play a significant role in pediatric GP in CF [42].

In general, vomiting was the most frequent presenting symptom, followed by abdominal pain, nausea, weight loss, early satiety, and bloating [41]. Infants and younger children are more likely to experience vomiting, while adolescents predominantly manifest symptoms of nausea and abdominal pain [46]. The diagnosis of GP is often challenging in pediatric patients, given the lack of normative data and poor standardization of testing in this population [40].

GP compromises patients' quality of life, increases financial burden, and significantly worsens nutrition status [47,48]. It often leads to food aversion, causing oral caloric intake restriction [49]. Adult studies have shown that patients with GP often have a diet deficient in calories, fat, protein, and several vitamins and minerals, on assessment [50]. A large proportion of patients consumed less than 60% of their daily energy requirements [50]. Wassem S. et al. reported a pediatric retrospective cohort of 239 patients with GP, with 27% of them experiencing weight loss [41]. While no study has focused explicitly on CF patients, there is a concern that GP will interfere with patients' adherence to supplements and oral medication, specifically pancreatic enzyme replacement therapy (PERT), which further worsens malabsorption and leads to nutritional deficiencies. As discussed in the GERD section, frequent usage of proton pump inhibitors also puts patients at greater risk of iron deficiency and small bowel bacterial overgrowth [51].

No specific pediatric treatment guidelines have been established for GP. GP-associated nutritional deficiency could arise due to inability to take in adequate intake or due to interventions required to treat GP, such as small feeds, tube feeding, and adverse effects of medications used for GP. Diet intervention, traditionally, may start with smaller and more frequent meals; however, it raises concern for possible insufficient caloric intake in the CF population, especially in the younger children [31]. A few studies in infants reported that feeding with medium-chain triglycerides and whey–hydrolysate formula improves gastric emptying [52–54]. Prokinetic agents are often required. Macrolide antibiotics, such as erythromycin and azithromycin, are widely used [40,55]. Amoxicillin/clavulanate, though mainly used for small bowel bacterial overgrowth, also seems to increase gastrointestinal motility [56]. Metoclopramide has a block-box warning by the FDA due to the risk of tardive dyskinesia, which has limited its use; however, it is often still utilized after partnering with patients and families regarding potential risks due to its effectiveness. Cisapride and domperidone, though effective, are not readily available in the United States due to increased risk of cardiac arrhythmias and death [57]. Other invasive treatment options for medically refractory GP, including pyloroplasty or gastric electrical stimulation, are beyond the scope of this paper and can be reviewed elsewhere [40]. Lastly, although with risks of complications, including food aversion, tube feeding (slow nasogastric or nasojejunal/gastrojejunal feeding) may be indicated to optimize patients' growth in all age groups [40].

6. Small Bowel Bacterial Overgrowth

Small intestine bacterial overgrowth (SIBO) has been reported in up to 40% of patients with CF [58,59]. Patients with CF have increased acidity of intestinal and biliary tract secretions due to decreased bicarbonate secretion by an injured exocrine pancreas and dysregulation of salts in biliary system due to the presence of dysfunction CFTR in cholangiocytes. Additionally, there is a slowing of intestinal motility, obstruction from inspissated intestinal secretions, and reduced gastric acid production, all of which can be associated with the development of SIBO. Another thought is that SIBO in CF is secondary to swallowed bacteria, which is higher in individuals with CF [60].

Symptoms of SIBO can overlap with gastrointestinal manifestations of CF and typically include abdominal discomfort, flatulence, and bloating [60]. Many symptoms of SIBO can be difficult to differentiate from abdominal symptoms in CF, so SIBO should be considered when abdominal symptoms persist after other diagnoses are ruled out. Gold-standard diagnosis of SIBO is with duodenal aspiration culture, although this is rarely performed due to its invasiveness. SIBO can also be diagnosed in the general population using a lactulose hydrogen breath test. However, as this test is not thought to be reliable in CF, diagnosis is most often made clinically with symptoms of gas, bloating, and/or increased steatorrhea in the setting of optimized PERT [59].

SIBO causes malnutrition through various pathogenesis, such as bacterial injury to gut epithelium, increased bacterial consumption of host nutrients, and decreased food intake because of abdominal discomfort [61,62]. Bacteria can directly affect gut epithelia and is associated with villous blunting and epithelial inflammation. It leads to decreased fatty acid production, increased gut permeability, and carbohydrate malabsorption due to reduced absorptive surface area, bacterial sugar degradation, and impaired brush-border enzyme activity [61,63,64]. Additionally, bacterial toxin production can directly impair carbohydrate and protein absorption.

Bile acid deconjugation by intraluminal bacterial contributes to inadequate micelle formation and enterocyte injury, resulting in steatorrhea and further micronutrient deficiency. In the general population, deficiencies of lipid-soluble vitamins, iron, thiamine, nicotinamide, and vitamin B12 have been found in patients with SIBO, though this has not yet been assessed specifically in children with CF [65]. Vitamin B12 deficiency may be secondary to bacterial B12 consumption and direct inhibition of normal B12 absorption. Bacteria may compete with the host consuming macronutrient and micronutrient consumption, leading to fewer nutrients for absorption. Villous blunting secondary to inflammation may also cause carbohydrate malabsorption [61].

Treatment goals include reduction in bacteria and amelioration of nutritional deficiencies. Antimicrobials are the first-line treatment [1] and can include metronidazole and rifaximin [61,66,67]. Amoxicillin/clavulanate, which can play a role in treating CF pulmonary exacerbations, also treats SIBO by stimulating contraction of the duodenum and increasing gut motility. One study showed that oral antibiotic therapy improves fat absorption and digestion in patients with SIBO [68]. Laxatives such as polyethylene glycol are another mainstay of SIBO treatment that works by increasing the excretion of bacteria [25,59]. One study found that inhaled ipratropium, a medication often used to treat pulmonary symptoms of CF, was protective against SIBO, though its anticholinergic effects on the GI tract are thought to favor SIBO through gut stasis [60].

7. Constipation and Distal Intestinal Obstruction Syndrome

Constipation is a significant chronic issue in CF patients, with a prevalence of approximately 47% [69]. Distal intestinal obstruction syndrome (DIOS) is an acute partial or complete obstruction with the potential for surgical intervention, with a varying incidence that ranges from 2.3 to 11.3 per 1000 patients per year [70]. The European Society for Paediatric Gastroenterology Hepatology and Nutrition (ESPHGAN) Cystic Fibrosis Working Group has defined constipation as a gradual fecal impaction presented with either abdominal pain/distension, decreased bowel movements, or increased fecal consistency, which all resolved after using laxatives. DIOS was defined as acute abdominal pain or distension complicated with signs of complete or incomplete ileocecum fecal obstruction [71].

Constipation and DIOS share some common pathophysiology. They may both result from the CFTR protein dysfunction, which may alter the intestinal fluid composition and induce thick mucus in the small bowel, leading to the anomalous intestinal milieu, biofilm formation, and dysbiosis.

Both are associated with a history of meconium ileus (MI) and inadequate fluid intake [69,72]. Though dysmotility is considered to be a potential cause for DIOS, objective motility assessments are lacking [31]. Additional risk factors for DIOS include more severe CF genotypes, history of DIOS, pancreatic insufficiency, and cystic-fibrosis-related diabetes (CFRD) [72]. There has been an inconsistent association between PERT with constipation and DIOS development [73,74].

Even with well-characterized clinical symptoms by the proposed criteria, the diagnosis remains challenging and requires careful history, physical exam, and possible imaging studies. Other differentials, such as SIBO, should be considered as significant symptom overlap.

While it is concerning that patients with chronic or recurrent abdominal symptoms have suboptimal nutrition intake, limited studies have described no association between constipation/DIOS and malnutrition. Lavie et al. performed a retrospective multicenter review on 350 patients with 20 years of follow-up; there was no statistical difference of Body Mass Index (BMI)/weight between the DIOS and control (non-DIOS) groups [75]. Mentessidou et al. conducted a 10-year retrospective review on 53 CF patients who operated on MI and found no growth impairment; however, the loss of follow-up may have confounded the result. A prospective cross-sectional study on 105 CF children by Stefano et al. also showed no statistical difference in malnutrition between the constipated and non-constipated groups [76]. Studies on the relationship between nutritional outcome and meconium ileus, an intraluminal obstruction over the distal ileum and ileocecal valve at birth caused by viscid meconium, have shown contradictory results. A further systemic review may help better characterize the association [77–79].

Although there is no clear evidence, exercise and appropriate fluid and fiber intake are recommended. Inadequate PERT, including poor adherence and under-dosing, is unlikely to play a role in DIOS but should be assessed as part of the overall dietetic review in CF. Proactive conservative treatment with laxatives, such as polyethylene glycol, is essential.

8. Other Gastrointestinal Enteropathies

With appropriate nutritional intervention, patients with refractory gastrointestinal symptoms or persistent malnutrition should be referred to a gastrointestinal specialist for further investigation, since other disorders may coexist with CF that can heavily affect nutrition.

Celiac disease is a gluten-induced T-cell-mediated systemic autoimmune disease in genetically predisposed populations. Celiac disease and CF share similar clinical presentations, including diarrhea, malnutrition, abdominal distension, abdominal pain, and liver involvement. Besides being two different entities, they do coexist. The true incidence of the coexistence is unclear, but current evidence suggests an up to three-fold increase in celiac disease in CF patients compared to those without CF. The prevalence of both celiac disease and CF was calculated as 1 in 2,000,000–1 in 5,900,000 [80]. A higher antigen load from increased intestinal permeability, inflammation, and pancreatic exocrine insufficiency secondary to CFTR dysfunction might be the leading cause [81]. Therefore, while a gluten-free diet is the standard treatment, ivacaftor has been proposed as a possible treatment strategy. Interestingly, there are cases found to have celiac disease while on ivacaftor [82]. Regardless, the clinician should have a high index of suspicion when patients with CF are not improving as expected with optimal nutrition supplementation.

Whether there is an increased prevalence of coexisting inflammatory bowel disease (IBD) remains controversial. Ongoing and persistent GI symptoms in these cases significantly contribute to decreasing ability to take adequate oral intake, as well as increased energy losses due to malabsorption. Additionally, gluten-free diets add another layer of barrier to food selection that could further restrict patients' caloric intake. Patients with CF and enteropathies also have increased metabolic demand, which contributes to persistence of malnutrition.

9. Functional Gastrointestinal Disorders

Patients with CF have chronic abdominal pain, with 60% of children and 36% of adults reported in a comparative study [83]. Another pilot prospective evaluation documented around 6% of the study population has recurrent abdominal pain [84]. Visceral hypersensitivity causing abdominal pain may be a plausible etiology in CF, given the presence of CFTR in neurons [85], and functional abdominal pain disorders (FAP) should be considered if the symptoms cannot be fully explained by another medical condition after appropriate evaluation. Currently, the true prevalence of FAP is unknown in patients with CF.

According to ROME IV criteria, functional gastrointestinal disorders (FGIDs) include functional nausea and vomiting disorders, functional abdominal pain disorders (FAP), and functional defecation disorders. FAP consists of functional dyspepsia, irritable bowel syndrome, abdominal migraine, and functional abdominal pain not otherwise specified. Therefore, an individual's symptoms vary and can include chest pain, heartburn, dysphagia, dyspepsia, epigastric or other abdominal pain, emesis, and changes in stool frequency or consistency. Chronic discomfort may cause poor appetite that compromises a child's nutritional status through missed meals and decreased food intake.

Pharmacological interventions include antispasmodics and mood stabilizers. Non-pharmacological therapies consist of hypnosis, cognitive-behavioral therapy, and other complementary therapies, such as mindfulness, and certain herbs, such as peppermint, chamomile, and caraway [86,87]. It is critical to emphasize the support from psychology in the care of these patients.

10. Disordered Eating

In the management of cystic fibrosis, there is a strong emphasis on high caloric and fat diet and maintaining or gaining weight. It is unclear how these may be associated with the development of eating disorders, such as anorexia nervosa and avoidant restrictive food intake disorder (ARFID). Consequences of CF, such as delayed puberty, low body weight, and emphasis on food management, may put these patients at risk of developing disordered eating habits. Additionally, psychosocial factors, including the demands of having a chronic illness, social isolation, and low self-esteem, may further increase risk of developing eating disorders [88].

In adults with CF, it has been reported that concerns about body image increases with declining nutritional status [89]. Disordered eating behaviors do not necessarily meet the criteria for an eating disorder, hence early recognition can help identify patients at risk. An eating disorder screening tool has been developed to identify disordered eating behaviors in patients with CF, though its validity has yet to be established [90]. In the CF population, there is an emphasis on maintaining optimal weight. Adolescent patients with chronic illness affecting diet and emphasis on maintaining weight have an increased risk of developing disordered eating behavior compared to their peers [91]. CF patients are no different when it comes to the risk of disordered eating. In fact, it has been shown that adolescents with CF have similar rates of eating disorders and self-esteem as their peers who do not have CF [90].

Some studies have shown that megestrol acetate, cyproheptadine, dronabinol, and mirtazapine may be used as appetite stimulants in patients with CF and anorexia nervosa [92]. It is important to consider eating disorders in the differential diagnosis in patients with CF who have unexplained weight loss and have risk factors for developing disordered eating for early identification and management.

11. Cystic-Fibrosis-Related Liver Disease

Cystic-fibrosis-related liver disease (CFLD), as an independent risk factor with a mortality rate of 2.5%, is the third leading cause of death in CF [93–95]. The term covers a full spectrum of diseases, including neonatal cholestasis, elevated liver enzymes, imaging abnormalities, histological changes, and gallbladder and biliary tract disorders. Severe CFLD is defined based on the presence of cirrhosis and portal hypertension [96]. The prevalence of CFLD ranges from <5% to 68% depending on the diagnostic criteria used [97,98]. The incidence of severe CFLD is 5–10% in the pediatric community [99,100]. Most of these arose in the first decade of life and developed with related complications during the second decade [95].

The pathogenesis of CFLD is not fully understood but seems to be multifactorial, including abnormal cholangiocyte function and altered biliary secretion, abnormal protein–protein interactions, abnormal innate inflammatory response, gut dysbiosis, genetics, and most recently proposed, obliterative portal venopathy [101,102]. Risk factors for CFLD include male sex, the presence of two severe mutations (loss-of-function CFTR), presence of SERPINA 1Z allele, exocrine pancreatic insufficiency, and CFRD [95,103].

Most patients are asymptomatic, and the diagnosis is challenging given the lack of consensus. ESPGHAN has proposed a diagnostic criterion with at least two of the following variables: (1) hepatomegaly or splenomegaly; (2) persistently elevated liver enzymes without other causes identified (>12 months, at least three consecutive labs); (3) abnormal liver ultrasound findings; and (4) abnormal liver histopathology, if indicated [96]. Studies have investigated numerous non-invasive biomarkers and imaging, hoping to predict or assess the severity of pediatric CFLD, such as GGT, aspartate aminotransferase to platelet ratio index (APRI), fibrosis-4 (Fib-4), liver ultrasound, transient elastography, MR elastography, and supersonic shear-wave elastography (SSWE) [95,104–107].

Patients with CFLD, especially those with a severe phenotype, are at risk of significant malnutrition, micronutrient deficiencies and worse lung function [108]. The underlying causes are complex, including but not limited to poor intake associated with anorexia, delayed gastric emptying, and early satiety from abdominal distension (ascites, organomegaly); increased energy expenditure; and maldigestion/malabsorption [109]. With unpredictable response with CFTR modulator therapy in CFRD, although not absolutely contraindicated, the hepatotoxicity potential may limit its usage in this patient population. [110,111].

Though guidelines and position papers were published, nutrition management should involve an experienced CF dietitian in the team [96,109]. In summary, patients with CFLD need higher daily calories up to 130 ~ 150% of the estimated requirement for age, preferably achieved by increasing the fat's proportion to 40–50% considering the risk of CFRD, with supplementation in medium-chain triglycerides and polyunsaturated fatty acids. It is essential to provide supplements to ensure adequate daily protein intake (3 g/kg/day), pancreatic enzymes to optimize fat absorption, and micronutrients, including fat-soluble vitamins and zinc, to prevent deficiency. Salt supplementation is common in CF patients but should be avoided for those with cirrhosis and portal hypertension due to the risk of developing or worsening ascites. Frequent monitoring of nutritional status is critical to evaluate the response and prevent vitamin toxicity. Enteral tube feeding is safe and may be indicated if the patient cannot achieve adequate intake.

12. Cystic-Fibrosis-Related Diabetes

CFRD is a common comorbidity in patients with CF and can be found in 19% of adolescents [112]. CFRD is not commonly observed in the prepubertal population, but 2% of children with CF have this diagnosis. Glucose abnormalities may be found in patients of all ages, including children as young as 3 months [113], and can reflect an increased risk of developing CFRD. Even young children with normal glucose levels had lower insulin secretion than controls after the age of two [113].

The etiology of CFRD is thought to be at least in part secondary to thick secretions leading to pancreatic obstruction, as well as due to progressive fibrosis and fatty infiltration resulting in islet cell damage [114] and pancreatic exocrine dysfunction [115]. Recent studies have found cytokine IL1-beta, which is thought to cause beta-cell apoptosis leading to insulin insufficiency in both type 1 and type 2 diabetes mellitus, in the islet cells of people with CFRD. Insulin insufficiency can lead to increased protein catabolism, which can result in malnutrition and weight loss. Early identification of CFRD is important, and the diagnostic method is similar to that of type 1 and type 2 diabetes mellitus (DM), except for HbA1C, which can be falsely low in patients with CF [116]. Screening for CFRD is recommended annually, starting at age 10 for all patients with CF, with a 2 hour oral glucose tolerance test [117].

Uncontrolled CFRD and CF-related pre-diabetes can worsen pulmonary function and have negative nutritional impacts. Studies have shown that individuals with CFRD have lower height and weight percentiles compared to their peers without diabetes [118]. A decline in growth velocity and weight can precede a diagnosis of CFRD [119]. Children with CFRD who have not yet reached peak height had lower BMI percentiles than their peers up to 2 years prior to diabetes diagnosis [120]. This phenomenon was thought to be secondary to insulin-insufficiency-triggered catabolism rather than hyperglycemia itself [121]. However, in patients with CF who do not have diabetes, weight loss can be associated with higher peak glucose levels and a higher proportion of time spent with elevated serum glucose levels [119].

Optimal glycemic control is critical in the management of CFRD, and management may differ from that for patients with DM. This is due to different nutritional needs, as well as a lower risk of cardiovascular disease. Rather than strict carbohydrate restriction recommended in general DM care, people with CFRD have higher calorie, protein, fat, and salt needs and may thus instead require careful carbohydrate counting [112,122]. In conjunction with nutritional therapies, insulin therapy is the first-line treatment to insulin insufficiency in people with CFRD and is thought to not only improve nutritional status but also lung function [117]. Oral diabetes agents, such as insulin secretagogues, metformin, and thiazolidinediones are generally not recommended in CFRD [123], though some early research suggests the insulin secretagogue repaglinide may have some utility in the treatment of early CFRD [124]. Furthermore, oral medications may be associated with gastrointestinal side effects (metformin and incretin mimetic agents) and decreased bone mineral density (thiazolidinediones) [125]. Lastly, though the role of CFTR in the beta cell is unknown, some research has shown improved insulin response to glucose with CFTR modulator therapy [126].

13. Renal Stone

Urolithiasis is common in adults but can also be seen in children with CF. Literature suggests that impaired fat absorption and resultant enteric hyperoxaluria may play a role in stone disease. Supersaturation of calcium and oxalate in the urine can lead to crystallization and stone formation [127]. Additionally, defects in renal calcium metabolism can lead to hypocalciuria and nephrocalcinosis [128]. One study found that enzyme supplementation was negatively associated with the presence of hyperoxaluria, though the amount of enzyme given did not seem to affect this [129].

The abdominal pain associated with urolithiasis can result in decreased food intake and appetite, further affecting nutritional status. Treatment of nephrolithiasis generally consists of hydration and analgesia. Long-term therapy can include low-oxalate diet, optimal PERT, pyridoxine supplementation, and adequate hydration [127].

14. Micronutrient Deficiency

Patients with CF have blunted fat absorption, leading to fat-soluble vitamin deficiencies even in patients with adequate pancreatic enzyme replacement. It is recommended that these vitamin levels are checked at diagnosis and annually, followed by appropriate supplementation [130].

14.1. Vitamin A

Vitamin A is essential for immunity and epithelial cell function. Studies have shown that up to 40% of patients with CF have Vitamin A deficiency [131]. It is crucial to remember that Vitamin A is an acute-phase reactant, so levels measured during acute illness may inaccurately suggest hypovitaminosis A [132]. Serum retinol is a commonly used marker of vitamin A status, though there are no universally available standard definitions for vitamin A deficiency. Other markers can include retinol-binding protein, as well as serum retinyl esters. A retinol-binding protein to retinol ratio of less than 0.8 may indicate vitamin A deficiency [133].

14.2. Vitamin D and Calcium

Vitamin D is important because it aids in calcium absorption and is critical for bone health. Contributors to vitamin D and calcium deficiency can include low intake, gastrointestinal malabsorption, and increased fecal calcium loss. Compared to the general population, children with CF have a higher prevalence of osteopenia and an increased risk of fractures [134,135]. It is recommended that calcium be assessed routinely. Increasing dietary consumption of calcium-rich foods can include dairy products, such as milk and cheese. Intervening early with calcium and vitamin D supplementation may address these risks.

14.3. Vitamin E

Vitamin E deficiency is a frequent finding in CF patients [136]. A compound of vitamin E, alpha–tocopherol, helps protect the body against oxidative damage. Vitamin E is best measured through a vitamin E to total cholesterol ratio. Vitamin E deficiency can result in neuromuscular degeneration, cognitive deficits, and eye problems. A child's vitamin E requirement may increase with higher levels of oxidative stress during a pulmonary CF exacerbation and with chronic respiratory infections.

14.4. Iron

Children with CF are at risk of iron deficiency. Multiple contributing factors include malabsorption, chronic inflammation or blood loss, and inadequate intake. Notably, ferritin, which is frequently used as a marker of iron status, is an acute phase reactant and thus may be falsely elevated in inflammatory states. Serum transferrin receptors are a good marker of iron stores that are not affected by inflammatory states; however, it may be difficult to test in a clinical setting. A simple marker can include checking hemoglobin and hematocrit annually or an iron panel containing ferritin and total iron-binding capacity [3].

14.5. Zinc

Zinc deficiency is common in patients with cystic fibrosis, and PERT can assist with zinc absorption. Symptoms of low zinc can include immunocompromise, poor growth, and poor appetite due to hypogeusia. However, zinc deficiency may be difficult to diagnose, as plasma zinc levels can be normal with true zinc deficiency. In patients with failure to thrive, six months of empiric zinc supplementation may be reasonable. Additionally, zinc can affect vitamin A, so it may be beneficial to supplement zinc in vitamin A deficiency settings that do not respond to vitamin A supplementation alone [3].

14.6. Essential Fatty Acids

Essential fatty acids are polyunsaturated fats that can be eventually metabolized from arachidonic acid (AA) into docosahexaenoic acid (DHA). This deficiency can be common in patients with CF, though it is not often symptomatic. Essential fatty acid deficiency should be considered in infants with failure to thrive. It is possible that in addition to fat malabsorption, patients with CF may have abnormal fatty acid metabolism. It is still unclear whether patients with CF should have supplementation with DHA, though foods rich in linolenic acid, such as cold-water fish and vegetable oils, can be sometimes recommended. Additionally, breastmilk contains DHA and may be beneficial for infants with CF [3,130].

15. Nutritional Impact of CF Therapy in the Era of HEMs

Patients with CF generally have lower body weight than age-matched peers. This may be due to increased energy expenditure secondary to increased work of breathing that leads to difficulty maintaining goal BMI. Improvements in CF management have resulted in overall improved nutritional status. In fact, there are now four times as many patients with CF who are overweight or obese compared to underweight [137].

Research suggests that CFTR modulators, such as elexacaftor/tezacaftor/ivacaftor, may increase weight in CF patients [138,139]. The mechanism behind this weight gain is thought to be multifactorial. Therapy can improve appetite, leading to increased food intake [12]. Improved CFTR function can lead to thinned mucus secretions and better nutrient absorption. Improved overall health may be correlated with fewer hospitalizations in which patients are required to fast. Additionally, elexacaftor/tezacaftor/ivacaftor therapy must be taken with dietary fat, which itself can increase calorie intake. Patients on adequate therapy may have improved respiratory function, which may correlate with decreased energy expenditure due to decreased respiratory muscle work. Therapy has also been associated with improved pancreatic exocrine function, which will theoretically improve intestinal nutrient absorption and thus nutritional status.

CFTR modulators have also been associated with increased high-density lipoprotein (HDL), low-density lipoprotein (LDL), and total cholesterol levels. This may be secondary to reduced systemic inflammation, as triglycerides are an acute-phase reactant. Inflammation and oxidative stress are linked to impaired glucose homeostasis, which may then theoretically improve with the use of CFTR modulators. Additionally, increased insulin secretion has been reported with CFTR modulator therapy [126].

Overall, data in adults suggest that the use of elexacaftor/tezacaftor/ivacaftor will increase the number of individuals who are overweight. Traditionally, nutrition recommendations for patients with CF who were generally undernourished and underweight in the past included high-fat and high-calorie diets. The Academy of Nutrition and Dietetics has altered their nutrition recommendations for patients with CF on HEM to reflect recommendations for the general population [137]. In conclusion, providers should monitor patients on CFTR modulator therapy for signs of overnutrition at every visit.

16. Conclusions

CF is a multisystemic chronic disease with many comorbidities playing a major role at the same time. This contributes significantly to existing nutritional challenges in this patient group and impacts nutritional management. CF providers have started seeing nutritional issues that are not considered typical in these patients, specifically, obesity. Multidisciplinary care from various providers (such as pulmonologists, gastroenterologists, nutritionists, otolaryngologists, endocrinologists) familiar with these issues is paramount to help manage the complex nutritional needs of patients with CF. CF is one of the many chronic diseases where the idea of co-production is truly needed, where providers, families, and patients all work interdependently to come together to establish common goals of care to improve care for CF patients.

Author Contributions: Conceptualization, D.P., A.S. and M.S.; writing—original draft preparation, A.S. and S.M.; writing review and editing, D.P., A.S., S.M. and M.S.; supervision, M.S. All authors have read and agreed to the published version of the manuscript.

Funding: This research received no external funding.

Institutional Review Board Statement: Not applicable.

Informed Consent Statement: Not applicable.

Data Availability Statement: Not applicable.

Conflicts of Interest: The authors declare no conflict of interest.

References

1. Ratchford, T.L.; Teckman, J.H.; Patel, D.R. Gastrointestinal pathophysiology and nutrition in cystic fibrosis. *Expert Rev. Gastroenterol. Hepatol.* **2018**, *12*, 853–862. [CrossRef] [PubMed]
2. Sullivan, J.S.; Mascarenhas, M.R. Nutrition: Prevention and management of nutritional failure in Cystic Fibrosis. *J. Cyst. Fibros.* **2017**, *16*, S87–S93. [CrossRef]
3. Borowitz, D.; Baker, R.D.; Stallings, V. Consensus report on nutrition for pediatric patients with cystic fibrosis. *J. Pediatr. Gastroenterol. Nutr.* **2002**, *35*, 246–259. [CrossRef] [PubMed]
4. Tipirneni, K.E.; Woodworth, B.A. Medical and Surgical Advancements in the Management of Cystic Fibrosis Chronic Rhinosinusitis. *Curr. Otorhinolaryngol. Rep.* **2017**, *5*, 24–34. [CrossRef] [PubMed]
5. Alvo, A.; Villarroel, G.; Sedano, C. Neonatal nasal obstruction. *Eur. Arch. Otorhinolaryngol.* **2021**, *278*, 3605–3611. [CrossRef]
6. Johansen, H.K.; Aanaes, K.; Pressler, T.; Nielsen, K.G.; Fisker, J.; Skov, M.; Hoiby, N.; von Buchwald, C. Colonisation and infection of the paranasal sinuses in cystic fibrosis patients is accompanied by a reduced PMN response. *J. Cyst. Fibros.* **2012**, *11*, 525–531. [CrossRef]
7. Safi, C.; Zheng, Z.; Dimango, E.; Keating, C.; Gudis, D.A. Chronic Rhinosinusitis in Cystic Fibrosis: Diagnosis and Medical Management. *Med. Sci.* **2019**, *7*, 32. [CrossRef]
8. Le, C.; McCrary, H.C.; Chang, E. Cystic Fibrosis Sinusitis. *Adv. Otorhinolaryngol.* **2016**, *79*, 29–37.
9. Illing, E.A.; Woodworth, B.A. Management of the upper airway in cystic fibrosis. *Curr. Opin. Pulm. Med.* **2014**, *20*, 623–631. [CrossRef]
10. Gramegna, A.; Contarini, M.; Aliberti, S.; Casciaro, R.; Blasi, F.; Castellani, C. From Ivacaftor to Triple Combination: A Systematic Review of Efficacy and Safety of CFTR Modulators in People with Cystic Fibrosis. *Int. J. Mol. Sci.* **2020**, *21*, 5882. [CrossRef]
11. Chang, E.H.; Tang, X.X.; Shah, V.S.; Launspach, J.L.; Ernst, S.E.; Hilkin, B.; Karp, P.H.; Abou Alaiwa, M.H.; Graham, S.M.; Hornick, D.B.; et al. Medical reversal of chronic sinusitis in a cystic fibrosis patient with ivacaftor. *Int. Forum. Allergy Rhinol.* **2015**, *5*, 178–181. [CrossRef] [PubMed]
12. Beswick, D.M.; Humphries, S.M.; Balkissoon, C.D.; Strand, M.; Vladar, E.K.; Lynch, D.A.; Taylor-Cousar, J.L. Impact of Cystic Fibrosis Transmembrane Conductance Regulator Therapy on Chronic Rhinosinusitis and Health Status: Deep Learning CT Analysis and Patient-reported Outcomes. *Ann. Am. Thorac. Soc.* **2022**, *19*, 12–19. [CrossRef] [PubMed]
13. Zemel, B.S.; Kawchak, D.A.; Cnaan, A.; Zhao, H.; Scanlin, T.F.; Stallings, V.A. Prospective evaluation of resting energy expenditure, nutritional status, pulmonary function, and genotype in children with cystic fibrosis. *Pediatr. Res.* **1996**, *40*, 578–586. [CrossRef] [PubMed]

14. Steinkamp, G.; Wiedemann, B. Relationship between nutritional status and lung function in cystic fibrosis: Cross sectional and longitudinal analyses from the German CF quality assurance (CFQA) project. *Thorax* **2002**, *57*, 596–601. [CrossRef]
15. Peterson, M.L.; Jacobs, D.R., Jr.; Milla, C.E. Longitudinal changes in growth parameters are correlated with changes in pulmonary function in children with cystic fibrosis. *Pediatrics* **2003**, *112 Pt 1*, 588–592. [CrossRef]
16. Millward, D.J. Nutrition, infection and stunting: The roles of deficiencies of individual nutrients and foods, and of inflammation, as determinants of reduced linear growth of children. *Nutr. Res. Rev.* **2017**, *30*, 50–72. [CrossRef]
17. Monteiro, R.; Azevedo, I. Chronic inflammation in obesity and the metabolic syndrome. *Mediat. Inflamm.* **2010**, *2010*, 289645. [CrossRef]
18. Welsh, P.; Polisecki, E.; Robertson, M.; Jahn, S.; Buckley, B.M.; de Craen, A.J.; Ford, I.; Jukema, J.W.; Macfarlane, P.W.; Packard, C.J.; et al. Unraveling the directional link between adiposity and inflammation: A bidirectional Mendelian randomization approach. *J. Clin. Endocrinol. Metab.* **2010**, *95*, 93–99. [CrossRef]
19. Heilbronn, L.K.; Noakes, M.; Clifton, P.M. Energy restriction and weight loss on very-low-fat diets reduce C-reactive protein concentrations in obese, healthy women. *Arterioscler. Thromb. Vasc. Biol.* **2001**, *21*, 968–970. [CrossRef]
20. Rosen, R.; Vandenplas, Y.; Singendonk, M.; Cabana, M.; DiLorenzo, C.; Gottrand, F.; Gupta, S.; Langendam, M.; Staiano, A.; Thapar, N.; et al. Pediatric Gastroesophageal Reflux Clinical Practice Guidelines: Joint Recommendations of the North American Society for Pediatric Gastroenterology, Hepatology, and Nutrition and the European Society for Pediatric Gastroenterology, Hepatology, and Nutrition. *J. Pediatr. Gastroenterol. Nutr.* **2018**, *66*, 516–554. [CrossRef]
21. Kelly, T.; Buxbaum, J. Gastrointestinal Manifestations of Cystic Fibrosis. *Dig. Dis. Sci.* **2015**, *60*, 1903–1913. [CrossRef] [PubMed]
22. Bolia, R.; Ooi, C.Y.; Lewindon, P.; Bishop, J.; Ranganathan, S.; Harrison, J. Ford K, van der Haak N, Oliver MR. Practical approach to the gastrointestinal manifestations of cystic fibrosis. *J. Paediatr. Child Health* **2018**, *54*, 609–619. [CrossRef]
23. Pauwels, A.; Blondeau, K.; Dupont, L.J.; Sifrim, D. Mechanisms of increased gastroesophageal reflux in patients with cystic fibrosis. *Am. J. Gastroenterol.* **2012**, *107*, 1346–1353. [CrossRef]
24. Mousa, H.M.; Woodley, F.W. Gastroesophageal reflux in cystic fibrosis: Current understandings of mechanisms and management. *Curr. Gastroenterol. Rep.* **2012**, *14*, 226–235. [CrossRef] [PubMed]
25. Ng, C.; Prayle, A. Gastrointestinal complications of cystic fibrosis. *Paediatr. Child Health* **2020**, *30*, 345–349. [CrossRef]
26. Cucchiara, S.; Santamaria, F.; Andreotti, M.R.; Minella, R.; Ercolini, P.; Oggero, V. Mechanisms of gastro-oesophageal reflux in cystic fibrosis. *Arch. Dis. Child* **1991**, *66*, 617–622. [CrossRef]
27. Assis, D.N.; Freedman, S.D. Gastrointestinal Disorders in Cystic Fibrosis. *Clin. Chest Med.* **2016**, *37*, 109–118. [CrossRef]
28. Gupta, S.K.; Hassall, E.; Chiu, Y.L.; Amer, F.; Heyman, M.B. Presenting symptoms of nonerosive and erosive esophagitis in pediatric patients. *Dig. Dis. Sci.* **2006**, *51*, 858–863. [CrossRef]
29. Duffield, R.A. Cystic fibrosis and the gastrointestinal tract. *J. Pediatr. Health Care* **1996**, *10*, 51–57. [CrossRef]
30. Tolone, S.; Savarino, E.; Docimo, L. Is there a role for high resolution manometry in GERD diagnosis? *Minerva Gastroenterol Dietol.* **2017**, *63*, 235–248. [CrossRef]
31. Henen, S.; Denton, C.; Teckman, J.; Borowitz, D.; Patel, D. Review of Gastrointestinal Motility in Cystic Fibrosis. *J. Cyst. Fibros.* **2021**, *20*, 578–585. [CrossRef] [PubMed]
32. Gifford, A.H.; Sanville, J.L.; Sathe, M.; Heltshe, S.L.; Goss, C.H. Use of proton pump inhibitors is associated with lower hemoglobin levels in people with cystic fibrosis. *Pediatr. Pulmonol.* **2021**, *56*, 2048–2056. [CrossRef]
33. Turk, H.; Hauser, B.; Brecelj, J.; Vandenplas, Y.; Orel, R. Effect of proton pump inhibition on acid, weakly acid and weakly alkaline gastro-esophageal reflux in children. *World J. Pediatrics* **2013**, *9*, 36–41. [CrossRef] [PubMed]
34. Dimango, E.; Walker, P.; Keating, C.; Berdella, M.; Robinson, N.; Langfelder-Schwind, E.; Levy, D.; Liu, X. Effect of esomeprazole versus placebo on pulmonary exacerbations in cystic fibrosis. *BMC Pulm. Med.* **2014**, *14*, 21. [CrossRef]
35. van der Doef, H.P.; Arets, H.G.; Froeling, S.P.; Westers, P.; Houwen, R.H. Gastric acid inhibition for fat malabsorption or gastroesophageal reflux disease in cystic fibrosis: Longitudinal effect on bacterial colonization and pulmonary function. *J. Pediatr.* **2009**, *155*, 629–633. [CrossRef] [PubMed]
36. Boesch, R.P.; Acton, J.D. Outcomes of fundoplication in children with cystic fibrosis. *J. Pediatr. Surg.* **2007**, *42*, 1341–1344. [CrossRef]
37. Sheikh, S.I.; Ryan-Wenger, N.A.; McCoy, K.S. Outcomes of surgical management of severe GERD in patients with cystic fibrosis. *Pediatr. Pulmonol.* **2013**, *48*, 556–562. [CrossRef]
38. Ng, J.; Friedmacher, F.; Pao, C.; Charlesworth, P. Gastroesophageal Reflux Disease and Need for Antireflux Surgery in Children with Cystic Fibrosis: A Systematic Review on Incidence, Surgical Complications, and Postoperative Outcomes. *Eur. J. Pediatr. Surg.* **2021**, *31*, 106–114. [CrossRef]
39. Davis, R.D.; Lau, C.L.; Eubanks, S.; Messier, R.H.; Hadjiliadis, D.; Steele, M.P.; Palmer, S.M. Improved lung allograft function after fundoplication in patients with gastroesophageal reflux disease undergoing lung transplantation. *J. Thorac. Cardiovasc. Surg.* **2003**, *125*, 533–542. [CrossRef]
40. Kovacic, K.; Elfar, W.; Rosen, J.M.; Yacob, D.; Raynor, J.; Mostamand, S.; Punati, J.; Fortunato, J.E.; Saps, M. Update on pediatric gastroparesis: A review of the published literature and recommendations for future research. *Neurogastroenterol. Motil.* **2020**, *32*, e13780. [CrossRef]
41. Waseem, S.; Islam, S.; Kahn, G.; Moshiree, B.; Talley, N.J. Spectrum of gastroparesis in children. *J. Pediatr. Gastroenterol. Nutr.* **2012**, *55*, 166–172. [CrossRef] [PubMed]

42. Corral, J.E.; Dye, C.W.; Mascarenhas, M.R.; Barkin, J.S.; Salathe, M.; Moshiree, B. Is Gastroparesis Found More Frequently in Patients with Cystic Fibrosis? A Systematic Review. *Scientifica* **2016**, *2016*, 1–11. [CrossRef] [PubMed]
43. Van Der Sijp, J.R.M.; Kamm, M.A.; Nightingale, J.M.D.; Britton, K.E.; Granowska, M.; Mather, S.J.; Akkermans, L.M.A.; Lennard-Jones, J.E. Disturbed gastric and small bowel transit in severe idiopathic constipation. *Dig. Dis. Sci.* **1993**, *38*, 837–844. [CrossRef] [PubMed]
44. Van Citters, G.W.; Lin, H.C. Ileal brake: Neuropeptidergic control of intestinal transit. *Curr. Gastroenterol. Rep.* **2006**, *8*, 367–373. [CrossRef]
45. Murphy, M.S.; Brunetto, A.L.; Pearson, A.D.J.; Ghatei, M.A.; Nelson, R.; Eastham, E.J.; Bloom, S.R.; Green, A.A. Gut hormones and gastrointestinal motility in children with cystic fibrosis. *Dig. Dis. Sci.* **1992**, *37*, 187–192. [CrossRef]
46. Rodriguez, L.; Irani, K.; Jiang, H.; Goldstein, A.M. Clinical presentation, response to therapy, and outcome of gastroparesis in children. *J. Pediatr. Gastroenterol. Nutr.* **2012**, *55*, 185–190. [CrossRef]
47. Parkman, H.P.; Wilson, L.A.; Yates, K.P.; Koch, K.L.; Abell, T.L.; McCallum, R.W.; Sarosiek, I.; Kuo, B.; Malik, Z.; Schey, R.; et al. Factors that contribute to the impairment of quality of life in gastroparesis. *Neurogastroenterol. Motil.* **2021**, *33*, e14087. [CrossRef]
48. Lu, P.L.; Moore-Clingenpeel, M.; Yacob, D.; Di Lorenzo, C.; Mousa, H.M. The rising cost of hospital care for children with gastroparesis: 2004–2013. *Neurogastroenterol. Motil.* **2016**, *28*, 1698–1704. [CrossRef]
49. Wytiaz, V.; Homko, C.; Duffy, F.; Schey, R.; Parkman, H.P. Foods provoking and alleviating symptoms in gastroparesis: Patient experiences. *Dig Dis Sci.* **2015**, *60*, 1052–1058. [CrossRef]
50. Parkman, H.P.; Yates, K.P.; Hasler, W.L.; Nguyan, L.; Pasricha, P.J.; Snape, W.J.; Farrugia, G.; Calles, J.; Koch, K.L.; Abell, T.L.; et al. Dietary Intake and Nutritional Deficiencies in Patients with Diabetic or Idiopathic Gastroparesis. *Gastroenterology* **2011**, *141*, 486–498.e7. [CrossRef]
51. Parrish, C.R. Nutrition concerns for the patient with gastroparesis. *Curr. Gastroenterol. Rep.* **2007**, *9*, 295–302. [CrossRef] [PubMed]
52. Siegel, M.; Krantz, B.; Lebenthal, E. Effect of fat and carbohydrate composition on the gastric emptying of isocaloric feedings in premature infants. *Gastroenterology* **1985**, *89*, 785–790. [CrossRef]
53. Tolia, V.; Lin, C.H.; Kuhns, L.R. Gastric emptying using three different formulas in infants with gastroesophageal reflux. *J. Pediatr. Gastroenterol. Nutr.* **1992**, *15*, 297–301. [CrossRef] [PubMed]
54. Pascale, J.A.; Mims, L.C.; Greenberg, M.G.; Alexander, J.B. Gastric response in low birth weight infants fed various formulas. *Biol. Neonate* **1978**, *34*, 150–154. [CrossRef]
55. Tonelli, A.R.; Drane, W.E.; Collins, D.P.; Nichols, W.; Antony, V.B.; Olson, E.L. Erythromycin improves gastric emptying half-time in adult cystic fibrosis patients with gastroparesis. *J. Cyst. Fibros.* **2009**, *8*, 193–197. [CrossRef]
56. Gomez, R.; Fernandez, S.; Aspirot, A.; Punati, J.; Skaggs, B.; Mousa, H.; Di Lorenzo, C. Effect of amoxicillin/clavulanate on gastrointestinal motility in children. *J. Pediatr. Gastroenterol. Nutr.* **2012**, *54*, 780–784. [CrossRef]
57. Schey, R.; Saadi, M.; Midani, D.; Roberts, A.C.; Parupalli, R.; Parkman, H.P. Domperidone to Treat Symptoms of Gastroparesis: Benefits and Side Effects from a Large Single-Center Cohort. *Dig. Dis. Sci.* **2016**, *61*, 3545–3551. [CrossRef]
58. Lisowska, A.; Wójtowicz, J.; Walkowiak, J. Small intestine bacterial overgrowth is frequent in cystic fibrosis: Combined hydrogen and methane measurements are required for its detection. *Acta Biochim. Pol.* **2009**, *56*, 631–634. [CrossRef]
59. Dorsey, J.; Gonska, T. Bacterial overgrowth, dysbiosis, inflammation, and dysmotility in the Cystic Fibrosis intestine. *J. Cyst. Fibros.* **2017**, *16* (Suppl. S2), S14–S23. [CrossRef]
60. Fridge, J.L.; Conrad, C.; Gerson, L.; Castillo, R.O.; Cox, K. Risk factors for small bowel bacterial overgrowth in cystic fibrosis. *J. Pediatr. Gastroenterol. Nutr.* **2007**, *44*, 212–218. [CrossRef]
61. Avelar Rodriguez, D.; Ryan, P.M.; Toro Monjaraz, E.M.; Ramirez Mayans, J.A.; Quigley, E.M. Small Intestinal Bacterial Overgrowth in Children: A State-Of-The-Art Review. *Front. Pediatr.* **2019**, *7*, 363. [CrossRef] [PubMed]
62. Sarker, S.A.; Ahmed, T.; Brüssow, H. Hunger and microbiology: Is a low gastric acid-induced bacterial overgrowth in the small intestine a contributor to malnutrition in developing countries? *Microb. Biotechnol.* **2017**, *10*, 1025–1030. [CrossRef]
63. Lappinga, P.J.; Abraham, S.C.; Murray, J.A.; Vetter, E.A.; Patel, R.; Wu, T.T. Small intestinal bacterial overgrowth: Histopathologic features and clinical correlates in an underrecognized entity. *Arch. Pathol. Lab Med.* **2010**, *134*, 264–270. [CrossRef] [PubMed]
64. Lewindon, P.J.; Robb, T.A.; Moore, D.J.; Davidson, G.P.; Martin, A.J. Bowel dysfunction in cystic fibrosis: Importance of breath testing. *J. Paediatr. Child Health* **1998**, *34*, 79–82. [CrossRef] [PubMed]
65. Elphick, D.A.; Chew, T.S.; Higham, S.E.; Bird, N.; Ahmad, A.; Sanders, D.S. Small bowel bacterial overgrowth in symptomatic older people: Can it be diagnosed earlier? *Gerontology* **2005**, *51*, 396–401. [CrossRef]
66. Furnari, M.; De Alessandri, A.; Cresta, F.; Haupt, M.; Bassi, M.; Calvi, A.; Haupt, R.; Bodini, G.; Ahmed, I.; Bagnasco, F.; et al. The role of small intestinal bacterial overgrowth in cystic fibrosis: A randomized case-controlled clinical trial with rifaximin. *J. Gastroenterol.* **2019**, *54*, 261–270. [CrossRef]
67. Burton, S.J.; Hachem, C.; Abraham, J.M. Luminal Gastrointestinal Manifestations of Cystic Fibrosis. *Curr. Gastroenterol. Rep.* **2021**, *23*, 4. [CrossRef] [PubMed]
68. Lisowska, A.; Pogorzelski, A.; Oracz, G.; Siuda, K.; Skorupa, W.; Rachel, M.; Cofta, S.; Piorunek, T.; Walkowiak, J. Oral antibiotic therapy improves fat absorption in cystic fibrosis patients with small intestine bacterial overgrowth. *J. Cyst. Fibros.* **2011**, *10*, 418–421. [CrossRef]
69. Van Der Doef, H.P.J.; Kokke, F.T.M.; Beek, F.J.A.; Woestenenk, J.W.; Froeling, S.P.; Houwen, R.H.J. Constipation in pediatric Cystic Fibrosis patients: An underestimated medical condition. *J. Cyst. Fibros.* **2010**, *9*, 59–63. [CrossRef]

70. Van Der Doef, H.P.J.; Kokke, F.T.M.; Van Der Ent, C.K.; Houwen, R.H.J. Intestinal Obstruction Syndromes in Cystic Fibrosis: Meconium Ileus, Distal Intestinal Obstruction Syndrome, and Constipation. *Curr. Gastroenterol. Rep.* **2011**, *13*, 265–270. [CrossRef]
71. Houwen, R.H.; van der Doef, H.P.; Sermet, I.; Munck, A.; Hauser, B.; Walkowiak, J.; Robberecht, E.; Colombo, C.; Sinaasappel, M.; Wilschanski, M.; et al. Defining DIOS and constipation in cystic fibrosis with a multicentre study on the incidence, characteristics, and treatment of DIOS. *J. Pediatr. Gastroenterol. Nutr.* **2010**, *50*, 38–42. [CrossRef] [PubMed]
72. Munck, A.; Alberti, C.; Colombo, C.; Kashirskaya, N.; Ellemunter, H.; Fotoulaki, M.; Houwen, R.; Robberecht, E.; Boizeau, P.; Wilschanski, M. International prospective study of distal intestinal obstruction syndrome in cystic fibrosis: Associated factors and outcome. *J. Cyst. Fibros.* **2016**, *15*, 531–539. [CrossRef] [PubMed]
73. Declercq, D.; Van Biervliet, S.; Robberecht, E. Nutrition and Pancreatic Enzyme Intake in Patients with Cystic Fibrosis with Distal Intestinal Obstruction Syndrome. *Nutr. Clin. Pract.* **2015**, *30*, 134–137. [CrossRef] [PubMed]
74. Baker, S.S.; Borowitz, D.; Duffy, L.; Fitzpatrick, L.; Gyamfi, J.; Baker, R.D. Pancreatic enzyme therapy and clinical outcomes in patients with cystic fibrosis. *J. Pediatr.* **2005**, *146*, 189–193. [CrossRef]
75. Lavie, M. Long-term follow-up of distal intestinal obstruction syndrome in cystic fibrosis. *World J. Gastroenterol.* **2015**, *21*, 318. [CrossRef]
76. Stefano, M.A.; Sandy, N.S.; Zagoya, C.; Duckstein, F.; Ribeiro, A.F.; Mainz, J.G.; Lomazi, E.A. Diagnosing constipation in patients with cystic fibrosis applying ESPGHAN criteria. *J. Cyst. Fibros.* **2021**. [CrossRef]
77. Tan, S.M.J.; Coffey, M.J.; Ooi, C.Y. Differences in clinical outcomes of paediatric cystic fibrosis patients with and without meconium ileus. *J. Cyst. Fibros.* **2019**, *18*, 857–862. [CrossRef]
78. Mentessidou, A.; Loukou, I.; Kampouroglou, G.; Livani, A.; Georgopoulos, I.; Mirilas, P. Long-term intestinal obstruction sequelae and growth in children with cystic fibrosis operated for meconium ileus: Expectancies and surprises. *J. Pediatric Surg.* **2018**, *53*, 1504–1508. [CrossRef]
79. Lai, H.-C.; Kosorok, M.R.; Laxova, A.; Davis, L.A.; Fitzsimmon, S.C.; Farrell, P.M. Nutritional Status of Patients with Cystic Fibrosis with Meconium Ileus: A Comparison with Patients without Meconium Ileus and Diagnosed Early Through Neonatal Screening. *Pediatrics* **2000**, *105*, 53–61. [CrossRef]
80. Chiaravalloti, G.; Baracchini, A.; Rossomando, V.; Ughi, C.; Ceccarelli, M. Celiac disease and cystic fibrosis: Casual association? *Minerva Pediatr.* **1995**, *47*, 23–26.
81. Emiralioglu, N.; Ademhan Tural, D.; Hizarcioglu Gulsen, H.; Ergen, Y.M.; Ozsezen, B.; Sunman, B.; Saltık Temizel, İ.; Yalcin, E.; Dogru, D.; Ozcelik, U.; et al. Does cystic fibrosis make susceptible to celiac disease? *Eur. J. Pediatrics* **2021**, *180*, 2807–2813. [CrossRef] [PubMed]
82. Hjelm, M.; Shaikhkhalil, A.K. Celiac Disease in Patients with Cystic Fibrosis on Ivacaftor: A Case Series. *J. Pediatr. Gastroenterol. Nutr.* **2020**, *71*, 257–260. [CrossRef] [PubMed]
83. Sermet-Gaudelus, I.; De Villartay, P.; De Dreuzy, P.; Clairicia, M.; Vrielynck, S.; Canoui, P.; Kirszenbaum, M.; Singh-Mali, I.; Agrario, L.; Salort, M.; et al. Pain in Children and Adults with Cystic Fibrosis: A Comparative Study. *J. Pain Symptom Manag.* **2009**, *38*, 281–290. [CrossRef]
84. Munck, A.; Pesle, A.; Cunin-Roy, C.; Gerardin, M.; Ignace, I.; Delaisi, B.; Wood, C. Recurrent abdominal pain in children with cystic fibrosis: A pilot prospective longitudinal evaluation of characteristics and management. *J. Cyst. Fibros.* **2012**, *11*, 46–48. [CrossRef]
85. Reznikov, L.R. Cystic Fibrosis and the Nervous System. *Chest* **2017**, *151*, 1147–1155. [CrossRef] [PubMed]
86. Lusman, S.S.; Grand, R. Approach to chronic abdominal pain in Cystic Fibrosis. *J. Cyst. Fibros.* **2017**, *16* (Suppl. S2), S24–S31. [CrossRef]
87. Rajindrajith, S.; Zeevenhooven, J.; Devanarayana, N.M.; Perera, B.J.C.; Benninga, M.A. Functional abdominal pain disorders in children. *Expert Rev. Gastroenterol. Hepatol.* **2018**, *12*, 369–390. [CrossRef]
88. Pinquart, M. Body image of children and adolescents with chronic illness: A meta-analytic comparison with healthy peers. *Body Image* **2013**, *10*, 141–148. [CrossRef]
89. Abbott, J.; Morton, A.M.; Musson, H.; Conway, S.P.; Etherington, C.; Gee, L.; Fitzjohn, J.; Webb, A.K. Nutritional status, perceived body image and eating behaviours in adults with cystic fibrosis. *Clin. Nutr.* **2007**, *26*, 91–99. [CrossRef]
90. Randlesome, K.; Bryon, M.; Evangeli, M. Developing a measure of eating attitudes and behaviours in cystic fibrosis. *J. Cyst. Fibros.* **2013**, *12*, 15–21. [CrossRef]
91. Quick, V.M.; Byrd-Bredbenner, C.; Neumark-Sztainer, D. Chronic illness and disordered eating: A discussion of the literature. *Adv. Nutr.* **2013**, *4*, 277–286. [CrossRef] [PubMed]
92. Nasr, S.Z.; Hurwitz, M.E.; Brown, R.W.; Elghoroury, M.; Rosen, D. Treatment of anorexia and weight loss with megestrol acetate in patients with cystic fibrosis. *Pediatr. Pulmonol.* **1999**, *28*, 380–382. [CrossRef]
93. Rowland, M.; Gallagher, C.; Gallagher, C.G.; Laoide, R.Ó.; Canny, G.; Broderick, A.M.; Drummond, J.; Greally, P.; Slattery, D.; Daly, L.; et al. Outcome in patients with cystic fibrosis liver disease. *J. Cyst. Fibros.* **2015**, *14*, 120–126. [CrossRef]
94. Rowland, M.; Gallagher, C.G.; O'Laoide, R.; Canny, G.; Broderick, A.; Hayes, R.; Greally, P.; Slattery, D.; Daly, L.; Durie, P.; et al. Outcome in cystic fibrosis liver disease. *Am. J. Gastroenterol.* **2011**, *106*, 104–109. [CrossRef] [PubMed]
95. Stonebraker, J.R.; Ooi, C.Y.; Pace, R.G.; Corvol, H.; Knowles, M.R.; Durie, P.R.; Ling, S.C. Features of Severe Liver Disease with Portal Hypertension in Patients with Cystic Fibrosis. *Clin. Gastroenterol. Hepatol.* **2016**, *14*, 1207–1215.e3. [CrossRef]

96. Debray, D.; Kelly, D.; Houwen, R.; Strandvik, B.; Colombo, C. Best practice guidance for the diagnosis and management of cystic fibrosis-associated liver disease. *J. Cyst. Fibros.* **2011**, *10* (Suppl. S2), S29–S36. [CrossRef]
97. Lamireau, T.; Monnereau, S.; Martin, S.; Marcotte, J.E.; Winnock, M.; Alvarez, F. Epidemiology of liver disease in cystic fibrosis: A longitudinal study. *J. Hepatol.* **2004**, *41*, 920–925. [CrossRef]
98. Dos Santos, A.L.M.; De Melo Santos, H.; Nogueira, M.B.; Távora, H.T.O.; De Lourdes Jaborandy Paim Da Cunha, M.; De Melo Seixas, R.B.P.; De Freitas Velloso Monte, L.; De Carvalho, E. Cystic Fibrosis: Clinical Phenotypes in Children and Adolescents. *Pediatric Gastroenterol. Hepatol. Nutr.* **2018**, *21*, 306. [CrossRef]
99. Colombo, C.; Battezzati, P.M.; Crosignani, A.; Morabito, A.; Costantini, D.; Padoan, R.; Giunta, A. Liver disease in cystic fibrosis: A prospective study on incidence, risk factors, and outcome. *Hepatology* **2002**, *36*, 1374–1382. [CrossRef]
100. Lindblad, A.; Glaumann, H.; Strandvik, B. Natural history of liver disease in cystic fibrosis. *Hepatology* **1999**, *30*, 1151–1158. [CrossRef]
101. Valamparampil, J.J.; Gupte, G.L. Cystic fibrosis associated liver disease in children. *World J. Hepatol.* **2021**, *13*, 1727–1742. [CrossRef] [PubMed]
102. Wu, H.; Vu, M.; Dhingra, S.; Ackah, R.; Goss, J.A.; Rana, A.; Quintanilla, N.; Patel, K.; Leung, D.H. Obliterative Portal Venopathy Without Cirrhosis Is Prevalent in Pediatric Cystic Fibrosis Liver Disease with Portal Hypertension. *Clin. Gastroenterol. Hepatol.* **2019**, *17*, 2134–2136. [CrossRef] [PubMed]
103. Bartlett, J.R. Genetic Modifiers of Liver Disease in Cystic Fibrosis. *JAMA* **2009**, *302*, 1076. [CrossRef]
104. Karnsakul, W.; Wasuwanich, P.; Ingviya, T.; Vasilescu, A.; Carson, K.A.; Mogayzel, P.J.; Schwarz, K.B. A longitudinal assessment of non-invasive biomarkers to diagnose and predict cystic fibrosis-associated liver disease. *J. Cyst. Fibros.* **2020**, *19*, 546–552. [CrossRef] [PubMed]
105. Leung, D.H.; Khan, M.; Minard, C.G.; Guffey, D.; Ramm, L.E.; Clouston, A.D.; Miller, G.; Lewindon, P.J.; Shepherd, R.W.; Ramm, G.A. Aspartate aminotransferase to platelet ratio and fibrosis-4 as biomarkers in biopsy-validated pediatric cystic fibrosis liver disease. *Hepatology* **2015**, *62*, 1576–1583. [CrossRef]
106. Calvopina, D.A.; Noble, C.; Weis, A.; Hartel, G.F.; Ramm, L.E.; Balouch, F.; Fernandez-Rojo, M.A.; Coleman, M.A.; Lewindon, P.J.; Ramm, G.A. Supersonic shear-wave elastography and APRI for the detection and staging of liver disease in pediatric cystic fibrosis. *J. Cyst. Fibros.* **2020**, *19*, 449–454. [CrossRef]
107. Loomba, R.; Adams, L.A. Advances in non-invasive assessment of hepatic fibrosis. *Gut* **2020**, *69*, 1343–1352. [CrossRef]
108. Singh, H.; Coffey, M.J.; Ooi, C.Y. Cystic Fibrosis-related Liver Disease is Associated with Increased Disease Burden and Endocrine Comorbidities. *J. Pediatr. Gastroenterol. Nutr.* **2020**, *70*, 796–800. [CrossRef]
109. Mouzaki, M.; Bronsky, J.; Gupte, G.; Hojsak, I.; Jahnel, J.; Pai, N.; Quiros-Tejeira, R.E.; Wieman, R.; Sundaram, S. Nutrition Support of Children with Chronic Liver Diseases: A Joint Position Paper of the North American Society for Pediatric Gastroenterology, Hepatology, and Nutrition and the European Society for Pediatric Gastroenterology, Hepatology, and Nutrition. *J. Pediatr. Gastroenterol. Nutr.* **2019**, *69*, 498–511. [CrossRef]
110. Lommatzsch, S.T.; Taylor-Cousar, J.L. The combination of tezacaftor and ivacaftor in the treatment of patients with cystic fibrosis: Clinical evidence and future prospects in cystic fibrosis therapy. *Ther. Adv. Respir. Dis.* **2019**, *13*, 175346661984442. [CrossRef]
111. Staufer, K. Current Treatment Options for Cystic Fibrosis-Related Liver Disease. *Int. J. Mol. Sci.* **2020**, *21*, 8586. [CrossRef] [PubMed]
112. Granados, A.; Chan, C.L.; Ode, K.L.; Moheet, A.; Moran, A.; Holl, R. Cystic fibrosis related diabetes: Pathophysiology, screening and diagnosis. *J. Cyst. Fibros.* **2019**, *18* (Suppl. S2), S3–S9. [CrossRef]
113. Yi, Y.; Norris, A.W.; Wang, K.; Sun, X.; Uc, A.; Moran, A.; Engelhardt, J.F.; Ode, K.L. Abnormal Glucose Tolerance in Infants and Young Children with Cystic Fibrosis. *Am. J. Respir. Crit. Care Med.* **2016**, *194*, 974–980. [CrossRef] [PubMed]
114. Löhr, M.; Goertchen, P.; Nizze, H.; Gould, N.S.; Gould, V.E.; Oberholzer, M.; Heitz, P.U.; Klöppel, G. Cystic fibrosis associated islet changes may provide a basis for diabetes. An immunocytochemical and morphometrical study. *Virchows Arch. A Pathol. Anat. Histopathol.* **1989**, *414*, 179–185. [CrossRef]
115. Kelly, A.; Moran, A. Update on cystic fibrosis-related diabetes. *J. Cyst. Fibros.* **2013**, *12*, 318–331. [CrossRef] [PubMed]
116. Lanng, S.; Hansen, A.; Thorsteinsson, B.; Nerup, J.; Koch, C. Glucose tolerance in patients with cystic fibrosis: Five year prospective study. *BMJ* **1995**, *311*, 655–659. [CrossRef] [PubMed]
117. Moran, A.; Brunzell, C. Clinical Care Guidelines for Cystic Fibrosis–Related Diabetes A position statement of the American Diabetes Association and a clinical practice guideline of the Cystic Fibrosis Foundation, endorsed by the Pediatric Endocrine Society. *Diabetes Care* **2010**, *33*, 2697–2708. [CrossRef]
118. Marshall, B.C.; Butler, S.M.; Stoddard, M.; Moran, A.M.; Liou, T.G.; Morgan, W.J. Epidemiology of cystic fibrosis-related diabetes. *J. Pediatr.* **2005**, *146*, 681–687. [CrossRef]
119. Hameed, S.; Morton, J.R.; Jaffé, A.; Field, P.I.; Belessis, Y.; Yoong, T.; Katz, T.; Verge, C.F. Early glucose abnormalities in cystic fibrosis are preceded by poor weight gain. *Diabetes Care* **2010**, *33*, 221–226. [CrossRef]
120. White, H.; Pollard, K.; Etherington, C.; Clifton, I.; Morton, A.M.; Owen, D.; Conway, S.P.; Peckham, D.G. Nutritional decline in cystic fibrosis related diabetes: The effect of intensive nutritional intervention. *J. Cyst. Fibros.* **2009**, *8*, 179–185. [CrossRef]
121. Moran, A.; Becker, D.; Casella, S.J.; Gottlieb, P.A.; Kirkman, M.S.; Marshall, B.C.; Slovis, B.; CFRD Consensus Conference Committee. Epidemiology, pathophysiology, and prognostic implications of cystic fibrosis-related diabetes: A technical review. *Diabetes Care* **2010**, *33*, 2677–2683. [CrossRef] [PubMed]

122. Kaminski, B.A.; Goldsweig, B.K.; Sidhaye, A.; Blackman, S.M.; Schindler, T.; Moran, A. Cystic fibrosis related diabetes: Nutrition and growth considerations. *J. Cyst. Fibros.* **2019**, *18* (Suppl. S2), S32–S37. [CrossRef] [PubMed]
123. Onady, G.M.; Stolfi, A. Insulin and oral agents for managing cystic fibrosis-related diabetes. *Cochrane Database Syst. Rev.* **2016**, *4*, Cd004730. [CrossRef]
124. Ballmann, M.; Hubert, D.; Assael, B.M.; Kronfeld, K.; Honer, M.; Holl, R.W. Open randomised prospective comparative multi-centre intervention study of patients with cystic fibrosis and early diagnosed diabetes mellitus. *BMC Pediatr.* **2014**, *14*, 70. [CrossRef] [PubMed]
125. Moran, A.; Pillay, K.; Becker, D.; Granados, A.; Hameed, S.; Acerini, C.L. ISPAD Clinical Practice Consensus Guidelines 2018: Management of cystic fibrosis-related diabetes in children and adolescents. *Pediatr. Diabetes* **2018**, *19* (Suppl. S27), 64–74. [CrossRef]
126. Bellin, M.D.; Laguna, T.; Leschyshyn, J.; Regelmann, W.; Dunitz, J.; Billings, J.; Moran, A. Insulin secretion improves in cystic fibrosis following ivacaftor correction of CFTR: A small pilot study. *Pediatr. Diabetes* **2013**, *14*, 417–421. [CrossRef]
127. Chidekel, A.S.; Dolan, T.F., Jr. Cystic fibrosis and calcium oxalate nephrolithiasis. *Yale J. Biol. Med.* **1996**, *69*, 317–321.
128. Kianifar, H.R.; Talebi, S.; Khazaei, M.; Alamdaran, A.; Hiradfar, S. Predisposing factors for nephrolithiasis and nephrocalcinosis in cystic fibrosis. *Iran J. Pediatr.* **2011**, *21*, 65–71.
129. Hoppe, B.; Hesse, A.; Brömme, S.; Rietschel, E.; Michalk, D. Urinary excretion substances in patients with cystic fibrosis: Risk of urolithiasis? *Pediatr. Nephrol.* **1998**, *12*, 275–279. [CrossRef]
130. Feranchak, A.P.; Sontag, M.K.; Wagener, J.S.; Hammond, K.B.; Accurso, F.J.; Sokol, R.J. Prospective, long-term study of fat-soluble vitamin status in children with cystic fibrosis identified by newborn screen. *J. Pediatr.* **1999**, *135*, 601–610. [CrossRef]
131. Solomons, N.W.; Wagonfeld, J.B.; Rieger, C.; Jacob, R.A.; Bolt, M.; Horst, J.V.; Rothberg, R.; Sandstead, H. Some biochemical indices of nutrition in treated cystic fibrosis patients. *Am. J. Clin. Nutr.* **1981**, *34*, 462–474. [CrossRef] [PubMed]
132. Duggan, C.; Colin, A.A.; Agil, A.; Higgins, L.; Rifai, N. Vitamin A status in acute exacerbations of cystic fibrosis. *Am. J. Clin. Nutr.* **1996**, *64*, 635–639. [CrossRef] [PubMed]
133. Saxby, N.P.C.; Kench, A.; King, S.; Crowder, T.; van der Haak, N.; The Australian and New Zealand Cystic Fibrosis Nutrition Guideline Authorship Group. *Nutrition Guidelines for Cystic Fibrosis in Australia and New Zealand*; Scott, C., Ed.; Bell, Thoracic Society of Australia and New Zealand: Sydney, Australia, 2017.
134. Gupta, S.; Mukherjee, A.; Khadgawat, R.; Kabra, M.; Lodha, R.; Kabra, S.K. Bone Mineral Density of Indian Children and Adolescents with Cystic Fibrosis. *Indian Pediatr.* **2017**, *54*, 545–549. [CrossRef] [PubMed]
135. Hahn, T.J.; Squires, A.E.; Halstead, L.R.; Strominger, D.B. Reduced serum 25-hydroxyvitamin D concentration and disordered mineral metabolism in patients with cystic fibrosis. *J. Pediatr.* **1979**, *94*, 38–42. [CrossRef]
136. Winklhofer-Roob, B.M.; Tuchschmid, P.E.; Molinari, L.; Shmerling, D.H. Response to a single oral dose of all-rac-alpha-tocopheryl acetate in patients with cystic fibrosis and in healthy individuals. *Am. J. Clin. Nutr.* **1996**, *63*, 717–721. [CrossRef]
137. McDonald, C.M.; Alvarez, J.A.; Bailey, J.; Bowser, E.K.; Farnham, K.; Mangus, M.; Padula, L.; Porco, K.; Rozga, M. Academy of Nutrition and Dietetics: 2020 Cystic Fibrosis Evidence Analysis Center Evidence-Based Nutrition Practice Guideline. *J. Acad. Nutr. Diet* **2021**, *121*, 1591–1636.e3. [CrossRef]
138. Bailey, J.; Rozga, M.; McDonald, C.M.; Bowser, E.K.; Farnham, K.; Mangus, M.; Padula, L.; Porco, K.; Alvarez, J.A. Effect of CFTR Modulators on Anthropometric Parameters in Individuals with Cystic Fibrosis: An Evidence Analysis Center Systematic Review. *J. Acad. Nutr. Diet* **2021**, *121*, 1364–1378.e2. [CrossRef]
139. Petersen, M.C.; Begnel, L.; Wallendorf, M.; Litvin, M. Effect of elexacaftor-tezacaftor-ivacaftor on body weight and metabolic parameters in adults with cystic fibrosis. *J. Cyst. Fibros.* **2021**, *in press*. [CrossRef]

Review

The Changing Landscape of Nutrition in Cystic Fibrosis: The Emergence of Overweight and Obesity

Julianna Bailey [1,*], Stefanie Krick [1,2] and Kevin R. Fontaine [3]

1. Division of Pulmonary, Allergy and Critical Care Medicine, University of Alabama at Birmingham, Birmingham, AL 35294, USA; skrick@uabmc.edu
2. Gregory Fleming James Cystic Fibrosis Research Center, University of Alabama at Birmingham, Birmingham, AL 35294, USA
3. Department of Health Behavior, School of Public Health, University of Alabama at Birmingham, Birmingham, AL 35294, USA; kfontai1@uab.edu
* Correspondence: JuliannaBailey@uabmc.edu; Tel.: +1-205-778-8667

Abstract: Cystic fibrosis has historically been characterized by malnutrition, and nutrition strategies have placed emphasis on weight gain due to its association with better pulmonary outcomes. As treatment for this disease has significantly improved, longevity has increased and overweight and obesity have emerged issues in this population. The effect of excess weight and adiposity on CF clinical outcomes is unknown but may produce similar health consequences and obesity-related diseases as those observed in the general population. This review examines the prevalence of overweight and obesity in CF, the medical and psychological impact, as well as the existing evidence for treatment in the general population and how this may be applied to people with CF. Clinicians should partner with individuals with CF and their families to provide a personalized, interdisciplinary approach that includes dietary modification, physical activity, and behavioral intervention. Additional research is needed to identify the optimal strategies for preventing and addressing overweight and obesity in CF.

Keywords: cystic fibrosis; nutrition; obesity; overweight; body composition; body mass index (BMI)

1. Introduction

Cystic Fibrosis (CF) is a rare, life-shortening multi-system organ disease that affects 30,000 people in the United States and 70,000 people worldwide [1]. Pulmonary failure is the main cause of death in this population; the heavy involvement of the gastrointestinal system creates significant nutritional impairments [2]. CF was initially called "Cystic fibrosis of the pancreas" due to the aggressive involvement of the GI system [3]. Historically, malnutrition and underweight have been the prevailing nutritional issue in people with CF (PwCF). Nutritional status has long been defined by body mass index (BMI) in this patient population due to epidemiological evidence that BMI is closely correlated with lung function and, ultimately, survival. Due to the positive correlation between BMI and lung function, the Cystic Fibrosis Foundation (CFF) has set BMI goals of at or above the 50th percentile for children and 22 kg/m^2 or higher for adult females and 23 kg/m^2 or higher for adult males with CF [4]. For this reason, and due to the history of malnutrition with CF, most nutrition interventions have focused on increasing BMI. A high-calorie, high-protein, high-fat diet is recommended for most PwCF, along with oral supplements and sometimes supplemental enteral feeds to help patients achieve the BMI goals associated with the best health outcomes [5]. Aggressive nutrition support has been recommended in pediatric patients with CF to avoid/prevent malnutrition, and to promote catch up growth. Moreover, adequate nutritional status has been associated with reduced pulmonary exacerbations and improved lung function [6].

However, overweight and obesity have emerged as an important issue in the CF population due to advancements in therapy and increased longevity, especially in recent years with the introduction of CF transmembrane conductance regulator (CFTR) modulator therapies [7–10]. There are limited data on overweight and obesity in CF, and few studies related to the impact of obesity on clinical outcomes in CF. The purpose of this review is to critically examine the literature on overweight and obesity in CF, and to make recommendations for clinical practice and future research in this area.

2. Overweight and Obesity in Cystic Fibrosis

BMI has been used to define overweight and obesity, with overweight classified as a BMI 25–29.9 kg/m^2 and obesity as BMI of $\geq$30 kg/m^2 in adults. In children, BMI percentiles are used to measure growth and 85–95th percentile indicates overweight with >95th percentile indicating obesity in children [11]. In the United States, 42.2% of adults are obese, with 9% being severely obese [12]. The prevalence of childhood obesity in the U.S. is 19.3% [13]. Obesity is associated with the development of many chronic and potentially life-threatening health conditions such as heart disease, cancer, and diabetes and virtually every organ system can be adversely affected by obesity. The medical cost associated with obesity is estimated to be USD 149 billion, making it both a serious public health crisis and economic issue in the general population [14]. While malnutrition has been the primary issue in CF due to increased work of breathing and malabsorption, obesity and overweight are becoming a growing concern [15]. Dyslipidemia and insulin resistance have been observed in the CF population; it is unknown how these factors affect risk for development of heart disease and diabetes, and how obesity may affect risk for development of comorbidities in CF due to the life-shortening nature of this disease [16–18].

Nutritional status has largely been defined by BMI in the CF population due to associations between BMI and pulmonary disease. Given the changing landscape of nutritional status in PwCF and the rise in overweight and obesity, body composition as a predictor of pulmonary and clinical outcomes has also become an area of interest. Further examination of body composition is merited given that the World Health Organization defines obesity as "an abnormal or excessive accumulation of fat that poses a risk to health" [19]. Research on obesity and body composition in the general population has demonstrated that body fat distribution, particularly increased upper body fat, visceral fat, and intramuscular fat, are important predictors of metabolic consequences commonly associated with obesity [20]. Evidence suggests that PwCF tend to have lower fat-free mass than healthy controls [21–27] and that lower fat-free mass is associated with lower lung function in CF [21,23,28–31]. Other body composition abnormalities including ectopic fat deposition with fatty replacement of the pancreas, central adiposity, visceral adiposity, and normal weight obesity have been observed in CF [28,29,32–35]. Few studies have explored the implications of increased abdominal fat and visceral fat specific to CF; however, one cross-sectional study of adults with CF found that high levels of visceral fat in CF were associated with decreased insulin sensitivity [36]. While associations between some clinical outcomes and body composition exist in the literature, more studies are needed to ascertain the role of body composition in nutritional assessment, and other clinical aspects of CF treatment including the management of overweight and obesity.

2.1. Prevalence

Several studies have assessed the prevalence of overweight and obesity in CF. A CF center-specific cross-sectional study in children at the Pittsburg CF Center found an increased rate of overweight and obesity in children with CF, with 15% of children qualifying as overweight and 8% of children were obese, based on BMI percentile [15]. Interestingly, 50% of the overweight children and 20% of the obese children suffered from malabsorption caused by CF-associated exocrine pancreatic insufficiency [15]. A recent single center study conducted in the Minnesota adult CF center found that >30% of patients were overweight or obese between the years of 2015–2017 [37]. A longitudinal analysis of the U.S. CF Foundation Patient Registry indicates that the prevalence of overweight and obesity increased over the past 20 years in both children and adults with CF. Between the years of 1998 and 2017, the percentage of overweight increased from 7% to 16% and the proportion of obese patients increased from 2% to 6%, which was noted to be a 345% increase in the number of PwCF who are obese due to CF population increase in the registry [38]. Increased overweight and obesity has also been observed in the UK CF population, and in a CF genotype typically associated with pancreatic insufficiency. An analysis of UK CF registry data for the year of 2002 attempted to determine the prevalence of overweight and obesity in children and adults with the most common CF mutation, DeltaF508 homozygous. Results indicated that 9% of PwCF and the DeltaF508 homozygous genetic mutation were overweight and 1% were obese [39]. While rates seem low compared to the obesity prevalence in the general population, this is significant because overweight and obesity are not expected in patients who are homozygous for DeltaF508 as this genotype has historically been associated with more severe nutritional deficiencies [40]. A cross-sectional observational study of 68 adults and children with CF in a Greek CF center found that 13.2% of patients with CF were either overweight or obese [35]. Analysis of a single CF center in Spain found that 6% of children with CF were overweight and 1% were obese, and that overweight/obese status did not provide improved pulmonary function [18]. Likewise, in Italy, a single site study of adults with CF found that 22% were either overweight or obese [41]. In addition to cross-sectional studies found in the literature, a longitudinal cohort study in Toronto of 909 adults with CF and that found that the percentage of overweight and obese CF patients increased from 7% to 18.4% between 1985 and 2011, similar to trends observed in the U.S. CF Patient Registry [8,38]. Prevalence of overweight and obesity in CF from cross-sectional and longitudinal studies in international CF samples is presented in Table 1. Given early data from weight gain on new CFTR modulators that are available to 90% of the CF population, we can expect these numbers to rise even further [9,42–44].

Table 1. Studies examining prevalence of overweight and obesity in CF.

Reference	Population	N	% Overweight	% Obese
Flume et al., 2019 [38]	United States CF Registry Data, Age ≥ 2 years	42,988 across study duration from 1998–2017	16% in 2017	6% in 2017
González Jiménez et al., 2017 [18]	Spain, 12 hospitals Age 4–57 years	451	6%	1%
Gramegna et al., 2021 [41]	Italian, multi-center Adults Age 32–45 years	321	20%	2%
Hanna et al., 2015 [15]	United States, single CF Center Children Age 2–18	226	15%	8%
Harindhanavudhi et al., 2020 [37]	United States, single CF Center, Minnesota Adults	484	25.6%	6.6%
Kastner-Cole et al., 2005 [39]	United Kingdom CF Registry Data Adults and children with F508del mutation	1869 children 1181 adults	9% in adults and children	1% in adults and children
Panagopoulou et al., 2014 [35]	Greece, single CF Center Age 2–38 years	68	6%	7%
Stephenson et al., 2013 [8]	Toronto single CF Center Adults	651	18.4%	3.3%
White et al., 2015 [45]	Australia, children from 16 inpatient hospitals	832	8.8%	9.9%

2.2. Etiology

2.2.1. Genetics

Multiple factors may contribute to the decreasing proportion of malnutrition and increasing prevalence of overweight and obesity in CF. There is evidence to suggest that some people are genetically predisposed to development of overweight and obesity [46]. In the CF population, some studies have found that obesity occurs more frequently in patients with less severe genetic mutations and patients who are pancreatic sufficient [8,39]. A US CF Registry-based study conducted by Flume and colleagues found that the prevalence of obesity was higher among individuals with CF who had class IV or class V CF causing genetic mutations, which are typically less severe mutations. However, a recent single CF center study in Minnesota found that overweight and obesity were present even in patients with genotypes associated with more severe disease [37], and a UK CF Registry-based study found that overweight and obesity were present in patients homozygous for F508del, which is known to present a more severe phenotype and is associated with pancreatic insufficiency [39].

2.2.2. Dietary Causes

The unrestricted high-calorie, high-fat diet that is often called the "legacy CF diet" likely contributes to positive energy balance that promotes excessive weight gain. PwCF have long been advised and behaviorally conditioned to consume an unrestricted high-calorie diet with an emphasis on weight gain since childhood and have received vastly different messaging around nutrition than the general population [4]. Indeed, studies of diet quality in children with CF have shown intake high in saturated fat, trans fat, and total calories but low in nutrient-dense foods such as fruits and vegetables [47]. A similar pattern has been observed in adults with CF in a cross-sectional study that demonstrated intake high in added sugars, refined grains, trans-fatty acids, and low intakes of whole grains and dietary fiber [36]. Further, poor diet quality was associated with high levels of visceral adipose tissue in this study [36]. Until recently, CF nutrition guidelines recommended an unrestricted high-calorie diet, with little focus on nutrient density or diet quality. As clinical care and treatments for CF have improved over the past decade, the high-calorie CF diet may have had led to positive energy balance and excessive weight gain in a subset of PwCF. Extended life expectancy is prompting a focus on diet quality to promote optimal health outcomes and prevention of other chronic lifestyle diseases in CF.

2.2.3. CFTR Modulator Therapy

Highly effective drugs that treat CF at the cellular level, known as CFTR modulators, are now available for up to 90% of the population with CF. CFTR modulators have been shown to improve pulmonary function, quality of life, and in some cases, cause weight gain and increase BMI. A recent systematic review concluded that the effect of CFTR modulators on anthropometric measurements is dependent on genetic mutation and modulator formulation [48]. Ivacaftor, the first CFTR modulator approved, was available to roughly 6% of the CF population who have gating mutations and was shown to increase weight and linear growth in children with CF [49–51]. In a long-term study of the clinical effects of ivacaftor, Guimbellot et al. observed that the proportion of overweight increased from 16% to 25% in adults and from 9% to 18% in pediatric patients over 5.5 years on the drug [10]. The rate of obesity remained stable in pediatric patients but increased slightly from 8% to 11% in adults [10]. A new highly effective CFTR modulator, elexacaftor-tezacaftor-ivacaftor (ETI), is now available to up to 90% of the CF population and has been shown to increase BMI in adolescents and adults with CF by 0.9–1.1 kg/m^2 in phase III clinical trials [43,44]. A single-center retrospective study recently confirmed weight gain in adults on ETI in the real-world setting, with an average BMI increase of 1.5 kg/m^2 after 12 months on ETI [9].

Few studies have explored mechanisms of weight gain on CFTR modulators, but decreased resting energy expenditure, increased caloric intake, and improved intestinal absorption are postulated to play a role in adults with gating mutations taking ivacaftor only [52]. One study of dietary changes on ivacaftor demonstrated that American participants increased their fat intake significantly, while Italian participants increased both total calorie and fat intake while taking ivacaftor [53]. This may be due to drug package instructions that advise patients to take the drug with "a fat containing food" twice per day while the quantity of fat that should be consumed for optimal drug efficacy is unknown. Some evidence also indicates that children with CF who took ivacaftor showed improvement, and in some cases, reversal of exocrine pancreatic insufficiency [54–56]. Improved pancreatic function likely plays a role in weight gain in children who take ivacaftor.

It is important to understand whether weight gain on CFTR modulators is due to fat mass or lean body mass accrual. There are little data on body composition changes on CFTR modulators, but one study of adults who took ivacaftor found an increase in both fat mass and fat-free mass as measured by DEXA after 3 months on the drug [52]. Another study found an increase in fat mass using bioelectrical impedance analysis after 28 days on ivacaftor [57]. In a long-term open label extension study of ivacaftor, King et al. found that both weight and fat mass significantly increased after 6 months of treatment and, at two years on the drug, 64% of weight gained was fat mass. Additionally, 25% of participants were classified as overweight and 10% were obese by the end of the two-year study [58]. More studies of longer duration are needed to examine body composition and fat distribution changes on CFTR modulators in the setting of unintended weight gain and the development of overweight and obesity on these drugs.

2.2.4. Social Determinants of Health

Social determinants of health are defined as environmental conditions that affect health and quality of life outcomes and risks [59]. Limited access to healthy food correlates with other unmet needs that indicate poverty. Poverty and food insecurity are both associated with obesity in the general population [60,61]. Health disparities related to poverty or low socioeconomic status (SES) have been documented in CF and affect survival, CF-related clinical outcomes, and access to optimal healthcare for CF [62–64]. Low SES is associated with obesity in the general population but socioeconomic factor contributions to overweight and obesity in the CF population have not been explored. Food insecurity and limited access to healthy food could increase risk for overweight and obesity in the population due to reliance on high-calorie convenience foods that lack nutrient density. Additionally, PwCF spend an average of 2–3 h per day on a complex medical treatment regimen, which leaves little time for healthy meal planning, particularly in the setting of food insecurity that limits access to healthy food. It is reasonable to suspect that a similar association between obesity and poverty and food insecurity observed in the general population may exist in CF, but national CF Registry-based studies are needed to understand the role of social determinants of health in the development of overweight and obesity in the CF population. Additional research could provide data to create targeted interventions to prevent over-nutrition in subsets of the CF population experiencing socioeconomic hardship.

3. Consequences of Overweight and Obesity in Cystic Fibrosis

3.1. Lung Function

There is a strong positive association between BMI and lung function in the CF population [2]. A large Canadian CF Registry analysis found that while overweight and obesity were associated with an increase in lung function, the magnitude of the increase was significantly less in the obese group than in the adequate-weight reference group [8]. Emerging evidence suggests that overweight and obesity may not provide benefit with respect to pulmonary function and could even be detrimental. In a single-center cross-sectional study conducted by Hanna et al., no associations were observed with overweight/obese status and lung function, prompting the authors to conclude that overweight and/or obesity does not confer pulmonary benefits in CF [15]. Several other studies examining prevalence and outcomes of obesity in CF have found that overweight and obesity were not protective with respect to lung function [8,15,18,65]. Additionally, emerging evidence suggests that in children with CF who are pancreatic sufficient, a BMI > 85th percentile has a detrimental effect on pulmonary function [66]. Large national registry-based studies are needed to determine longitudinal associations between higher BMI and pulmonary outcomes in the CF population.

3.2. Cardiometabolic Parameters

Few studies have addressed the effect of obesity on other comorbidities in CF. Shorter life expectancy has typically not allowed for patients with CF to live to the age where obesity-associated comorbidities such as heart disease, cancer, and type 2 diabetes would emerge. Additionally, at least one study estimates that up to 50% of adults with CF develop CF-related diabetes, which is a unique disease entity distinct from both type 1 and type 2 diabetes, but with features of both insulin deficiency and insulin resistance [16]. The endocrine abnormalities inherent to CF make understanding the association between obesity and development of diabetes more difficult to assess in this population. However, a cross-sectional study by Coderre and colleagues found that 13% of adults with CF had a BMI > 25 kg/m^2 and these overweight/obese patients had higher fasting insulin, total cholesterol, and LDL cholesterol than patients with CF at lower BMI. They also found that LDL cholesterol and insulin area under the curve were associated with lower lung function [67]. In a comparison of underweight, normal weight, and overweight PwCF to determine differences in insulin levels, insulin response, and prevalence of diabetes, 5.5% of patients were overweight and overweight patients had higher fasting insulin, higher insulin resistance (HOMA-IR), and higher insulin levels during an oral glucose tolerance test [65]. Authors concluded that overweight patients with CF may be at increased risk for the development of diabetes [65]. Dyslipidemia has also been observed in the CF population and studies have found a positive association between total cholesterol and triglyceride levels and BMI in the CF population [8,67–70]. In a recent study, ETI was shown to increase total cholesterol, HDL, and LDL in adult patients with CF-related diabetes. The implications of dyslipidemia in PwCF are not clear, and the role of dietary intervention in CF dyslipidemia is unknown [9]. More longitudinal research is needed to understand the association between overweight and obesity in CF and how this affects risk for the development of diabetes and cardiovascular disease.

3.3. Lung Transplantation

Lung transplantation is a treatment option for end-stage pulmonary diseases including CF. Large database evidence suggests that both malnutrition and overweight are associated with increased risk of death following a lung transplant. Currently, class II or III obesity (BMI $\geq$ 35 kg/m^2) is considered an absolute contraindication to lung transplant, and class I obesity (BMI 30–34.9 kg/m^2) is considered a relative contraindication. In a recent study of 546 lung transplant patients, patients who were overweight or obese had a significantly higher mortality rate. Of note, only two of the patients who had CF were overweight and none were obese. More research is needed to understand how overweight and obesity affect lung transplant outcomes in CF, and assessing body fat distribution may be especially important as abdominal and visceral fat can play a role in restrictive lung disease [71].

3.4. Weight Stigma and Psychological Consequences

Obesity is common in the general population but is a highly stigmatized condition in many settings including the medical environment. Evidence suggests that people with obesity experience prejudice, derogatory comments, and that in some cases, health care professionals hold negative perceptions of patients who are overweight or obese. Patient experience of weight stigma can also lead to increased stress and subsequent avoidance of clinical care [72]. A recent systematic review of the impact of weight stigma in overweight and obese individuals found that weight stigma was positively associated with cortisol levels, diabetes risk, eating disturbances, depression, anxiety, body image issues, and was negatively associated with self-esteem [73]. In the general population, weight stigma has been associated with decreased motivation to change eating patterns and less healthy eating behaviors, which could lead to negative health consequences [74]. While there are no studies related to weight stigma in overweight and obese PwCF, there is a growing body of evidence related to body image disturbances and even eating disorders in this population [75–77]. Clinicians should therefore take a sensitive and empathetic approach when having discussions about overweight and obesity, and be cognizant of the psychological consequences of weight stigma. Additionally, it would be beneficial to include the CF team social worker and/or psychologist early on in the care of these patients to address the mental health impact of overweight and obesity.

4. Treatment Strategies

Management of overweight and obesity in the general population typically involves weight loss given evidence that a reduction of 5–10% body weight can reduce the risk of cardiovascular disease, improve lipid panels, decrease blood pressure, and decrease risk of diabetes [78,79]. Even a sustained weight reduction of as little as 3–5% has been associated with some positive health benefits and reduced health risk [80]. Weight loss in overweight and obesity can be effectively achieved by creating negative energy balance through dietary modification to reduce energy intake paired with physical activity to increase energy expenditure, and supported by behavioral counseling to facilitate lifestyle change [81,82]. Studies show that this approach can produce a 3–5% body weight loss that is associated with an improvement in health parameters [83]. Other markers of health can include cardiometabolic parameters such as lipid panels, blood pressure, diet quality, eating behaviors, hemoglobin A1c (in diabetic patients), body composition parameters, and quality of life. When lifestyle modification is unsuccessful, medication and even surgical management may be necessary.

To our knowledge, no studies exist regarding treatment strategies for overweight and obesity in in CF. Studies conducted thus far suggest some associations between overweight/obesity and comorbidities in CF, worse lung transplant outcomes, dyslipidemia, and increased diabetes risk with higher fasting insulin levels. Therefore, further work is needed to better understand the etiology, consequences, and treatment and prevention strategies for PwCF who are overweight and obese. It is important to have a full evaluation by all members of the multidisciplinary CF care team when a patient has unintended weight gain that results in overweight or obesity. Causes of weight gain such as medications and other disease processes should be ruled out before pursuing interventions for overweight and obesity management. While research is conducted to identify the optimal approach to overweight and obesity in CF, it is reasonable to utilize evidence-based methods for the general population presented in this section.

4.1. Dietary Intake Recommendations

The CF Foundation has not formally updated CF Nutrition Guidelines since obesity and overweight emerged as a significant issue or since highly effective CFTR modulators recently became available for most of the CF population. However, several organizations have created evidence-based guidelines in recent years, which are presented in Table 2. Most recently, the Academy of Nutrition and Dietetics (AND) published guidance based on a systematic review that suggests that there is no evidence to support that PwCF have any benefit from consuming a dietary pattern outside of what is recommended for the general population [84]. This guideline also emphasizes just as some patients with CF experience under-nutrition, some patients may require caloric reduction with a focus on improving diet quality given that undesired weight gain, overweight, and obesity have been reported in a subset of the CF population, especially with CFTR modulator use. Both the Australia and New Zealand CF Nutrition Guideline and the AND CF Nutrition Guideline emphasize taking an individualized approach to nutrition care based on patients' genetic mutations, clinical status, laboratory values, personal health goals, nutrition status, culture, and food preferences [84,85]. The AND guideline suggests a diet rich in whole grains, fruits, vegetables, seafood, legumes, lean protein, nuts and beans, and low-fat dairy as these foods have been associated with positive health outcomes in the general population [84].

Table 2. Summary of cystic fibrosis nutrition guidelines.

Guideline Reference	Total Calorie Intake	Macro-Nutrient Balance	Overweight/Obesity
Cystic Fibrosis Foundation, 2008 [4]	110–120% of estimated energy requirements for general population	35–40% of calories from fat 40–45% of calories from carb 20% of calories from protein	N/A
ESPGAN, 2016 [86]	110–120% of energy requirements for same age healthy children and adults	20% of calories from protein; notes lack of evidence for recommending macro-balance	"We suggest adjusting energy intake upward to achieve normal growth and nutritional status while avoiding obesity" [86]
Thoracic Society of Australia New Zealand, 2018 [85]	110–200% of the population-based energy requirements emphasizing frequent RD assessment and individualized energy intake goals	Upper limit of 25% of calories from protein	High BMI in Pediatrics: Overweight: BMI 85th to <95th BMI Obese: >95th percentile High BMI in Adults: ≥ 27 kg/m^2 AND/OR unintentional weight gain from previously acceptable BMI of >5 kg within a year [85]
Academy of Nutrition and Dietetics CF Nutrition Guideline, 2020 [84]	110–200% of the population-based energy requirements emphasizing RD frequent assessment and individualized energy intake goals	"Macronutrients in same percentage distribution as is recommended for the typical, age-matched population"	"For individuals with CF who are overweight or obese, it is reasonable for the RDN or international equivalent to advise an age-appropriate diet that emphasizes foods associated with positive health outcomes in the general population, with energy needs adjusted to achieve or maintain normal growth (pediatrics) or BMI status (adults)." [84]

In the general population, a deficit of 500 calories per day can produce weight loss of one pound per week [87]. Negative energy balance may be achieved through diet alone or through a combination of diet and exercise. Evidence suggests that a variety of dietary patterns can be effective for calorie reduction including low carbohydrate, low fat, portion control, and the Mediterranean diet pattern with calorie restriction [81,82,88]. Other dietary trends for weight loss are gaining popularity in the general population, including time-restricted plans such as intermittent fasting; however, there is currently no evidence to suggest efficacy in the CF population. Given that the best predictor of weight loss is adherence to a chosen dietary plan, it is important to select the diet best suited for the individual in order to promote the sustainability of dietary patterns and lifestyle changes [89]. PwCF have been counseled to eat a high-calorie diet since early childhood, so clinicians should recognize that shifting a dietary intake will take time and often trial and error to find the pattern that works best for each individual. Small incremental changes may be useful to build confidence and avoid overwhelming patients [90]. Table 3 presents suggestions for modifications to the legacy CF diet adapted from the Academy of Nutrition and Dietetics Evidence Based Guideline for Nutrition in Cystic Fibrosis and the Dietary Guidelines for Americans with the goal of reducing calories and increasing nutrient dense foods and foods known to promote health and reduce the risk of chronic disease.

Table 3. Dietary modifications for a generally healthful eating pattern.

Food Group	Recommended Foods	Foods to Consider Reducing
Carbohydrates	Whole grains (at least half of grains should be whole grains). • Whole wheat bread, buns, rolls, tortillas, and crackers • Whole wheat pasta • Brown rice • Wild rice • Quinoa • Oats • Barley • Whole grain cereals	Refined Grains • White breads • White rice • Biscuits • Cakes Added sugars • Soda • Sweetened Coffee and tea • Fruit drink and lemonade • Many breakfast cereals • Granola bars • Deserts and candy
Fats	Unsaturated Fats • Vegetable oils • Nuts, nut butters • Fish • Avocado	Saturated fats • Butter and stick margarine • Heavy Cream • Cream cheese
Protein	Lean Meats • Poultry • Eggs • Beans, peas, lentils • Seafood • Soy products • Nuts and Seeds	Fatty meats • Beef ribs • Sausage • Processed meats • Fried meats
Fruits	Whole Fruits • Berries • Melons • Citrus Fruits (oranges, grapefruit, limes) • Other fruits (apples, banana, pears, apricot, etc.)	Juices that are not 100% fruit juice
Vegetables	Vegetables of all types including • Dark green vegetables (spinach, kale, broccoli) • Red and Orange vegetables (peppers, squash, carrots) • Starchy vegetables (potatoes, corn, yucca, jicama) • Beans, peas, lentils • Other vegetables (sprouts, cauliflower, asparagus, eggplant, etc.)	Fried vegetables
Dairy	Fat-free low-fat dairy • Skim milk • 1% (low fat) milk • Low-fat or fat-free yogurt • Cheeses • Non-dairy alternatives such as soy milk	Full-fat dairy: • Whole milk • Full-fat yogurt • Full-fat cheeses

Additional resources on following for healthful diet plan:
www.choosemyplate.gov (accessed on 14 December 2021)
www.dietaryguidelines.gov (accessed on 14 December 2021)

Adapted from data found in Academy of Nutrition and Dietetics CF Nutrition Guidelines and the Dietary Guidelines for Americans.

In considering dietary change recommendations for PwCF as overweight and obesity emerge, it should be noted that malnutrition still exists in subsets of the CF population, particularly in those who are not eligible to take CFTR modulators due to their genetic mutations and in patients with advanced lung disease. PwCF who also experience malnutrition will likely continue to require a nutrient-dense high-calorie diet to optimize nutrition status. The broad spectrum of nutrition status observed in CF currently highlights the need for highly individualized nutrition therapy. The CF Care Team Registered Dietitian should work carefully with patients in co-producing customized nutrition care plans to help patients meet personal goals and optimize health outcomes, based on individual clinical status, cultural considerations, and food preferences. Emphasis should also be placed on screening for food insecurity and other social determinants of health that could impact patient's ability to afford healthy foods. Registered Dietitians can assist patients in planning healthful meals and eating patterns that are affordable and should work in tandem with the CF Social Worker and other care team members to provide resources to improve food access for patients experiencing food insecurity.

4.2. Physical Activity

While diet is a crucial element in the management of overweight and obesity, a multidisciplinary lifestyle intervention is recommended [91]. Exercise is an important component of comprehensive lifestyle plans to treat overweight, obesity, and associated comorbidities. Regular physical activity has been shown to assist in creating negative energy balance for weight loss and improve cardiometabolic risk factors [92]. In the general population, it is recommended to participate in 150 min per week of moderate physical activity or 75 min per week of vigorous activity. To promote weight loss in the absence of dietary changes, 225–420 min of exercise is recommended. However, physical activity paired with reduced caloric intake is recommended as this strategy is more likely to promote weight loss [93]. Research has also consistently demonstrated that physical activity is necessary for weight loss maintenance in the general population [80]. Aside from weight loss, regular physical activity is associated with reduced risk of cardiovascular disease and diabetes in obese individuals, independent of weight loss [92]. In the CF population, physical activity, particularly strength training, is an important component of exercise plans in order to promote preservation and accrual of lean body mass [94], which is associated with improved pulmonary function in this population. Additionally, exercise in CF also provides additional benefits beyond weight management, including increased aerobic and pulmonary capacity and airway clearance [95]. Further, exercise can provide mental health benefits that could combat anxiety and depression associated with overweight and obesity [96]. The CF Care Team Physical Therapist is an invaluable resource in creating customized exercise plans to help patients meet their personal health, weight, and body composition goals and that synergize with the nutrition care plans provided by Registered Dietitians.

4.3. Behavioral Interventions

Comprehensive lifestyle interventions for overweight and obesity include dietary intervention, physical activity, and behavioral approaches to lifestyle change [97]. A systematic review that defined successful weight loss interventions for obesity as ≥5% initial body weight loss maintained at 12 months found that 92% of successful interventions included a behavioral component [82]. Intensive behavioral therapy (IBT) for obesity that includes nutritional guidance and exercise has been shown to produce significant weight loss, lower blood glucose levels, decrease waist circumference, and decrease blood pressure. Additionally, IBT decreased the development of diabetes by 50% in patients who had elevated blood glucose at baseline [98]. Many behavioral change theories, models, and strategies provide an evidence-based approach to changing behaviors related to energy balance in the treatment of overweight and obesity. Cognitive behavioral therapy (CBT) and motivational interviewing (MI) are two modalities of behavioral change that have been used with success in obesity treatment [81,99]. CBT involves such skills as self-monitoring, goal setting, problem solving, and stimulus control [81]. MI is an approach that emphasizes partnership for a client-driven lifestyle change intervention [100]. Many different behavior change models exist and a variety of combinations of behavioral techniques and strategies can be used to facilitate behavior change.

The intensity of the comprehensive lifestyle intervention seems to have an impact on the efficacy of behavioral interventions for the management of overweight and obesity. Research has shown that 14 sessions in a 6-month time frame led to more weight loss than a less frequent intervention of 12 or less sessions in 6 months [97]. It should be noted that most patients with CF see their care team in the clinic once every 3 months, and that more frequent multidisciplinary visits that provide comprehensive lifestyle intervention-focused on management of overweight and obesity may be warranted. Telehealth and home-based interdisciplinary lifestyle interventions may be an option for increasing frequency of visits centered around treatment of overweight and obesity, particularly for patients who live far away from their CF care centers.

Successful behavioral interventions targeting BMI in children with CF have been conducted, although these interventions are typically focused on normalizing growth and promoting weight gain. These interventions in the pediatric CF population have involved, with strong parental support and participation, implementing a behavioral intervention in combination with nutrition education, and this has been found to be more effective than nutrition education alone [101]. These behavioral nutrition interventions focus on positive reinforcement and promoting positive mealtime behaviors to prevent growth decline in children with CF. It is possible that these behavioral intervention strategies could potentially be used to prevent the development of overweight and obesity in children with CF.

Only one behavioral nutrition intervention conducted in adults with CF was identified in the review of the literature. This randomized controlled trial was a Social Cognitive Theory home-based behavioral intervention called "Eat Well with CF" and had a duration of 10 weeks. While this intervention was successful in improving self-efficacy scores related to nutritional self-management, there was no significant change in BMI or quality of life at the end of the intervention [102]. Self-efficacy is believed to be crucial for facilitating and maintaining behavioral changes [100]. While this intervention did not produce a significant change in BMI, it is possible that the improvement in self-efficacy could be an early indicator of behavioral change, especially given the short duration of the intervention [102]. It is also possible that early changes in body composition occurred but were not captured by measuring BMI only. Home-based behavioral interventions that involve both nutrition and physical activity components could be an avenue for increasing the frequency and intensity of multidisciplinary weight management interventions, especially given that telehealth is growing in popularity, with several studies documenting a high level of patient satisfaction and positive experience of telehealth care in CF [103,104].

4.4. Weight-Neutral Approaches

While weight loss is the most common goal in the treatment of overweight and obesity due to associations with improved cardiometabolic outcomes, it is unknown if the benefits associated with a 5–10% body weight loss in the general population translate to PwCF. Given that normal weight obesity and lean body mass depletion have been documented in CF [29], it is unclear if weight loss interventions for overweight and obese individuals with CF will result in predominantly fat loss and could cause further reduction in lean body mass stores. Body image issues and eating disorders have been documented in CF, and this population deals with an intense focus on weight patterns and BMI throughout their lifespan, due to the association between BMI and pulmonary function and guideline recommendations based on BMI and growth [77]. There is also some evidence that suggests that the influence of eating experiences in childhood are sustained into adulthood in CF [99]. For these reasons, some individuals may instead prefer to focus on other markers of health as outcomes, rather than weight, when designing a plan to address overweight or obesity and related health consequences. The Health at Every Size® (HAES®) approach is a weight-inclusive model that focuses on improving physical, behavioral, and psychological parameters rather than promoting weight loss. This method also places emphasis on body acceptance and improving the individual's relationship with food [105,106]. HAES® has been gaining popularity in the general population and has been studied regarding overweight and obesity. A recent systematic review of randomized controlled trials in overweight and obese individuals found that the HAES® approach is associated with improved quality of life, improved cardiovascular endpoints, increased physical activity, reduction in disordered eating, reduction in binging as well as improved diet quality [107]. Despite the weight-neutral approach utilized, some studies have demonstrated reduction in BMI, waist circumference, and fat loss as a result of the HAES® interventions [108]. While there are no data on the efficacy of HAES® in CF, clinicians should be aware of this approach as individuals with CF who have a history of body image disturbances or disordered eating may prefer a weight-neutral approach to addressing the health consequences associated with the development of overweight and obesity. More research is needed on comprehensive health behavior interventions, including weight-neutral approaches to improve cardiometabolic outcomes and quality of life in the CF population.

4.5. Medical and Surgical Treatment

When lifestyle approaches are ineffective, pharmacological approaches are recommended to treat overweight and obesity in the general population when BMI is >30 kg/m^2 or >27 kg/m^2 with comorbid conditions in adults [109]. Several drugs are approved by the FDA to treat obesity and include orlistat, lorcaserin, liraglutide, topiramate/phentermine, naltrexone/bupropion, and newly approved semaglutide [110]. In addition, prescription appetite suppressants are available and include phentermine, benzphetamine, diethylpropion, and phendimetrazine. It should be noted that appetite suppressants are only approved for short-term use of 12 weeks or less in the general population. Orlistat, Liraglutide, and setmelanotide are the only drugs approved for use in adolescents. However, setmelanotide is only approved for three specific rare genetic conditions which do not include CF. None of these drugs have been studied in CF and their use should be carefully considered by the medical team, weighing the risks and benefits with patients and their families. Of note, use of Orlistat may not be ideal in PwCF who also have pancreatic insufficiency as this could compound malabsorption and lead to vitamin and essential fatty deficiencies in this population that is already at high risk of micronutrient deficiencies.

A variety of bariatric procedures are available for the management of obesity and include surgical (gastric bypass, gastric banding, sleeve gastrectomy) and non-surgical (intragastric-balloon) options. Bariatric procedures are considered in the general population when BMI is $\geq$40 kg/m^2 or BMI $\geq$ 35 kg/m^2 with one or more severe obesity-related complications that can be improved with weight loss [109]. There currently is no evidence on outcomes regarding bariatric procedures to treat obesity in CF, and this could lead to micronutrient deficiencies and worsened gastrointestinal issues, particularly in patients who have exocrine pancreatic insufficiency. Given what is known about nutrition risk in the CF population, bariatric surgery may not be an ideal option for PwCF until more evidence is available.

5. Discussion

Nutrition status as measured by BMI is closely linked with pulmonary outcomes and survival in CF [4]. As life expectancy continues to increase in this population with the use of CFTR modulators, non-traditional nutrition issues have emerged, and overweight/obesity have become areas of interest in CF clinical care and research. Studies indicate that overnutrition is increasing in CF, and this has been linked to insulin resistance and other hormonal disturbances, which are postulated to play a role in development of diabetes and the trend of increased central fat distribution observed in this population [65,67]. Additionally, central adiposity, visceral adiposity, and normal weight obesity have been documented in CF and could have negative health consequences [28,33–35].

Additional perspective is needed to determine if excess weight and adipose tissue distribution in CF are causally related to pulmonary and cardiometabolic outcomes. Specifically, additional studies are needed to determine the interplay between central adiposity and CF-related endocrine, metabolic, and pulmonary outcomes. Given the association between fat-free mass and lung function in CF, future research should also focus on exploration of the causes and predictors of decreased fat-free mass so that interventions can be developed to promote accrual of lean body mass and reduction in central adiposity, regardless of BMI status, in PwCF. While weight loss is a common goal of obesity intervention success, clinicians should also track body composition changes to ensure that lean body mass is preserved and optimized during weight reduction.

The care team should be cognizant of the stigma associated with overweight and obesity and the psychological impact of undesired weight gain. It should be noted that some individuals with CF may prefer weight-neutral interventions that place emphasis on health-promoting behaviors and repairing their relationship with food rather than weight loss as their main goal. Collaborating with the CF care team Social Worker and/or Psychologist is advised, both for support around these issues and assistance with behavioral strategies to promote optimal health and wellbeing during weight management interventions.

There is limited evidence on the optimal diet for PwCF, especially in the setting of new highly effective CFTR modulators and the development of overweight and obesity in this population. Caloric deficit is necessary for weight loss, and many dietary approaches have been shown to promote weight reduction in the general population. The sustainability of any intervention relies on an individual's ability to adhere to the chosen regimen. It is therefore important to take an individualized approach to nutrition care and weight management plans, and it is the role of the CF care team Registered Dietitian to partner closely with PwCF and their families to determine a customized nutrition care plan that is the best fit for patient preferences, culture, and clinical status. Registered Dietitians should also closely collaborate with the CF team Physical Therapist and Social Worker/Psychologist to create a comprehensive lifestyle approach to address overweight and obesity that works for each patient's individual needs. When these comprehensive lifestyle interventions are unsuccessful, medical management of obesity may be considered in overweight and obese individuals with CF. Given the paucity of evidence regarding weight loss drugs and bariatric surgery in CF, clinicians should carefully weigh the risks and benefits of medical management considering patient and family goals and only pursue those interventions in highly specialized centers.

Additional research is needed to determine the optimal dietary pattern for PwCF, particularly in the setting of overweight and obesity. Research should also explore the efficacy of different dietary patterns as well as weight-neutral approaches in improving cardiometabolic outcomes independent of weight loss in PwCF who are overweight or obese. Qualitative studies that explore patient and family attitudes and perceptions of their weight status and body composition, as well as what would make effective weight management interventions in CF, are also warranted. Using patients and family voices combined with emerging scientific evidence to guide interdisciplinary behavioral nutrition interventions for weight management is recommended.

6. Conclusions

A comprehensive interdisciplinary approach to lifestyle change, including nutrition care plans, enjoyable customized exercise, and behavioral strategies, is necessary to address the new challenge of overweight and obesity in CF. The CF Care Team should take a sensitive, highly individualized approach to enhance the sustainability of lifestyle interventions to address overweight and obesity, as well as partnership in care. Expanding the research on overweight and obesity in CF is necessary to determine the impact on cardiometabolic, pulmonary, and quality of life outcomes, and to determine optimal behavioral nutrition interventions for clinical practice as longevity continues to increase in this population.

Author Contributions: Conceptualization, J.B., S.K. and K.R.F.; writing—original draft preparation, J.B.; writing—review and editing, S.K., K.R.F. and J.B.; supervision, K.R.F. All authors have read and agreed to the published version of the manuscript.

Funding: This research received no external funding.

Institutional Review Board Statement: Not applicable.

Informed Consent Statement: Not applicable.

Conflicts of Interest: The authors declare no conflict of interest.

References

1. Rowe, S.M.; Miller, S.; Sorscher, E.J. Cystic Fibrosis. *N. Engl. J. Med.* **2005**, *352*, 1992–2001. [CrossRef] [PubMed]
2. Cystic Fibrosis Foundation Patient Registry: 2020 Annual Data Report. Available online: https://www.cff.org/medical-professionals/patient-registry (accessed on 12 March 2021).
3. Wells, G.D.; Heale, L.; Msc, J.E.S.; Wilkes, D.L.; Atenafu, E.; Coates, A.L.; Ratjen, F. Assessment of body composition in pediatric patients with cystic fibrosis. *Pediatr. Pulmonol.* **2008**, *43*, 1025–1032. [CrossRef] [PubMed]
4. Stallings, V.A.; Stark, L.J.; Robinson, K.A.; Feranchak, A.P.; Quinton, H. Evidence-Based Practice Recommendations for Nutrition-Related Management of Children and Adults with Cystic Fibrosis and Pancreatic Insufficiency: Results of a Systematic Review. *J. Am. Diet. Assoc.* **2008**, *108*, 832–839. [CrossRef] [PubMed]
5. Jelalian, E.; Stark, L.J.; Reynolds, L.; Seifer, R. Nutrition intervention for weight gain in cystic fibrosis: A meta analysis. *J. Pediatr.* **1998**, *132*, 486–492. [CrossRef]
6. Shepherd, R.W.; Holt, T.L.; Thomas, B.J.; Kay, L.; Isles, A.; Francis, P.J.; Ward, L.C. Nutritional rehabilitation in cystic fibrosis: Controlled studies of effects on nutritional growth retardation, body protein turnover, and course of pulmonary disease. *J. Pediatr.* **1986**, *109*, 788–794. [CrossRef]
7. Litvin, M.; Yoon, J.C. Nutritional excess in cystic fibrosis: The skinny on obesity. *J. Cyst. Fibros.* **2019**, *19*, 3–5. [CrossRef] [PubMed]
8. Stephenson, A.L.; Mannik, L.A.; Walsh, S.; Brotherwood, M.; Robert, R.; Darling, P.B.; Nisenbaum, R.; Moerman, J.; Stanojevic, S. Longitudinal trends in nutritional status and the relation between lung function and BMI in cystic fibrosis: A population-based cohort study. *Am. J. Clin. Nutr.* **2013**, *97*, 872–877. [CrossRef] [PubMed]
9. Petersen, M.C.; Begnel, L.; Wallendorf, M.; Litvin, M. Effect of elexacaftor-tezacaftor-ivacaftor on body weight and metabolic parameters in adults with cystic fibrosis. *J. Cyst. Fibros.* **2021**. [CrossRef] [PubMed]
10. Guimbellot, J.S.; Baines, A.; Paynter, A.; Heltshe, S.L.; Van Dalfsen, J.; Jain, M.; Rowe, S.M.; Sagel, S.D. Long term clinical effectiveness of ivacaftor in people with the G551D CFTR mutation. *J. Cyst. Fibros.* **2021**, *20*, 213–219. [CrossRef] [PubMed]
11. Society for Adolescent Health and Medicine Preventing and Treating Adolescent Obesity: A Position Paper of the Society for Adolescent Health and Medicine. *J. Adolesc. Health* **2016**, *59*, 602–606. [CrossRef] [PubMed]
12. Hales, C.M.; Carroll, M.D.; Fryar, C.D.; Ogden, C.L. Prevalence of Obesity and Severe Obesity Among Adults: United States, 2017–2018. *NCHS Data Brief* **2020**, *360*, 1–8.
13. Fryar, C.D.; Carroll, M.D.; Afful, J. Prevalence of overweight, obesity, and severe obesity among children and adolescents aged 2–19 years: United States, 1963–1965 through 2017–2018. *NCHS Health E-Stats* **2020**. Available online: https://www.cdc.gov/nchs/data/hestat/obesity-child-17-18/obesity-child.htm (accessed on 5 January 2022).
14. Kim, D.D.; Basu, A. Estimating the Medical Care Costs of Obesity in the United States: Systematic Review, Meta-Analysis, and Empirical Analysis. *Value Health* **2016**, *19*, 602–613. [CrossRef] [PubMed]
15. Hanna, R.M.; Weiner, D.J. Overweight and obesity in patients with cystic fibrosis: A center-based analysis. *Pediatr. Pulmonol.* **2015**, *50*, 35–41. [CrossRef] [PubMed]
16. Kelly, A.; Moran, A. Update on cystic fibrosis-related diabetes. *J. Cyst. Fibros.* **2013**, *12*, 318–331. [CrossRef] [PubMed]
17. Worgall, T.S. Lipid metabolism in cystic fibrosis. *Curr. Opin. Clin. Nutr. Metab. Care* **2009**, *12*, 105–109. [CrossRef]
18. Jiménez, D.G.; Muñoz-Codoceo, R.; Garriga-García, M.; Molina-Arias, M.; Álvarez-Beltrán, M.; García-Romero, R.; Martínez-Costa, C.; Meavilla-Olivas, S.M.; Peña-Quintana, L.; Gutiérrez, S.G.; et al. Excess weight in patients with cystic fibrosis: Is it always beneficial? *Nutr. Hosp.* **2017**, *34*, 578. [CrossRef]
19. World Health Organization. Available online: https://www.who.int/health-topics/obesity#tab=tab_1 (accessed on 23 October 2021).
20. Booth, A.; Magnuson, A.; Foster, M. Detrimental and protective fat: Body fat distribution and its relation to metabolic disease. *Horm. Mol. Biol. Clin. Investig.* **2014**, *17*, 13–27. [CrossRef]
21. King, S.J.; Nyulasi, I.B.; Strauss, B.J.G.; Kotsimbos, T.; Bailey, M.; Wilson, J.W. Fat-free mass depletion in cystic fibrosis: Associated with lung disease severity but poorly detected by body mass index. *Nutrition* **2010**, *26*, 753–759. [CrossRef]
22. King, S.; Wilson, J.; Kotsimbos, T.; Bailey, M.; Nyulasi, I. Body composition assessment in adults with cystic fibrosis: Comparison of dual-energy X-ray absorptiometry with skinfolds and bioelectrical impedance analysis. *Nutrition* **2005**, *21*, 1087–1094. [CrossRef]
23. Sheikh, S.; Zemel, B.S.; Stallings, V.A.; Rubenstein, R.C.; Kelly, A. Body Composition and Pulmonary Function in Cystic Fibrosis. *Front. Pediatr.* **2014**, *2*, 33. [CrossRef] [PubMed]
24. Ahmad, A.; Ahmed, A.; Patrizio, P. Cystic fibrosis and fertility. *Curr. Opin. Obstet. Gynecol.* **2013**, *25*, 167–172. [CrossRef] [PubMed]

25. Tashjian, A.H.; Gagel, R.F. Teriparatide [Human PTH(1-34)]: 2.5 Years of Experience on the Use and Safety of the Drug for the Treatment of Osteoporosis. *J. Bone Miner. Res.* **2005**, *21*, 354–365. [CrossRef] [PubMed]
26. Hauschild, D.B.; Barbosa, E.; Moreira, E.A.M.; Neto, N.L.; Platt, V.B.; Filho, E.P.; Wazlawik, E.; Moreno, Y.M.F. Nutrition Status Parameters and Hydration Status by Bioelectrical Impedance Vector Analysis Were Associated with Lung Function Impairment in Children and Adolescents with Cystic Fibrosis. *Nutr. Clin. Pract.* **2016**, *31*, 378–386. [CrossRef] [PubMed]
27. Reix, P.; Bellon, G.; Braillon, P. Bone mineral and body composition alterations in paediatric cystic fibrosis patients. *Pediatr. Radiol.* **2010**, *40*, 301–308. [CrossRef]
28. Moriconi, N.; Kraenzlin, M.; Muller, B.; Keller, U.; Nusbaumer, C.P.G.; Stöhr, S.; Tamm, M.; Puder, J.J. Body Composition and Adiponectin Serum Concentrations in Adult Patients with Cystic Fibrosis. *J. Clin. Endocrinol. Metab.* **2006**, *91*, 1586–1590. [CrossRef] [PubMed]
29. Alvarez, J.A.; Ziegler, T.R.; Millson, E.C.; Stecenko, A.A. Body composition and lung function in cystic fibrosis and their association with adiposity and normal-weight obesity. *Nutrition* **2015**, *32*, 447–452. [CrossRef]
30. Ionescu, A.A.; Nixon, L.S.; Luzio, S.; Lewis-Jenkins, V.; Evans, W.D.; Stone, M.D.; Owens, D.R.; Routledge, P.A.; Shale, D.J. Pulmonary Function, Body Composition, and Protein Catabolism in Adults with Cystic Fibrosis. *Am. J. Respir. Crit. Care Med.* **2002**, *165*, 495–500. [CrossRef]
31. Calella, P.; Calella, P.; Valerio, G.; Valerio, G.; Brodlie, M.; Brodlie, M.; Donini, L.M.; Donini, L.M.; Siervo, M.; Siervo, M. Cystic fibrosis, body composition, and health outcomes: A systematic review. *Nutrition* **2018**, *55–56*, 131–139. [CrossRef]
32. Soyer, P.; Spelle, L.; Pelage, J.-P.; Dufresne, A.-C.; Rondeau, Y.; Gouhiri, M.H.; Scherrer, A.; Rymer, R. Cystic Fibrosis in Adolescents and Adults: Fatty Replacement of the Pancreas—CT Evaluation and Functional Correlation. *Radiology* **1999**, *210*, 611–615. [CrossRef]
33. Haroun, D.; Wells, J.C.K.; Lau, C.; Hadji-Lucas, E.; Lawson, M.S. Assessment of obesity status in outpatients from three disease states. *Acta Paediatr.* **2006**, *95*, 970–974. [CrossRef] [PubMed]
34. Chaves, C.R.M.D.M.; Da Cunha, A.L.P.; Da Costa, A.C.; Costa, R.D.S.S.D.; Lacerda, S.V. Estado nutricional e distribuição de gordura corporal em crianças e adolescentes com Fibrose Cística. *Cienc. Saude Coletiva* **2015**, *20*, 3319–3328. [CrossRef]
35. Panagopoulou, P.; Fotoulaki, M.; Nikolaou, A.; Nousia-Arvanitakis, S.; Maria, F. Prevalence of malnutrition and obesity among cystic fibrosis patients. *Pediatr. Int.* **2014**, *56*, 89–94. [CrossRef] [PubMed]
36. Bellissimo, M.P.; Zhang, I; Ivie, E.A.; Tran, P.H.; Tangpricha, V.; Hunt, W.R.; Stecenko, A.A.; Ziegler, T.R.; Alvarez, J.A. Visceral adipose tissue is associated with poor diet quality and higher fasting glucose in adults with cystic fibrosis. *J. Cyst. Fibros.* **2019**, *18*, 430–435. [CrossRef]
37. Harindhanavudhi, T.; Wang, Q.; Dunitz, J.; Moran, A.; Moheet, A. Prevalence and factors associated with overweight and obesity in adults with cystic fibrosis: A single-center analysis. *J. Cyst. Fibros.* **2019**, *19*, 139–145. [CrossRef] [PubMed]
38. Flume, P.A.; Fernandez, G.S.; Schechter, M.S.; Fink, A. Prevalence of obesity in people with cystic fibrosis over a 20-year period. *Pediatr. Pulmonol.* **2019**, *54*, 263.
39. Kastner-Cole, D.; Palmer, C.N.; Ogston, S.A.; Mehta, A.; Mukhopadhyay, S. Overweight and Obesity in ΔF508 Homozygous Cystic Fibrosis. *J. Pediatr.* **2005**, *147*, 402–404. [CrossRef]
40. Dray, X.; Kanaan, R.; Bienvenu, T.; Desmazes-Dufeu, N.; Dusser, D.; Marteau, P.; Hubert, D. Malnutrition in adults with cystic fibrosis. *Eur. J. Clin. Nutr.* **2004**, *59*, 152–154. [CrossRef]
41. Gramegna, A.; Aliberti, S.; Contarini, M.; Savi, D.; Sotgiu, G.; Majo, F.; Saderi, L.; Lucidi, V.; Amati, F.; Pappaletera, M.; et al. Overweight and obesity in adults with cystic fibrosis: An Italian multicenter cohort study. *J. Cyst. Fibros.* **2022**, *21*, 111–114. [CrossRef]
42. Bailey, J.; Rozga, M.; McDonald, C.M.; Bowser, E.K.; Farnham, K.; Mangus, M.; Padula, L.; Porco, K.; Alvarez, J.A. Effect of CFTR Modulators on Anthropometric Parameters in Individuals with Cystic Fibrosis: An Evidence Analysis Center Systematic Review. *J. Acad. Nutr. Diet.* **2020**, *121*, 1364–1378.e2. [CrossRef]
43. Heijerman, H.G.M.; McKone, E.F.; Downey, D.G.; Van Braeckel, E.; Rowe, S.M.; Tullis, E.; Mall, M.A.; Welter, J.J.; Ramsey, B.W.; McKee, C.M.; et al. Efficacy and safety of the elexacaftor plus tezacaftor plus ivacaftor combination regimen in people with cystic fibrosis homozygous for the F508del mutation: A double-blind, randomised, phase 3 trial. *Lancet* **2019**, *394*, 1940–1948. [CrossRef]
44. Middleton, P.G.; Mall, M.A.; Dřevínek, P.; Lands, L.C.; McKone, E.F.; Polineni, D.; Ramsey, B.W.; Taylor-Cousar, J.L.; Tullis, E.; Vermeulen, F.; et al. Elexacaftor–Tezacaftor–Ivacaftor for Cystic Fibrosis with a Single Phe508del Allele. *N. Engl. J. Med.* **2019**, *381*, 1809–1819. [CrossRef] [PubMed]
45. White, M.; Dennis, N.; Ramsey, R.; Barwick, K.; Graham, C.; Kane, S.; Kepreotes, H.; Queit, L.; Sweeney, A.; Winderlich, J.; et al. Prevalence of malnutrition, obesity and nutritional risk of Australian paediatric inpatients: A national one-day snapshot. *J. Paediatr. Child Health* **2015**, *51*, 314–320. [CrossRef] [PubMed]
46. Singh, R.K.; Kumar, P.; Mahalingam, K. Molecular genetics of human obesity: A comprehensive review. *Comptes Rendus. Biol.* **2017**, *340*, 87–108. [CrossRef]
47. McDonald, C.M.; Bowser, E.K.; Farnham, K.; Alvarez, J.A.; Padula, L.; Rozga, M. Dietary Macronutrient Distribution and Nutrition Outcomes in Persons with Cystic Fibrosis: An Evidence Analysis Center Systematic Review. *J. Acad. Nutr. Diet.* **2021**, *121*, 1574–1590.e3. [CrossRef]
48. Bailey, J.; Garcia, L.; Rutland, S.; Oates, G. Prevalence and correlates of overweight and obesity in a national cohort of children and adolescents with cystic fibrosis. *Pediatr. Pulmonol.* **2020**, *56*, 144.
49. Stalvey, M.S.; Pace, J.; Niknian, M.; Higgins, M.N.; Tarn, V.; Davis, J.; Heltshe, S.L.; Rowe, S.M. Growth in Prepubertal Children With Cystic Fibrosis Treated With Ivacaftor. *Pediatrics* **2017**, *139*, e20162522. [CrossRef]

50. Borowitz, D.; Lubarsky, B.; Wilschanski, M.; Munck, A.; Gelfond, D.; Bodewes, F.A.J.A.; Schwarzenberg, S.J. Nutritional Status Improved in Cystic Fibrosis Patients with the G551D Mutation After Treatment with Ivacaftor. *Am. J. Dig. Dis.* **2016**, *61*, 198–207. [CrossRef]
51. Davies, J.C.; Wainwright, C.E.; Canny, G.J.; Chilvers, M.A.; Howenstine, M.S.; Munck, A.; Mainz, J.G.; Rodriguez, S.; Li, H.; Yen, K.; et al. Efficacy and Safety of Ivacaftor in Patients Aged 6 to 11 Years with Cystic Fibrosis with a G551D Mutation. *Am. J. Respir. Crit. Care Med.* **2013**, *187*, 1219–1225. [CrossRef]
52. Stallings, V.A.; Sainath, N.; Oberle, M.; Bertolaso, C.; Schall, J.I. Energy Balance and Mechanisms of Weight Gain with Ivacaftor Treatment of Cystic Fibrosis Gating Mutations. *J. Pediatr.* **2018**, *201*, 229–237.e4. [CrossRef]
53. Sainath, N.N.; Schall, J.; Bertolaso, C.; McAnlis, K.; Stallings, V.A. Italian and North American dietary intake after ivacaftor treatment for Cystic Fibrosis Gating Mutations. *J. Cyst. Fibros.* **2019**, *18*, 135–143. [CrossRef] [PubMed]
54. Rosenfeld, M.; Cunningham, S.; Harris, W.T.; Lapey, A.; Regelmann, W.E.; Sawicki, G.S.; Southern, K.W.; Chilvers, M.; Higgins, M.; Tian, S.; et al. An open-label extension study of ivacaftor in children with CF and a CFTR gating mutation initiating treatment at age 2–5 years (KLIMB). *J. Cyst. Fibros.* **2019**, *18*, 838–843. [CrossRef] [PubMed]
55. Davies, J.C.; Cunningham, S.; Harris, W.T.; Lapey, A.; Regelmann, W.E.; Sawicki, G.S.; Southern, K.W.; Robertson, S.; Green, Y.; Cooke, J.; et al. Safety, pharmacokinetics, and pharmacodynamics of ivacaftor in patients aged 2–5 years with cystic fibrosis and a CFTR gating mutation (KIWI): An open-label, single-arm study. *Lancet Respir. Med.* **2016**, *4*, 107–115. [CrossRef]
56. Gould, M.J.; Smith, H.; Rayment, J.H.; Machida, H.; Gonska, T.; Galante, G.J. CFTR modulators increase risk of acute pancreatitis in pancreatic insufficient patients with cystic fibrosis. *J. Cyst. Fibros.* **2021**. [CrossRef]
57. Edgeworth, D.; Keating, D.; Ellis, M.; Button, B.; Williams, E.; Clark, D.; Tierney, A.; Heritier, S.; Kotsimbos, T.; Wilson, J. Improvement in exercise duration, lung function and well-being in G551D-cystic fibrosis patients: A double-blind, placebo-controlled, randomized, cross-over study with ivacaftor treatment. *Clin. Sci.* **2017**, *131*, 2037–2045. [CrossRef] [PubMed]
58. King, S.J.; Tierney, A.C.; Edgeworth, D.; Keating, D.; Williams, E.; Kotsimbos, T.; Button, B.M.; Wilson, J.W. Body composition and weight changes after ivacaftor treatment in adults with cystic fibrosis carrying the G551 D cystic fibrosis transmembrane conductance regulator mutation: A double-blind, placebo-controlled, randomized, crossover study with open-label extension. *Nutrition* **2021**, *85*, 111124. [CrossRef]
59. US Department of Health and Human Services. Social Determinants of Health. Available online: https://health.gov/healthypeople/objectives-and-data/social-determinants-health (accessed on 26 July 2021).
60. Franklin, B.; Jones, A.; Love, D.; Puckett, S.; Macklin, J.; White-Means, S. Exploring Mediators of Food Insecurity and Obesity: A Review of Recent Literature. *J. Community Health* **2011**, *37*, 253–264. [CrossRef]
61. Dinour, L.M.; Bergen, D.; Yeh, M.-C. The Food Insecurity–Obesity Paradox: A Review of the Literature and the Role Food Stamps May Play. *J. Am. Diet. Assoc.* **2007**, *107*, 1952–1961. [CrossRef]
62. Quon, B.S.; Psoter, K.; Mayer-Hamblett, N.; Aitken, M.L.; Li, C.I.; Goss, C.H. Disparities in Access to Lung Transplantation for Patients with Cystic Fibrosis by Socioeconomic Status. *Am. J. Respir. Crit. Care Med.* **2012**, *186*, 1008–1013. [CrossRef]
63. Schechter, M.S.; Shelton, B.J.; Margolis, P.A.; Fitzsimmons, S.C. The Association of Socioeconomic Status with Outcomes in Cystic Fibrosis Patients in the United States. *Am. J. Respir. Crit. Care Med.* **2001**, *163*, 1331–1337. [CrossRef]
64. Oates, G.; Schechter, M.S. Socioeconomic status and health outcomes: Cystic fibrosis as a model. *Expert Rev. Respir. Med.* **2016**, *10*, 967–977. [CrossRef] [PubMed]
65. Jiménez, D.G.; García, C.B.; Crespo, M.R.; Martín, J.D.; Quirós, M.A.; González, S.H.; Aguirre, A.S.; Otero, J.G. Resistencia insulínica en pacientes pediátricos con fibrosis quística y sobrepeso. *An. Pediatr.* **2012**, *76*, 279–284. [CrossRef] [PubMed]
66. Madde, A.; Okoniewski, W.; Sanders, D.B.; Ren, C.L.; Weiner, D.J.; Forno, E. Nutritional status and lung function in children with pancreatic-sufficient cystic fibrosis. *J. Cyst. Fibros.* **2021**. [CrossRef] [PubMed]
67. Coderre, L.; Fadainia, C.; Belson, L.; Belisle, V.; Ziai, S.; Mailhot, G.; Berthiaume, Y.; Rabasa-Lhoret, R. LDL-cholesterol and insulin are independently associated with body mass index in adult cystic fibrosis patients. *J. Cyst. Fibros.* **2012**, *11*, 393–397. [CrossRef] [PubMed]
68. Rhodes, B.; Nash, E.F.; Tullis, E.; Pencharz, P.B.; Brotherwood, M.; Dupuis, A.; Stephenson, A. Prevalence of dyslipidemia in adults with cystic fibrosis. *J. Cyst. Fibros.* **2010**, *9*, 24–28. [CrossRef] [PubMed]
69. Georgiopoulou, V.V.; Denker, A.; Bishop, K.L.; Brown, J.M.; Hirsh, B.; Wolfenden, L.; Sperling, L. Metabolic abnormalities in adults with cystic fibrosis. *Respirology* **2010**, *15*, 823–829. [CrossRef]
70. Nowak, J.K.; Szczepanik, M.; Wojsyk-Banaszak, I.; Mądry, E.; Wykrętowicz, A.; Krzyżanowska-Jankowska, P.; Drzymała-Czyż, S.; Nowicka, A.; Pogorzelski, A.; Sapiejka, E.; et al. Cystic fibrosis dyslipidaemia: A cross-sectional study. *J. Cyst. Fibros.* **2019**, *18*, 566–571. [CrossRef] [PubMed]
71. Mafort, T.T.; Rufino, R.; Costa, C.H.; Lopes, A.J. Obesity: Systemic and pulmonary complications, biochemical abnormalities, and impairment of lung function. *Multidiscip. Respir. Med.* **2016**, *11*, 28. [CrossRef]
72. Phelan, S.M.; Burgess, D.J.; Yeazel, M.W.; Hellerstedt, W.L.; Griffin, J.M.; Van Ryn, M. Impact of weight bias and stigma on quality of care and outcomes for patients with obesity. *Obes. Rev.* **2015**, *16*, 319–326. [CrossRef]
73. Wu, Y.-K.; Berry, D.C. Impact of weight stigma on physiological and psychological health outcomes for overweight and obese adults: A systematic review. *J. Adv. Nurs.* **2017**, *74*, 1030–1042. [CrossRef]
74. Vartanian, L.R.; Porter, A.M. Weight stigma and eating behavior: A review of the literature. *Appetite* **2016**, *102*, 3–14. [CrossRef] [PubMed]
75. Tierney, S. Body image and cystic fibrosis: A critical review. *Body Image* **2012**, *9*, 12–19. [CrossRef] [PubMed]
76. Helms, S.W.; Christon, L.M.; Dellon, E.P.; Prinstein, M. Patient and Provider Perspectives on Communication About Body Image with Adolescents and Young Adults with Cystic Fibrosis. *J. Pediatr. Psychol.* **2017**, *42*, 1040–1050. [CrossRef] [PubMed]

77. Darukhanavala, A.; Merjaneh, L.; Mason, K.; Le, T. Eating disorders and body image in cystic fibrosis. *J. Clin. Transl. Endocrinol.* **2021**, *26*, 100280. [CrossRef] [PubMed]
78. Pi-Sunyer, X.; Blackburn, G.; Brancati, F.L.; Bray, G.A.; Bright, R.; Clark, J.M.; Curtis, J.M.; Espeland, M.A.; Foreyt, J.P.; Graves, K.; et al. Reduction in weight and cardiovascular disease risk factors in individuals with type 2 diabetes: One-year results of the look AHEAD trial. *Diabetes Care* **2007**, *30*, 1374–1383. [CrossRef] [PubMed]
79. Wing, R.R. Long-term Effects of a Lifestyle Intervention on Weight and Cardiovascular Risk Factors in Individuals with Type 2 Diabetes Mellitus. *Arch. Intern. Med.* **2010**, *170*, 1566–1575. [CrossRef] [PubMed]
80. Donnelly, J.E.; Blair, S.N.; Jakicic, J.M.; Manore, M.M.; Rankin, J.W.; Smith, B.K.; American College of Sports Medicine. American College of Sports Medicine Position Stand. Appropriate Physical Activity Intervention Strategies for Weight Loss and Prevention of Weight Regain for Adults. *Med. Sci. Sports Exerc.* **2009**, *41*, 459–471. [CrossRef] [PubMed]
81. Raynor, H.; Champagne, C.M. Position of the Academy of Nutrition and Dietetics: Interventions for the Treatment of Overweight and Obesity in Adults. *J. Acad. Nutr. Diet.* **2016**, *116*, 129–147. [CrossRef] [PubMed]
82. Ramage, S.; Farmer, A.; Eccles, K.A.; McCargar, L. Healthy strategies for successful weight loss and weight maintenance: A systematic review. *Appl. Physiol. Nutr. Metab.* **2014**, *39*, 1–20. [CrossRef] [PubMed]
83. Jensen, M.D.; Ryan, D.; Apovian, C.M.; Ard, J.; Comuzzie, A.G.; Donato, K.A.; Hu, F.B.; Hubbard, V.S.; Jakicic, J.M.; Kushner, R.F.; et al. 2013 AHA/ACC/TOS Guideline for the Management of Overweight and Obesity in Adults. *J. Am. Coll. Cardiol.* **2013**, *63*, 2985–3023. [CrossRef] [PubMed]
84. McDonald, C.M.; Alvarez, J.A.; Bailey, J.; Bowser, E.K.; Farnham, K.; Mangus, M.; Padula, L.; Porco, K.; Rozga, M. Academy of Nutrition and Dietetics: 2020 Cystic Fibrosis Evidence Analysis Center Evidence-Based Nutrition Practice Guideline. *J. Acad. Nutr. Diet.* **2020**, *121*, 1591–1636.e3. [CrossRef] [PubMed]
85. Saxby, N.P.C.; Kench, A.; King, S.; Crowder, T.; van der Hank, N.; The Australian and New Zealand Cystic Fibrosis Nutrition Guideline Authorship Group. *Nutrition Guidelines for Cystic Fibrosis in Australia and New Zealand*; Bell, S.C., Ed.; Thoracic Society of Australia and New Zealand: Sydney, Australia, 2017.
86. Turck, D.; Braegger, C.P.; Colombo, C.; Declercq, D.; Morton, A.; Pancheva, R.; Robberecht, E.; Stern, M.; Strandvik, B.; Wolfe, S.; et al. ESPEN-ESPGHAN-ECFS guidelines on nutrition care for infants, children, and adults with cystic fibrosis. *Clin. Nutr.* **2016**, *35*, 557–577. [CrossRef]
87. Antonetti, V.W. The equations governing weight change in human beings. *Am. J. Clin. Nutr.* **1973**, *26*, 64–71. [CrossRef]
88. Shai, I.; Schwarzfuchs, D.; Henkin, Y.; Shahar, D.R.; Witkow, S.; Greenberg, I.; Golan, R.; Fraser, D.; Bolotin, A.; Vardi, H.; et al. weight loss with a low-carbohydrate, Mediterranean, or low-fat diet. *N. Engl. J. Med.* **2008**, *359*, 229–241. [CrossRef] [PubMed]
89. Hebestreit, H.; Lands, L.C.; Alarie, N.; Schaeff, J.; Karila, C.; Orenstein, D.M.; Urquhart, D.S.; Hulzebos, E.H.J.; Stein, L.; Schindler, C.; et al. Effects of a partially supervised conditioning programme in cystic fibrosis: An international multi-centre randomised controlled trial (ACTIVATE-CF): Study protocol. *BMC Pulm. Med.* **2018**, *18*, 31. [CrossRef]
90. Satter, E. Eating Competence: Nutrition Education with the Satter Eating Competence Model. *J. Nutr. Educ. Behav.* **2007**, *39*, S189–S194. [CrossRef] [PubMed]
91. Laddu, D.; Dow, C.; Hingle, M.; Thomson, C.; Going, S. A Review of Evidence-Based Strategies to Treat Obesity in Adults. *Nutr. Clin. Pract.* **2011**, *26*, 512–525. [CrossRef] [PubMed]
92. Swift, D.L.; McGee, J.E.; Earnest, C.; Carlisle, E.; Nygard, M.; Johannsen, N.M. The Effects of Exercise and Physical Activity on Weight Loss and Maintenance. *Prog. Cardiovasc. Dis.* **2018**, *61*, 206–213. [CrossRef]
93. Johns, D.J.; Hartmann-Boyce, J.; Jebb, S.A.; Aveyard, P. Diet or Exercise Interventions vs Combined Behavioral Weight Management Programs: A Systematic Review and Meta-Analysis of Direct Comparisons. *J. Acad. Nutr. Diet.* **2014**, *114*, 1557–1568. [CrossRef]
94. Prévotat, A.; Godin, J.; Bernard, H.; Perez, T.; Le Rouzic, O.; Wallaert, B. Improvement in body composition following a supervised exercise-training program of adult patients with cystic fibrosis. *Respir. Med. Res.* **2019**, *75*, 5–9. [CrossRef]
95. Ding, S.; Zhong, C. Exercise and Cystic Fibrosis. *Adv. Exp. Med. Biol.* **2020**, *1228*, 381–391. [CrossRef] [PubMed]
96. Ruegsegger, G.; Booth, F.W. Health Benefits of Exercise. *Cold Spring Harb. Perspect. Med.* **2017**, *8*, a029694. [CrossRef] [PubMed]
97. Jensen, M.D.; Ryan, D.H.; Apovian, C.M.; Jamy, D.A.; Comuzzie, A.G.; Donato, K.A.; Hu, F.B.; Hubbard, V.S.; Jakicic, J.M.; Kushner, R.F.; et al. 2013 AHA/ACC/TOS Guideline for the Management of Overweight and Obesity in Adults: A report of the American College of Cardiology/American Heart Association Task Force on Practice Guidelines and The Obesity Society. *Circulation* **2014**, *129* (Suppl. 2), S102–S138. [CrossRef] [PubMed]
98. Yao, A. Screening for and Management of Obesity in Adults: U.S. Preventive Services Task Force Recommendation Statement: A Policy Review. *Ann. Med. Surg.* **2013**, *2*, 18–21. [CrossRef]
99. Barrett, J.; Slatter, G.; Whitehouse, J.L.; Nash, E.F. Perception, experience and relationship with food and eating in adults with cystic fibrosis. *J. Hum. Nutr. Diet.* **2021**. [CrossRef]
100. Hardcastle, S.J.; Taylor, A.H.; Bailey, M.P.; Harley, R.A.; Hagger, M.S. Effectiveness of a motivational interviewing intervention on weight loss, physical activity and cardiovascular disease risk factors: A randomised controlled trial with a 12-month post-intervention follow-up. *Int. J. Behav. Nutr. Phys. Act.* **2013**, *10*, 40. [CrossRef]
101. Powers, S.W.; Mitchell, M.J.; Patton, S.R.; Byars, K.C.; Jelalian, E.; Mulvihill, M.M.; Hovell, M.F.; Stark, L.J. Mealtime behaviors in families of infants and toddlers with cystic fibrosis. *J. Cyst. Fibros.* **2005**, *4*, 175–182. [CrossRef]
102. Watson, H.; Bilton, D.; Truby, H. A Randomized Controlled Trial of a New Behavioral Home-Based Nutrition Education Program, "Eat Well with CF," in Adults with Cystic Fibrosis. *J. Am. Diet. Assoc.* **2008**, *108*, 847–852. [CrossRef]

103. Jaclyn, D.; Andrew, N.; Ryan, P.; Julianna, B.; Christopher, S.; Nauman, C.; Powers, M.; Gregory, S.S.; George, M.S. Patient and family perceptions of telehealth as part of the cystic fibrosis care model during COVID-19. *J. Cyst. Fibros.* **2021**, *20*, e23–e28. [CrossRef]
104. Solomon, G.M.; Bailey, J.; Lawlor, J.; Scalia, P.; Sawicki, G.S.; Dowd, C.; Sabadosa, K.A.; Van Citters, A. Patient and family experience of telehealth care delivery as part of the CF chronic care model early in the COVID-19 pandemic. *J. Cyst. Fibros.* **2021**, *20*, 41–46. [CrossRef]
105. Bacon, L.; Aphramor, L. Weight Science: Evaluating the Evidence for a Paradigm Shift. *Nutr. J.* **2011**, *10*, 9. [CrossRef]
106. Tylka, T.L.; Annunziato, R.A.; Burgard, D.; Daníelsdóttir, S.; Shuman, E.; Davis, C.; Calogero, R.M. The Weight-Inclusive versus Weight-Normative Approach to Health: Evaluating the Evidence for Prioritizing Well-Being over Weight Loss. *J. Obes.* **2014**, *2014*, 983495. [CrossRef]
107. Ulian, M.D.; Aburad, L.; Oliveira, M.S.D.S.; Poppe, A.C.M.; Sabatini, F.; Perez, I.; Gualano, B.; Benatti, F.; Pinto, A.J.; Roble, O.J.; et al. Effects of health at every size® interventions on health-related outcomes of people with overweight and obesity: A systematic review. *Obes. Rev.* **2018**, *19*, 1659–1666. [CrossRef]
108. Ulian, M.D.; Benatti, F.B.; De Campos-Ferraz, P.L.; Roble, O.J.; Unsain, R.F.; Sato, P.; Brito, B.C.; Murakawa, K.A.; Modesto, B.T.; Aburad, L.; et al. The Effects of a "Health at Every Size®"-Based Approach in Obese Women: A Pilot-Trial of the "Health and Wellness in Obesity" Study. *Front. Nutr.* **2015**, *2*, 34. [CrossRef]
109. Mechanick, J.I.; Apovian, C.; Brethauer, S.; Garvey, W.T.; Joffe, A.M.; Kim, J.; Kushner, R.F.; Lindquist, R.; Pessah-Pollack, R.; Seger, J.; et al. Clinical Practice Guidelines for the Perioperative Nutrition, Metabolic, and Nonsurgical Support of Patients Undergoing Bariatric Procedures—2019 Update: Cosponsored by American Association of Clinical Endocrinologists/American College of Endocrinology, The Obesity Society, American Society for Metabolic and Bariatric Surgery, Obesity Medicine Association, and American Society of Anesthesiologists. *Obesity* **2020**, *28*, O1–O58. [CrossRef]
110. Ammori, B.J.; Skarulis, M.C.; Soran, H.; Syed, A.A.; Eledrisi, M.; Malik, R.A. Medical and surgical management of obesity and diabetes: What's new? *Diabet. Med. J. Br. Diabet. Assoc.* **2020**, *37*, 203–210. [CrossRef]

Review

Pancreatic Enzyme Replacement Therapy in Cystic Fibrosis

Peter N. Freswick [1,*], Elizabeth K. Reid [2] and Maria R. Mascarenhas [2]

1. Helen DeVos Children's Hospital, Grand Rapids, MI 49503, USA
2. Children's Hospital of Philadelphia, Philadelphia, PA 19104, USA; reide1@chop.edu (E.K.R.); mascarenhas@chop.edu (M.R.M.)
* Correspondence: peter.freswick@helendevoschildrens.org

Abstract: While typically considered a pulmonary disease, cystic fibrosis patients develop significant nutritional complications and comorbidities, especially those who are pancreatic insufficient. Clinicians must have a high suspicion for cystic fibrosis among patients with clinical symptoms of pancreatic insufficiency, and pancreatic enzymatic replacement therapy (PERT) must be urgently initiated. PERT presents a myriad of considerations for patients and their supporting dieticians and clinicians, including types of administration, therapy failures, and complications.

Keywords: cystic fibrosis; pancreatic insufficiency; PERT; pancreatic enzymes; nutrition

1. Introduction

Cystic fibrosis (CF) is the most common life-shortening autosomal recessive disorder in North America, effecting ~30,000 people in the United States alone [1]. CF is caused by various mutations in the gene that codes for the cystic fibrosis transmembrane conductance regulator (CFTR) gene which encodes a cyclic adenosine monophosphate regulated chloride channel, responsible for chloride and bicarbonate secretion across epithelia cells. Abnormal chloride transport through the CFTR leads to viscous sodium bicarbonate-depleted fluid, and ultimately can lead to pancreatic insufficiency with subsequent malnutrition and malabsorption, fat-soluble vitamin deficiencies, and progressive obstructive lung disease [2].

While many now consider CF primarily a pulmonary disease, CF was initially recognized by Dr Dorothy Andersen in 1938 as a distinct diagnosis for patients with failure to thrive [3]. She labeled the disease "cystic fibrosis of the pancreas" based off her autopsy findings of children who died of malnutrition. As such, the lack of pancreatic function was the initial defining characteristic of CF with many children succumbing to malnutrition before CF lung disease progressed. Now, patients with CF are quickly tested for pancreatic sufficiency and are thus characterized as "pancreatic sufficient" (PS) or "pancreatic insufficient" (PI) [4].

The pancreas consists of the exocrine pancreas that produces pancreatic enzymes and the endocrine pancreas that produces insulin. In this review, the term pancreatic insufficiency refers to inadequate production of pancreatic enzymes, bicarbonate and fluid by the pancreas resulting in maldigestion and malabsorption. Approximately 85% of CF patients have evidence of maldigestion due to pancreatic insufficiency requiring treatment [5]. CF patients experience progressive and substantial pancreatic injury early in life, even in utero, thus often developing PI in infancy [4].

While first recognized in 1938, clinicians had a significant breakthrough in understanding CF disease pathology in 1989 when Lap-Chee Tsui and colleagues cloned the CF gene. From this discovery, scientists have defined six classes of mutations CFTR gene, classes I–VI, stratified in decreasing severity. Such a genotype dichotomy has helped predict a patient's likelihood of PI: patients with two copies of class IV, V, or VI mutations tend to be PS, whereas those with two copies of class I, II, or III mutations tend to be PI [5].

Aggressive nutritional therapy is paramount to improved clinical outcomes for patients with CF. With the development of comprehensive North American CF care centers, clinicians could recognize varying outcomes between care centers. Corey et al. compared various metrics between Boston and Toronto's CF care centers in 1988 finding that Toronto's patients had significantly better survival despite having similar FEV_1 values. Possibly explaining this drastic difference, patients in Toronto had drastically better weight for height as Toronto advocated for a high-fat, high-calorie diet with aggressive pancreatic enzyme replacement therapy (PERT), whereas Boston advocated for a low-fat, high-calorie diet with less emphasis on PERT [6]. A subsequent prospective study validated Corey et al.'s findings, demonstrating that higher weight percentile in early childhood predicts improved FEV_1 and survival at 18 years old [7]. Given the advent of the newborn screen (NBS) [8] and a focused interest in therapeutic nutrition and PERT in CF children beginning in infancy, median World Health Organization height percentiles in US children with CF younger than 2 years old increased from less than 20% in 1993 to 44% in 2019 [9]. Leung et al., compared a historic cohort of infants from 1994 to1995 to their prospective cohort concluding that the initiation of a universal NBS, aggressive PERT, and early nutritional therapy improved the nutritional status of CF infants [10].

Given numerous studies demonstrating aggressive nutrition's vital impact on health in the CF population and the subsequent use of PERT, this review will explore current updates on the clinical presentation of patients with PI CF, the effects of PERT, various nuances to consider, and special considerations of PERT therapy in the age of CFTR modulators.

2. Clinical Considerations

2.1. Presentation

Patients with malabsorption present classically with malodorous oily stools that can be difficult to flush, chronic diarrhea, failure to thrive, weight loss, bloating, and dyspepsia [11]. All causes of malabsorption, not just pancreatic insufficiency, will present similarly, and thus a broad differential diagnosis should be considered when approaching an infant or child with malabsorption symptoms [4,11,12] (see Table 1). Regardless, anytime a pediatric patient presents with failure to thrive, a full differential diagnosis must be considered.

Table 1. Causes of childhood malabsorption.

Intestinal	Extra-Intestinal
Crohn's disease Celiac disease Small intestinal bacterial overgrowth (SIBO) Infectious diarrhea (*Giardia, Cryptosporidium*) Brush border enzyme deficiencies Short bowel syndrome Acrodermatitis enteropathica	Cystic fibrosis Zollinger-Ellison syndrome Gastroparesis Chronic cholestasis Schwachman-Diamond syndrome Johanson-Blizzard syndrome Pearson syndrome Jeune syndrome Pancreatic aplasia Cholestatic liver disease

Before universal newborn screening was available in the USA, cystic fibrosis would present with meconium ileus, rectal prolapse, failure to thrive and weight loss, respiratory difficulties, and steatorrhea, among others [2]. As of 2010, all 50 states are performing CF screening at birth [13]. Screening typically entails measuring an infant's immune-reactive trypsinogen (IRT) serum level, occasionally coupled with DNA analysis for common CFTR mutations. Should an infant screen positive, he/she should be immediately referred for a confirmation sweat chloride test [14]. However, some infants with CF will not screen positive [15,16], thus clinicians must always remain vigilant to the diagnosis of CF in all infants with concerning symptoms.

2.2. Diagnosis

Since newborn screening has been in place, infants often present for the initial CF center visit at 2 weeks of age. For infants with identified CFTR mutations associated with PI or poor initial weight gain, PERT should be initiated as soon as possible, even before a patient has been officially diagnosed as PI. Caregivers may bring a stool sample to be sent for fecal elastase (FE) and PERT administration may be demonstrated and started at the initial appointment. Taking PERT will not affect a FE test result, and PERT can be discontinued if the test result is negative for PI.

Early diagnosis of PI and intervention leads to improved nutritional status [10], regardless of age [4]. PI testing can prove difficult, as testing only easily detects severe PI and there is no gold standard for PI diagnosis or severity [17]. PI testing is either considered indirect or direct.

Indirect testing evaluates the secondary fecal effects due to lack of pancreatic enzymes [17]. The historical gold standard has been the 72 h fecal fat test, with a coefficient of fat absorption normal if $\geq 85\%$ if the patient is less than 6 months of age and $\geq 93\%$ if 6 months or older [18]. However, 72 h tests are unpleasant and time consuming. An acid steatocrit (AS) test has been proposed as an alternative test. Walkowiak et al., compared the AS among CF patient without or with mild steatorrhea, but unfortunately found that AS did not reflect the fecal fat excretion in these CF patients [19]. However, a FE > 100 micrograms/gram stool has a 99% negative predictive value for pancreatic insufficiency. FE is easier to obtain than a 72 h fecal fat and, as proven to be a valid screening method for PI, has become the standard test for PI [20]. In general, indirect tests are only accurate for advanced stages of PI [17].

Direct tests measure the secreted pancreatic enzymes and bicarbonate [17]. Cholecystokinin stimulates pancreatic enzyme secretion and secretin stimulates pancreatic bicarbonate secretion. Both Dreiling tubes (nasal tube with gastric and duodenal ports to decompress the stomach and measure duodenal secretions, respectively) and endoscopy have been utilized for direct exocrine pancreatic testing [4]. However, among patients with chronic pancreatitis, direct PI testing was similar between endoscopic and Dreiling tube methods, and the endoscopic method eases the performance of these tests [21].

2.3. Subsequent Testing

After the diagnosis of PI in a patient with CF, it is important to note that the fecal elastase values may fluctuate significantly in the first year of life. O'Sullivan et al. following 61 CF patients obtained at least 8 fecal elastase levels initially obtained at age < 3.5 months and the final level obtained at age ≥ 9 months old. They found significant variability in fecal elastase testing across all initial fecal elastase groups. In total, 27% of those with an initial fecal elastase between 50–200 micrograms/gram had at least one fecal elastase > 200 micrograms/gram. Furthermore, 15% of those with an initial fecal elastase > 200 micrograms/gram had a fecal elastase < 200 micrograms/gram by the end of the first year [22].

Given this variability, clinicians should always consider the evolution of a patient's pancreatic exocrine function. Thus, guidelines now recommend that PS CF children and adults undergo an annual assessment of pancreatic function by fecal elastase measurement, and more frequent tests are needed for poor growth or inadequate nutritional status [23].

3. Treatment

3.1. Types of PERT

PI is treated with PERT capsules which contain pancreatic extract (lipase, protease, and amylase) to replace the missing endogenous pancreatic enzymes. All the current FDA-approved products are derived from porcine origin. Patients who follow religious or cultural preference to avoid pork products may be given a medical dispensation. Non-porcine enzymes are present in clinical trials; however, there is no effective alternative currently available. There are over the counter preparations available which are sold as digestive aids. These are only available in very low doses and are not efficacious to treat PI.

There are three formulations: enteric-coated, non-enteric coated, and a lipase enzyme cartridge. There are different brands of PERT which come in multiple sizes and dosing strengths based on lipase units (See Figure 1).

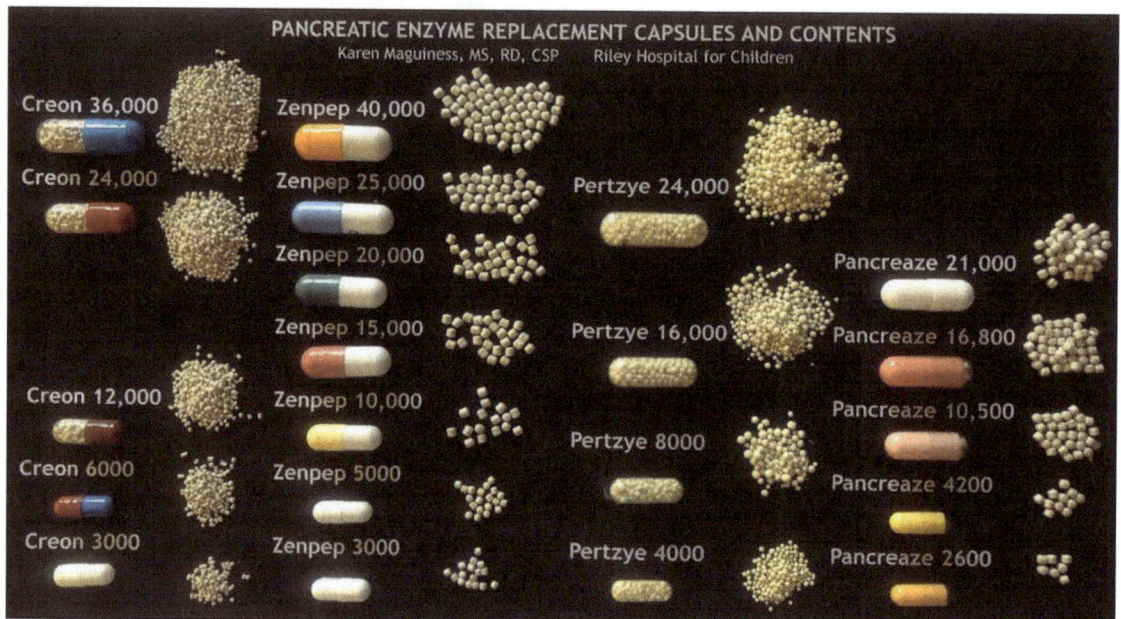

Figure 1. PERT Capsules and Contents. Reprinted with permission from Karen Maguiness, MS, RD, CSP, Riley Hospital for Children.

Products for oral use are generally in capsule form containing microspheres or microtablets with a pH-sensitive coating protecting the enzymes from gastric acid and allow activation in the more alkaline environment of the duodenum.

3.2. Oral PERT Administration

The capsules are administered by either swallowing whole or opening the capsules and sprinkling the contents in a small amount of acidic food such as applesauce. The enzymes beads or microtablets must be swallowed whole; they should not be chewed or crushed.

PERT may be administered to infants by sprinkling the capsule contents in a small amount of applesauce and offering right before breast or bottle feeding. The infant's mouth should be checked for retained beads and mucosal irritation. Starting a skin barrier cream is recommended because some of the beads may pass through the immature intestine and cause perianal irritation.

PERT dose can be determined by patient weight or by the fat content of the meal or snack. PERT dosing should be individualized (see Table 2); PI CF patients should be started on the lowest effective dose and then have the dose titrated based on weight gain and gastrointestinal symptoms to the lowest effective dose, not to exceed 2500 lipase units/kg/meal and a total of 10,000 lipase units/kg/day. The number after an enzyme name denotes lipase units per capsule multiplied by 1000. Several companies provide varying capsule dosages with varying bead sizes, best illustrated by the excellent image prepared by Karen Maguiness, MS, RD and used with permission (see Figure 1). Some young infants may transiently exceed this upper limit of dosing recommendations due to the need for frequent PERT dosing with on demand feeding and should be monitored by an experienced clinician [24]. Optimal enzyme dosing is associated with typical rates of weight gain and growth determined by genetic potential and evidence of some improvement of indirect markers of maldigestion such as vitamins levels. Doses are adjusted based on clinical symptoms of poor weight gain and/or fat malabsorption by increasing 1 capsule per dose not to exceed 2500 lipase units/kg/meal. Another consideration is tailoring the enzyme dose to the specific foods by dosing per fat gram. Always consider consultation with a nutrition or PERT expert when dosing enzymes to determine the best method for use and for dose titration recommendations.

Table 2. PERT Dosing Recommendations.

Patient Age	Age-Based Dosing Recommendations	Focus Guidance About Administration	Titration
PERT for Oral Feeds—Use Enteric-Coated Formulation			
Premature and full-term infants, <12 months	Initiate when taking >60 mL per feed (formula/breastmilk). Starting dose: 3000 lipase units/feed Range: 1000–2500 lipase units/kg/feed Max: 10,000 lipase units/kg/day	Open capsule, sprinkle enzyme beads on a small amount of applesauce Administer at start of feed Give by mouth even if portion of feed is enteral Never give via feeding tube (clogs tube) Check infant's mouth for retained beads and mucosal irritation Start skin barrier cream and monitor for perianal irritation	Increase by 1 capsule per dose based on clinical symptoms of malabsorption and/or poor weight gain Max dose may transiently exceed 10,000 lipase units per kg/day due to frequency of infant feedings
Children and Adolescents	Starting dose 1–4 years 1000 lipase units/kg/meal Titrate to max 2500 lipase units/kg/meal ≥4 years 500 lipase units/kg/meal Titrate to max 2500 lipase units/kg/meal Range: 500–2500 units/kg/meal Max: 10,000 lipase units/kg/day Snack dose: Half of the meal dose	Give capsule by mouth Give by mouth even if portion of feed is enteral If unable to swallow capsule: open capsule, sprinkle enzyme beads on a small amount of applesauce Meals lasting longer than 30 min: split dose and administer halfway through meals	Increase by 1 capsule per dose based on clinical symptoms of malabsorption and/or poor weight gain. Consult CF RD or another PERT expert and dose as needed
PERT for Tube Feeding—Bolus Feeds			
Enteric-Coated Enzyme	Weight based Starting dose Children: 500–1000 lipase units/kg/feed Range: 500–2500 lipase units/kg/feed Max: 10,000 lipase units/kg/day	Only use if patient is able to take enzymes by mouth Give at start of feed Give at start of feed Start at lower end of dosing range Initiate when taking ≥60 mL per feed Never crush or chew enzymes	
	Grams of fat based Typical dose: 1800–2200 lipase units/g of fat Range: 500–4000 lipase units/g of fat	Dose enzymes based on total grams of fat in the formula per RD recommendations Non-CF: start at lower end of dosing range	
Non-Enteric-Coated Enzyme (Viokace®)	Grams of fat based Typical dose: 1800–2200 lipase units/g of fat Range: 500–4000 lipase units/g of fat	Use ONLY if patient is unable to take enteric-coated enzymes by mouth Crush Viokace and add to formula Lipase units that come from Viokace do NOT count towards total max dose per day of 10,000 units lipase/kg/day Round to the nearest 1/2 tablet of Viokace. Dosage options are either 10,440 or 20,880 lipase units per tablet Non-CF: start at lower end of dosing range	

Table 2. Cont.

Patient Age	Age-Based Dosing Recommendations	Focus Guidance About Administration	Titration
PERT for Tube Feeding—Continuous/Overnight Feeds			
RELiZORB™	Patient is ≥5 years of age Starting dose: 1 cartridge per 500 mL formula Max: 2 cartridges/24 h period Optimal rate of feed: 24–120 mL/h	Preferred method for continuous tube feeds Avoid fiber containing and blenderized formulas, can clog cartridge Minimum tube feeding rate is 24 mL/h If rate is lower than 24 mL/h, use crushed (Viokace) method below	
Non-Enteric-Coated Enzyme (Viokace®)	Grams of fat based Typical dose: 1800–2200 lipase units/g of fat Range: 500–4000 lipase units/g of fat	Crush Viokace and add to formula Initiate when taking ≥15 mL per hour (formula/breast milk)If unable to use RELiZORB, this method provides the next best option Lipase units that come from Viokace do NOT count towards total max dose per day of 10,000 units lipase/kg/day Round to the nearest 1/2 tablet of Viokace. Dosage options are either 10,440 or 20,880 lipase units per tablet Non-CF: start at lower end of dosing range	
Enteric-Coated Enzyme	Given orally only Dose based on weight or grams of fat in tube feeding formula Weight based Range: 500–2500 lipase units/kg/feed Max: 10,000 lipase units/kg/day Grams of fat based Typical dose: 1800–2200 lipase units/g of fat Range: 500–4000 lipase units/g of fat	This option is for patients >2 years on overnight feeds who can take oral enzymes when RELiZORB or Viokace are not available Use a meal dose of enzymes at start of tube feeds. Patients may need an additional half dose at the end of the feed Do not use for 24 h continuous feeds Initiate when receiving >15 mL per hour (formula/breast milk) Do not recommend enteric-coated enzymes through any enteral tube due to risk for clogging Non-CF: start at lower end of dosing range	

Copied with permission from Children's Hospital of Philadelphia's Clinical Pathway for PERT in Children with or at Risk for Exocrine PI [25].

3.3. PERT Administration for Bolus and Continuous Enteral Feeding

Providing PERT for patients with PI receiving continuous enteral feeding has been challenging. Until recently, a common practice has been to provide oral PERT capsules at the beginning and end of the tube feeding. This off label practice is not practical from a physiologic sense to provide a bolus of PERT which lasted 45–60 min during a continuous infusion of nutrients often over 4–8 h or more. Some but not all patients tolerate this method and achieve desired weight gain. However, some may have ongoing abdominal pain, bloating, bowel urgency and other undesirable side effects and often results in decreased patient adherence.

The PERT dose for bolus enteral feeding can be based on the weight, or more accurately determined based on the grams of fat in the formula. A typical dose would be calculated at approximately 2000 lipase units per gram of fat in the formula, with a range between 500–4000 lipase units/gram of fat (see Table 2 PERT dosing recommendations). PERT may be administered orally at the start of the bolus feed with enteric-coated capsules either swallowed whole or opened and sprinkled on acidic fruit sauce. For instances when PERT cannot be swallowed, a non-enteric-coated tablet may be carefully crushed and added to the volume of formula for the bolus. The dose would be based on the grams of fat in the formula, typically at 2000 lipase units per gram of fat and rounded to the nearest 1/2 tablet size within a range of 500–4000 lipase units per gram of fat (see Table 2 PERT dosing recommendations).

The first FDA-approved device for patients with fat malabsorption ages 5 years and older receiving continuous enteral feeding is a lipase only cartridge (Relizorb) placed in-line with the tube through which the enteral feeding flows. The lipase inside the cartridge continuously breaks down long-chain triglycerides into absorbable components throughout the entire duration of the enteral feeding which then flow into the patient. One cartridge is recommended for 500 mL of formula at flow rates between 10–120 mL per hour. Cartridges can be connected in tandem to accommodate volumes between 500–1000 mL at rates between 24–120 mL per hour (see Table 3). Several studies report good results with improved weight gain, improved essential fatty acid profiles, and fat-soluble vitamins as well as improved gastrointestinal symptoms [26,27]. There is a list of compatible formulas on the company website with corresponding fat hydrolysis data which are updated as more information becomes available (relizorb.com). Note there are no data for infant formulas or breast milk as the cartridge has not been FDA approved in these age groups although there may be case reports.

Table 3. Enzyme Types and Considerations.

Enzyme Type	Considerations

Enteric-Coated	Capsules containing enteric-coated beads or microtablets
	Coating protects enzymes from gastric acid, allows activation in duodenum
	Administration:
	• Give by mouth at the start of feeds/meals/snacks/beverages.
	   ○ Swallowed whole or opened, and contents sprinkled on a small amount of applesauce.
	   ○ Give even if portion of their feeding is via an enteral tube.
	• Do NOT:
	   ○ Administer via feeding tube, will clog tube.
	   ○ Crush or chew enzyme beads.

Non-Enteric-Coated (Viokace®)	Powdered tablets Most often used for patients on tube feedings Administration: • Crush and add to enteral formula to pre-digest nutrients in feeding bag prior to enteral tube administration. • Do NOT give orally
Lipase Cartridge (RELiZORB™)	Enzyme cartridge only containing enzyme lipase For patients on continuous tube feeds only Administration: • Cartridge is connected in-line with the enteral tube feeding set. • Enteral formula flows through the cartridge and fat is digested in the formula. Refer to RELiZORB™ manufacturer's data sheet for compatible formulas

Copied with permission from Children's Hospital of Philadelphia's Clinical Pathway for PERT in Children with or at Risk for Exocrine PI [25].

Another off-label method used for pancrelipase coverage during continuous enteral feeding when unable to use the Relizorb cartridge is Viokace, a non-enteric-coated pancrelipase tablet which must be crushed and added to the formula before infusion (see Table 3). This method is preferable to giving oral enteric-coated pancrelipase before overnight formula infusion and in cases when the Relizorb cartridge is not compatible with the prescribed enteral regimen. The Viokace tablets are crushed carefully to a fine powder and added to the formula prior to infusion. Care must be taken to protect eyes and skin and not breathe in the powder. Dosing is based on grams of fat in the formula or breast milk (see Table 2).

3.4. Considerations for Treatment Failure

Poor growth is common among the CF population, even in clinics with robust dietician support. Numerous considerations must be entertained. A primary cause of poor growth is poor caloric intake, especially common given the many symptoms that CF patients have (abdominal distension, abdominal pain, constipation, etc.). Of course, other causes of malabsorption and other factors besides PI that may affect fat malabsorption and/or weight loss must be considered as well, such as constipation or celiac disease (see Table 1 for a more complete differential diagnosis). Among CF patients with mild PI CF-associated liver disease, Drzymala-Czyz et al., found that ursodeoxycholic acid supplementation enhanced fat absorption [28].

Enzyme failure must also be carefully considered; for common causes of enzyme failure see Table 4. As CF patients have increased gastrointestinal acidity [29], and acidity may decrease enzyme activity efficacy [30] and/or precipitate bile acids in the CF intestine [31] (and thus decreased micelle formation and fat absorption), adding acid-suppressing agents may improve fat absorption [32]. However, acid-suppression therapies (most commonly proton pump inhibitors) are still being evaluated and other adjustments might be considered first. In general, an acid suppression therapy trial is typically limited with clear endpoints such as weight gain [4].

Table 4. Troubleshooting PERT Failure.

	PERT Considerations

Timing of PERT administration	• Enzymes should be administered prior to eating all meals and snacks.
	   ○ Food should be eating within 45–60 min of enzyme dose to ensure appropriate enzyme activity.
	• Slow eaters, gastroparesis, or fat eaten at end of meal.
	   ○ Consider splitting dose, taking 1/2 at start of meal and 1/2 dose partway through the meal.
	• If most food is eaten in 1 sitting.
	   ○ Dose adjustment may be difficult. Consider spreading food and fat intake over course of day

Type of food eaten	• Match enzyme dose to food eaten. ○ Consider dosing enzymes based on fat grams.

Storage	• Keep lid tightly closed on enzyme container.
	• Enzymes should be kept at room temperature (59–86 degrees Fahrenheit).
	   ○ Heat destroys enzymes: do not store on top of refrigerator or toaster oven, keep out of hot cars.
	   ○ Cold temperatures can harm enzymes, do not refrigerate.

Expiration date	• Enzymes degrade overtime so always check the expiration date.

3.5. PERT Complications

Despite PERT's vital component to any PI CF patient treatment regimen, PERT has side effects that must be considered by the clinician. Fibrosing colonopathy is a well-known complication of PERT associated with high PERT doses [33]. FitzSimmons et al.'s case-controlled study from 1997 demonstrated that higher doses of PERT were associated with fibrosing colonopathy, finding that affected patients had a daily dose of PERT 2.5 times higher than the unaffected control group [34]. However, even the control group had an daily average PERT intake of 18,000 units lipase/kg/day [34], far higher than the current US care guidelines [18]. Thus, it has been argued that the maximal PERT dose of 10,000 units lipase/kg/day is not evidence based, and clinicians can consider higher doses especially in infants as patient's needs necessitate [33].

4. PERT Efficacy

While PERT efficacy could be measured by weight changes over time and PI symptom improvement, these endpoints are lagging and subjective. Thus, nearly all studies for PERT efficacy utilize coefficient of fat absorption (CFA) over 72 h to measure improvement in steatorrhea, and coefficient of nitrogen absorption (CNA) to quantify changes in azotorrhea. While CFA is an especially accurate and precise measurement for steatorrhea in otherwise healthy individuals, CFA is less accurate and precise among patients with CF. However, CFA remains the primary outcome measure to investigate PERT efficacy [35].

Accurate measurement of fecal fat is not only important for diagnosing PI, but also PERT dosing. Overall, 72 h fecal collections for CFA are arduous and difficult for patients. Caras et al. demonstrated that the mean calculated CFA of three random stool samples over 3 days was as sensitive to predict a percentage fat < 30% as a 72 h CFA [36]. While still difficult to obtain, these random stool samples would be more easily obtained than a strict 72 h fecal collection. Clinically, random stool collections are not used to titrate PERT dosing.

A recent metanalysis on PERT's efficacy in patients with chronic pancreatitis showed that PERT significantly improved CFA and CNA, and among randomized controlled trials PERT improved GI symptoms and decreased fecal weight, fecal fat, and nitrogen excretion [37]. A Cochrane review recently published noted that there are no high-quality trials comparing PERT to placebo in individuals with CF. However, when comparing enteric-coated microspheres (ECM) to enteric-coated tablets (ECT), ECM has superior outcomes regarding abdominal pain, stool frequency, and fecal fat excretion suggesting ECM's significant efficacy among patients with CF. Interestingly, there are no outcome differences among the various formulations of ECMs [38].

5. PERT and Highly Effective Modulator Therapies

Severe PI has been thought to be irreversible, and CF-related PI is often present from birth or develops in most instances within the first year of life. The introduction of highly effective modulator therapies (HEMTs) has demonstrated that pancreatic function is more dynamic than previously thought. HEMTs include ivacaftor (IVA) for gating mutations such as G551D and IVA/tezacaftor/elexacaftor for at least a heterozygous F508del mutation [39]. In some cases, long-term use of HEMTs (especially IVA for gating mutations) may reverse the PI in some PI CF patients as seen in the following studies.

In the ARRIVAL study, 19 children aged 12 to <24 months with at least one gating mutation were placed on CFTR modulator IVA for 24 weeks [40]. After the 24-week trial, the patient's fecal elastase improved from a mean of 182.2 ug/g to 326.9 ug/g and immunoreactive trypsinogen improved from a mean of 1154.9 ng/mL to 505.4 ng/mL. Eleven children were considered pancreatic insufficient at baseline (all 11 had baseline values < 50 μg/g). Nine of these children had both baseline and week-24 fecal elastase values; six of these nine children had fecal elastase > 200 μg/g at week 24, a value consistent with normal pancreatic function.

In the 24-week KIWI trial of IVA given to older children ages 2–5 years, a mean increase in FE of 99.8 ug/g was observed and sustained throughout the open-label extension KLIMB study. Subjects were followed from week 24 to 84 although no further significant changes were noted in FE. For the 18 subjects who had FE measurements completed at baseline and week 84, one had a FE > 200 ug/g at baseline and five had FE > 200 ug/g at week 84 [41].

Thus, it seems that HEMTs can improve, and possibly rescue, pancreatic exocrine function in some patients, even in older patients according to these studies. There is significant variability in fecal elastase response among the patients in the KIWI and ARRIVAL trials, suggesting that a subset of CF PI patients may have enough residual acinar activity that can be rescued. Given this variability, approaches to PERT therapy for patients on HEMT are currently highly variable and not standardized. As of this time, we are in the early days of HEMT use and have yet to fully appreciate the effects of long-term therapy on exocrine pancreatic function.

6. Conclusions

The CF population presents unique considerations for their physicians and dieticians caring for them. Improved nutrition for the CF patient is paramount for improved clinical outcomes and survivability, and PERT is essential for PI CF patients. We have explored the nuances, considerations, administrative techniques, and complications of PERT in this review. As PI CF patients have unique disease pathology, the providers and dieticians that care for CF patients must have a robust knowledge of PERT.

Funding: This research received no external funding.

Institutional Review Board Statement: Not applicable.

Informed Consent Statement: Not applicable.

Data Availability Statement: Not applicable.

Conflicts of Interest: The authors declare no conflict of interest.

References

1. Comeau, A.M.; Accurso, F.J.; White, T.B.; Campbell, P.W.; Hoffman, G.; Parad, R.B.; Wilfond, B.S.; Rosenfeld, M.; Sontag, M.K.; Massie, J.; et al. Guidelines for Implementation of Cystic Fibrosis Newborn Screening Programs: Cystic Fibrosis Foundation Workshop Report. *Pediatrics* **2007**, *119*, e495–e518. [CrossRef] [PubMed]
2. Atlas, A.; Rosh, J. *Pediatric Gastroenterology and Liver Disease*; Elsevier: Amsterdam, The Netherlands, 2016; pp. 999–1015.
3. Andersen, D.H. Cystic Fibrosis of the Pancreas and its Relation to Celiac Disease. *Am. J. Dis. Child.* **1938**, *2*, 344–399. [CrossRef]
4. Singh, V.K.; Schwarzenberg, S.J. Pancreatic insufficiency in Cystic Fibrosis. *J. Cyst. Fibros.* **2017**, *16*, S70–S78. [CrossRef] [PubMed]
5. Wilschanski, M.; Durie, P.R. Pathology of pancreatic and intestinal disorders in cystic fibrosis. *J. R. Soc. Med.* **1998**, *91*, 40–49. [CrossRef]
6. Corey, M.; McLaughlin, F.J.; Williams, M.; Levison, H. A comparison of survival, growth, and pulmonary function in patients with cystic fibrosis in Boston and Toronto. *J. Clin. Epidemiol.* **1988**, *41*, 583–591. [CrossRef]
7. Yen, E.H.; Quinton, H.; Borowitz, D. Better Nutritional Status in Early Childhood Is Associated with Improved Clinical Outcomes and Survival in Patients with Cystic Fibrosis. *J. Pediatrics* **2013**, *162*, 530–535.e1. [CrossRef]
8. Farrell, P.M.; Lai, H.J.; Li, Z.; Kosorok, M.R.; Laxova, A.; Green, C.G.; Collins, J.; Hoffman, G.; Laessig, R.; Rock, M.J.; et al. Evidence on Improved Outcomes with Early Diagnosis of Cystic Fibrosis Through Neonatal Screening: Enough is Enough! *J. Pediatrics* **2005**, *147*, S30–S36. [CrossRef]
9. 2019 Patient Registry Annual Data Report. 2019. Available online: https://www.cff.org/Research/Researcher-Resources/Patient-Registry/2019-Patient-Registry-Annual-Data-Report.pdf (accessed on 22 January 2022).
10. Leung, D.H.; Heltshe, S.L.; Borowitz, D.; Gelfond, D.; Kloster, M.; Heubi, J.E.; Stalvey, M.; Ramsey, B.W.; for the Baby Observational and Nutrition Study (BONUS) Investigators of the Cystic Fibrosis Foundation Therapeutics Development Network. Effects of Diagnosis by Newborn Screening for Cystic Fibrosis on Weight and Length in the First Year of Life. *JAMA Pediatrics* **2017**, *171*, 546. [CrossRef]
11. Pietzak, M.M.; Thomas, D.W. Childhood Malabsorption. *Pediatrics Rev.* **2003**, *24*, 195–206. [CrossRef]
12. Clark, R.; Johnson, R. Malabsorption Syndromes. *Nurs. Clin.* **2018**, *53*, 361–374. [CrossRef]
13. Hoch, H.; Sontag, M.K.; Scarbro, S.; Juarez-Colunga, E.; McLean, C.; Kempe, A.; Sagel, S.D. Clinical outcomes in U.S. infants with cystic fibrosis from 2001 to 2012. *Pediatric Pulmonol.* **2018**, *53*, 1492–1497. [CrossRef] [PubMed]
14. Southern, K.W.; Munck, A.; Pollitt, R.; Travert, G.; Zanolla, L.; Dankert-Roelse, J.; Castellani, C.; on behalf of the ECFS CF Neonatal Screening Working Group. A survey of newborn screening for cystic fibrosis in Europe. *J. Cyst. Fibros.* **2007**, *6*, 57–65. [CrossRef] [PubMed]
15. Rock, M.J.; Levy, H.; Zaleski, C.; Farrell, P.M. Factors accounting for a missed diagnosis of cystic fibrosis after newborn screening. *Pediatric Pulmonol.* **2011**, *46*, 1166–1174. [CrossRef] [PubMed]
16. Coffey, M.J.; Whitaker, V.; Gentin, N.; Junek, R.; Shalhoub, C.; Nightingale, S.; Hilton, J.; Wiley, V.; Wilcken, B.; Gaskin, K.J.; et al. Differences in Outcomes between Early and Late Diagnosis of Cystic Fibrosis in the Newborn Screening Era. *J. Pediatrics* **2017**, *181*, 137–145.e1. [CrossRef] [PubMed]
17. Keller, J.; Aghdassi, A.A.; Lerch, M.M.; Mayerle, J.V.; Layer, P. Tests of pancreatic exocrine function—Clinical significance in pancreatic and non-pancreatic disorders. *Best Pract. Res. Clin. Gastroenterol.* **2009**, *23*, 425–439. [CrossRef]
18. Borowitz, D.; Robinson, K.A.; Rosenfeld, M.; Davis, S.D.; Sabadosa, K.A.; Spear, S.L.; Michel, S.H.; Parad, R.B.; White, T.B.; Farrell, P.M.; et al. Cystic Fibrosis Foundation Evidence-Based Guidelines for Management of Infants with Cystic Fibrosis. *J. Pediatrics* **2009**, *155*, S73–S93. [CrossRef]
19. Walkowiak, J.; Lisowska, A.; Blask-Osipa, A.; Drzymala-Czyz, S.; Sobkowiak, P.; Cichy, W.; Breborowicz, A.; Herzig, K.; Radzikowski, A. Acid Steatocrit Determination Is Not Helpful In Cystic Fibrosis Patients Without or With Mild Steatorrhea. *Pediatric Pulmonol.* **2010**, *45*, 249–254. [CrossRef]
20. Beharry, S.; Ellis, L.; Corey, M.; Marcon, M.; Durie, P. How useful is fecal pancreatic elastase 1 as a marker of exocrine pancreatic disease? *J. Pediatrics* **2002**, *141*, 84–90. [CrossRef]
21. Stevens, T.; Conwell, D.L.; Zuccaro, G.; Lente, F.V.; Lopez, R.; Purich, E.; Fein, S. A prospective crossover study comparing secretin-stimulated endoscopic and Dreiling tube pancreatic function testing in patients evaluated for chronic pancreatitis. *Gastrointest. Endosc.* **2008**, *67*, 458–466. [CrossRef]
22. O'Sullivan, B.P.; Baker, D.; Leung, K.G.; Reed, G.; Baker, S.S.; Borowitz, D. Evolution of Pancreatic Function during the First Year in Infants with Cystic Fibrosis. *J. Pediatrics* **2013**, *162*, 808–812.e1. [CrossRef]
23. Turck, D.; Braegger, C.P.; Colombo, C.; Declercq, D.; Morton, A.; Pancheva, R.; Robberecht, E.; Stern, M.; Strandvik, B.; Wolfe, S.; et al. ESPEN-ESPGHAN-ECFS guidelines on nutrition care for infants, children, and adults with cystic fibrosis. *Clin. Nutr.* **2016**, *35*, 557–577. [CrossRef] [PubMed]
24. Borowitz, D.S.; Grand, R.J.; Durie, P.R.; the Consensus Committee. Use of pancreatic enzyme supplements for patients with cystic fibrosis in the context of fibrosing colonopathy. *J. Pediatrics* **1995**, *127*, 681–684. [CrossRef]
25. Padula, L.; Brownwell, J.; Reid, E.; Jansma, B.; Mascarenhas, M.; Sadgwar, S.; Shanley, L.; McKnight-Menci, H.; Maqbool, A. Childrens Hospital of Philadelphia Clinical Pathway for Pancreatic Enzyme Replacement Therapy (PERT) in Children with or at Risk for Exocrine Pancreatic Insufficiency (EPI). 2019. Available online: https://www.chop.edu/clinical-pathway/initiating-pancreatic-enzyme-replacement-therapy-pert-clinical-pathway (accessed on 22 March 2022).

26. Stevens, J.; Wyatt, C.; Brown, P.; Patel, D.; Grujic, D.; Freedman, S.D. Absorption and Safety With Sustained Use of RELiZORB Evaluation (ASSURE) Study in Patients With Cystic Fibrosis Receiving Enteral Feeding. *J. Pediatric Gastroenterol. Nutr.* **2018**, *67*, 527–532. [CrossRef] [PubMed]
27. Freedman, S.; Orenstein, D.; Black, P.; Brown, P.; McCoy, K.; Stevens, J.; Grujic, D.; Clayton, R. Increased Fat Absorption From Enteral Formula Through an In-line Digestive Cartridge in Patients With Cystic Fibrosis. *J. Pediatric Gastroenterol. Nutr.* **2017**, *65*, 97–101. [CrossRef]
28. Drzymala-Czyz, S.; Jonczyk-Potoczna, K.; Lisowska, A.; Stajgis, M.; Walkowiak, J. Supplementation of Ursodeoxycholic acid improves fat digestion and absorption in cystic fibrosis patients with mild liver involvement. *Eur. J. Gastroenterol. Hepatol.* **2016**, *28*, 645–649. [CrossRef]
29. Youngberg, C.A.; Berardi, R.R.; Howatt, W.F.; Hyneck, M.L.; Amidon, G.L.; Meyer, J.H.; Dressman, J.B. Comparison of gastrointestinal pH in cystic fibrosis and healthy subjects. *Digest. Dis. Sci.* **1987**, *32*, 472–480. [CrossRef]
30. Brady, M.S.; Garson, J.L.; Krug, S.K.; Kaul, A.; Rickard, K.A.; Caffrey, H.H.; Fineberg, N.; Balistreri, W.F.; Stevens, J.C. An Enteric-Coated High-Buffered Pancrelipase Reduces Steatorrhea in Patients with Cystic Fibrosis: A Prospective, Randomized Study. *J. Am. Diet. Assoc.* **2006**, *106*, 1181–1186. [CrossRef]
31. Borowitz, D.; Durie, P.R.; Clarke, L.L.; Werlin, S.L.; Taylor, C.J.; Semler, J.; Lisle, R.C.D.; Lewindon, P.; Lichtman, S.M.; Sinaasappel, M.; et al. Gastrointestinal Outcomes and Confounders in Cystic Fibrosis. *J. Pediatric Gastroenterol. Nutr.* **2005**, *41*, 273–285. [CrossRef]
32. Proesmans, M.; Boeck, K.D. Omeprazole, a proton pump inhibitor, improves residual steatorrhoea in cystic fibrosis patients treated with high dose pancreatic enzymes. *Eur. J. Pediatrics* **2003**, *162*, 760–763. [CrossRef]
33. Borowitz, D.; Gelfond, D.; Maguiness, K.; Heubi, J.E.; Ramsey, B. Maximal daily dose of pancreatic enzyme replacement therapy in infants with cystic fibrosis: A reconsideration. *J. Cyst. Fibros.* **2013**, *12*, 784–785. [CrossRef]
34. FitzSimmons, S.C.; Burkhart, G.A.; Borowitz, D.; Grand, R.J.; Hammerstrom, T.; Durie, P.R.; Lloyd-Still, J.D.; Lowenfels, A.B. High-Dose Pancreatic-Enzyme Supplements and Fibrosing Colonopathy in Children with Cystic Fibrosis. *N. Engl. J. Med.* **1997**, *336*, 1283–1289. [CrossRef] [PubMed]
35. Borowitz, D.; Konstan, M.W.; O'Rourke, A.; Cohen, M.; Hendeles, L.; Murray, F.T. Coefficients of Fat and Nitrogen Absorption in Healthy Subjects and Individuals with Cystic Fibrosis. *J. Pediatric Pharmacol. Ther.* **2007**, *12*, 47–52. [CrossRef] [PubMed]
36. Caras, S.; Boyd, D.; Zipfel, L.; Sander-Struckmeier, S. Evaluation of Stool Collections to Measure Efficacy of PERT in Subjects With Exocrine Pancreatic Insufficiency. *J. Pediatric Gastroenterol. Nutr.* **2011**, *53*, 634–640. [CrossRef] [PubMed]
37. De la Iglesia-García, D.; Huang, W.; Szatmary, P.; Baston-Rey, I.; Gonzalez-Lopez, J.; Prada-Ramallal, G.; Mukherjee, R.; Nunes, Q.M.; Domínguez-Muñoz, J.E.; Sutton, R.; et al. Efficacy of pancreatic enzyme replacement therapy in chronic pancreatitis: Systematic review and meta-analysis. *Gut* **2017**, *66*, 1354–1355. [CrossRef]
38. Somaraju, U.R.R.; Solis-Moya, A. Pancreatic enzyme replacement therapy for people with cystic fibrosis. *Cochrane Database Syst. Rev.* **2020**, *8*, CD008227.
39. Dave, K.; Dobra, R.; Scott, S.; Saunders, C.; Matthews, J.; Simmonds, N.J.; Davies, J.C. Entering the era of highly effective modulator therapies. *Pediatric Pulmonol.* **2021**, *56*, S79–S89. [CrossRef]
40. Rosenfeld, M.; Cunningham, S.; Harris, W.T.; Lapey, A.; Regelmann, W.E.; Sawicki, G.S.; Southern, K.W.; Chilvers, M.; Higgins, M.; Tian, S.; et al. An open-label extension study of ivacaftor in children with CF and a CFTR gating mutation initiating treatment at age 2–5 years (KLIMB). *J. Cyst. Fibros.* **2019**, *18*, 838–843. [CrossRef]
41. Rosenfeld, M.; Wainwright, C.E.; Higgins, M.; Wang, L.T.; McKee, C.; Campbell, D.; Tian, S.; Schneider, J.; Cunningham, S.; Davies, J.C.; et al. Ivacaftor treatment of cystic fibrosis in children aged 12 to <24 months and with a CFTR gating mutation (ARRIVAL): A phase 3 single-arm study. *Lancet Respir. Med.* **2018**, *6*, 545–553. [CrossRef]

Article

Experiences and Perspectives of Individuals with Cystic Fibrosis and Their Families Related to Food Insecurity

Montserrat A. Corbera-Hincapie [1,*], Samar E. Atteih [1], Olivia M. Stransky [2], Daniel J. Weiner [1], Iris M. Yann [1] and Traci M. Kazmerski [1,2]

1 Department of Pediatrics, University of Pittsburgh School of Medicine, Pittsburgh, PA 15213, USA; atteihse@upmc.edu (S.E.A.); daniel.weiner@chp.edu (D.J.W.); iris.yann@chp.edu (I.M.Y.); traci.kazmerski@chp.edu (T.M.K.)
2 Center for Innovative Research on Gender Health Equity (CONVERGE), Pittsburgh, PA 15224, USA; ols25@pitt.edu
* Correspondence: mac572@pitt.edu; Tel.: +1-786-412-1300

Abstract: Food insecurity (FI) rates among people with cystic fibrosis (CF) are significantly higher than in the general US population. This study explored the experiences and perceptions of adults and parents of children with CF surrounding FI. We recruited parents of children with CF ages 0–18 years and adults with CF ages 18 years and older from a large, accredited U.S. CF care center and the Cystic Fibrosis Foundation Community Voice to participate in a qualitative study using semi-structured telephone interviews to explore experiences and preferences related to food insecurity. Two coders independently reviewed each transcript to apply the codebook and identify any emerging codes using an ongoing, iterative process to identify central themes. We interviewed 20 participants (six adults with CF and 14 parents of children with CF) and identified five major themes: (1) FI in CF is influenced by a variety of factors, ranging from nutritional demands to competing financial barriers; (2) FI impacts CF health outcomes; (3) Open patient-provider communication around FI is vital; (4) FI screening and discussions should be normalized in CF care; (5) Comprehensive FI resources are vital. FI is an important topic that should routinely be addressed with the CF care team to destigmatize and encourage individuals to be more forthcoming about their FI status. Results from this study will inform future larger investigations on the impact of FI on CF health and aid in the design and planning of targeted interventions and advocacy efforts.

Keywords: cystic fibrosis; food insecurity; experiences

Citation: Corbera-Hincapie, M.A.; Atteih, S.E.; Stransky, O.M.; Weiner, D.J.; Yann, I.M.; Kazmerski, T.M. Experiences and Perspectives of Individuals with Cystic Fibrosis and Their Families Related to Food Insecurity. *Nutrients* **2022**, *14*, 2573. https://doi.org/10.3390/nu14132573

Academic Editors: Maria R. Mascarenhas and Jessica Alvarez

Received: 26 May 2022
Accepted: 16 June 2022
Published: 21 June 2022

Publisher's Note: MDPI stays neutral with regard to jurisdictional claims in published maps and institutional affiliations.

Copyright: © 2022 by the authors. Licensee MDPI, Basel, Switzerland. This article is an open access article distributed under the terms and conditions of the Creative Commons Attribution (CC BY) license (https://creativecommons.org/licenses/by/4.0/).

1. Introduction

Food insecurity (FI), defined as a household condition in which access to adequate food is limited by lack of money and other resources, is an important health concern that has the potential to impact on children and families in a dramatic way [1]. In 2020, 10.5 percent of households in the United States (US) were food insecure, and nearly 15 percent of households with children were food insecure [2]. FI, especially in households with children, has since been exacerbated by the COVID-19 pandemic [3]. Furthermore, individuals living with chronic illnesses in the US have an increased prevalence of FI [4].

FI rates among people with cystic fibrosis (CF) are significantly higher than the general US population [5]. Ideal nutritional status, defined as body mass index (BMI) at or above the 50th percentile for children and adolescents aged 2 to 20 years or BMI of at least 22 and 23 for females and males respectively, has been found to correlate with improved long-term linear growth, pulmonary function, and long-term survival in CF. As such, the CF Foundation (CFF) recommends an intake of at least 500 calories per day over the standard daily requirement with fat intake composing about 35–40% of those calories [6–8]. Given that people with CF have increased caloric demands, FI may contribute to the inability to achieve and maintain appropriate weight gain and lead to malnutrition in CF and is a nutritional risk factor for the CF population [9–12].

The individual and family experience of FI in CF, where diet and nutrition are so heavily valued and monitored, has not been previously studied. We explore the lived experiences of both adults with CF and the parents of children with CF, who have reported experiencing FI on routine screening. We investigated perceptions surrounding FI screening, patient-provider interactions, and emotions surrounding disclosure with the overarching goal of improving identification and provision of resources to those experiencing FI in CF.

2. Materials and Methods

2.1. Study Participants

We recruited parents or primary caregivers of children with CF ages 0–18 years and adults with CF ages 18 years and older using two methods. First, we recruited participants from a large, accredited U. S. CF care center during outpatient visits between November 2020 and May 2021 who reported experiencing FI on verbal screening using a validated screening tool. We also recruited participants who self-reported FI in June 2021 using a national convenience sample from the CFF Community Voice (CV), a virtual avenue for people with CF and their family members to share their experiences and perspectives on CF research and care [13]. We used pseudonyms to represent study participants.

2.2. Interviews and Data Collection

The principal investigator (MACH), a pediatric gastroenterologist and health services researcher, conducted the interviews in a private space using a secure telephone line. The interviewer was not directly involved in the participants' clinical care. We collected participant demographic information, including age, race, and gender of the person(s) with CF and structured interviews by key open-ended questions. The interviews explored the participants' experiences surrounding FI, their understanding of the impact of FI on their or their child's health, their experiences discussing FI with the CF team and other healthcare providers, their coping strategies for addressing FI, and their perspectives on what resources were needed to better support those with CF facing FI (Box 1).

The institutional review board (IRB protocol #20080145) approved the study. Patient consent was waived due to the study's anonymous low-risk nature. We compensated each participant with a $50 gift card. All interviews were conducted in English and audio recorded. Interviews lasted between 25 and 60 min. We reached thematic saturation after the 10th interview, suggesting that our total sample was adequate for establishing important themes as they relate to FI in CF.

A co-investigator (TMK), a CF physician and health services researcher with extensive experience on qualitative research, served as a resource during the entire data collection process.

2.3. Data Analysis

We transcribed the interviews and used Dedoose software to facilitate data management and coding. The primary investigator (MACH) developed an initial codebook based on the interview guide. MACH and a second coder (SA), a pediatric pulmonologist, then independently reviewed each transcript to apply the initial codebook and identify any emerging codes. The coders reviewed their coding together and defined any new codes after the first few transcripts were independently coded. This approach was used in an ongoing, iterative process. The coders applied the final codes to all transcripts, and the study team identified central themes and representative quotations [14–16].

Box 1. Key Interview Questions.

- Can you tell me more about your experiences with food insecurity?
- In the past 12 months, have you worried about paying the rent/mortgage, falling behind on utilities, facing transportation or health care barriers? Tell me more about this.
- How do you think food insecurity has affected your [or your child's] health?
- Did you ever bring up food insecurity with any health care provider? Tell me more about this.
- Should the CF team be asking about food insecurity? Why or why not?
- Has anyone on your CF care team (i.e., your doctor, nutritionist, nurse, social worker, etc.) ever talked to you about food insecurity? How did this conversation go?
- What resources would be helpful for you?
- Have you ever been offered resources during an office clinic? If so, which did you find the most helpful and why?
- How can your CF team better support people with CF with food insecurity?
- What advice do you have for other parents with children with CF or individuals with CF, who are experiencing food insecurity?
- What advice do you have for your CF team on how to make these discussions more comfortable?
- What coping strategies do you have related to food insecurity?
- What is the best way to screen for food insecurity? Tell me more about this.
- If you have questions about food insecurity that are specific to CF, where do you go for information?
- In the future, how would you want to hear about any information regarding food insecurity?

CF: cystic fibrosis.

3. Results

We interviewed 20 participants (six adults with CF and 14 parents of children with CF). We recruited 15 participants directly from clinic and five from CV. Table 1 includes participant demographic information. Major themes and subthemes with representative quotations are below.

3.1. Theme 1: FI in CF Is Influenced by A Variety of Factors, Ranging from the Nutritional Demands to Competing Financial Barriers

Many participants expressed significant concerns surrounding the expense of a CF diet, due to increased caloric requirements and increased need for expensive food items, such as proteins and fats. Half of the participants felt that they purchased lower quality food due to these considerations. "Avery", a parent of three boys with CF, discussed their family's monthly food budget and stated, "It's a tricky thing because I know they need to eat more, a lot more, but how much money can you put into a grocery budget . . . things that they need that are gonna be good for their diet are also the expensive things".

Most participants felt that there were several competing barriers that compounded the effects of FI, including basic living expenses (i.e., utilities, rent), single income households, healthcare costs, transportation barriers, and households with multiple children with CF. Participants expressed the difficulty of having to choose between food and competing costs. "Riley", a parent of two children with CF, shared "I don't make a whole lot of money, but I also don't qualify for food stamps. Everything that I buy I pay for myself. It gets hard on a single [parent]". "Morgan", age 38, shared their experiences with transportation costs and stated, "I do pulmonary rehabilitation at the hospital three times a week, so driving back and forth, that takes gas and a lot of things. I usually end up just barely making it, month to month". "Avery" shared that having three children with CF is "going to be like I'm feeding six boys . . . they need twice as much food as the average kid".

Most participants felt that COVID-19 was an external contributor to FI. "Emerson", a parent of two girls with CF, shared that due to COVID-19, their family "ended up having to pick between if we were paying bills or buying groceries because we were already rammed with credit card debt". The pandemic worsened FI for some families and led to FI in some families that had not experienced it previously.

Table 1. Characteristics of the CF patient sample.

Participant Characteristics	Mean (SD) or % (n)
People with CF $n = 6$	
Age	32 (range 18–38)
Gender	
Female	67% (4)
Male	33% (2)
Modulator use	
Yes	67% (4)
No	33% (2)
Race/Ethnicity	
White	67% (4)
Black	17% (1)
Multiracial (White and Native American)	16% (1)
Parents $n = 14$	
Age of children with CF *	10.3 (range 2–19)
Gender of children with CF	
Female	43% (9)
Male	52% (11)
Transmasculine	5% (1)
Gender of parents participating	
Female	100% (14)
Modulator use in children	
Yes	52% (11)
No	48% (10)
Race/Ethnicity	
White	95% (20)
Black	5% (1)
Households with multiple children with CF	43% (6)

*, 21 children total.

3.2. Theme 2: FI Impacts CF Health Outcomes

Nearly all participants felt that FI played a role in their own or their children's weight with an emphasis on being underweight. "Avery" reflected on how FI has impacted their children's weight by stating "we do try to maintain their health as best within our power, but I think they could probably do so much better with their weight if I had the means to give them what I wish I could as far as food".

Some participants felt that FI also played a role in their lung function and susceptibility to infections. "Dallas", age 37, reflected on how they "had more [lung] infections . . . getting sicker a lot more" when they were underweight due to FI. "Avery" said "If they could just have what they need to do better, and just nutrition . . . It's just everything rides on that 'cause they'll have less hospitalizations if they're nutritionally better. I just hope that this is somethin' that gets more talked about, and more taught to the professionals".

3.3. Theme 3: Open Patient-Provider Communication around FI Is Vital

Most participants felt that FI was associated with embarrassment and guilt. This emotional impact led to difficulty and fear of disclosing FI status and asking for help. "Riley" noted that "it does get embarrassing being asked . . . it's hard to say yes . . . that you're struggling . . . sometimes it's just hard for people to say that they are". "Charlie", a parent of two girls with CF, said that "there's a feeling like you're failing, or some desperation" when disclosing FI and accepting resources.

Almost half of the participants expressed difficulty in disclosing FI because they felt undeserving of help and that other families or patients could use FI resources more. "Avery" shared, "I don't want to put 'yes' on [the FI screener] knowing there's probably people who can't even get their kid one package of bacon. I feel like those resources should go to those people, but I don't know what you do about the people in the middle who don't qualify for SNAP [Supplemental Nutrition Assistance Program] but aren't quite able to get what they need".

Most participants felt strongly that overcoming this fear of asking for help and accepting FI resources were the best ways to advocate for loved ones or oneself. "Brooklyn", parent of a boy with CF, expressed, "[Don't be] scared to say something because [the CF providers] don't know what's going on unless you bring it to their attention. Don't be scared to speak up because this is your child's voice right now".

3.4. Theme 4: FI Screening and Discussions Should Be Normalized in CF Care

Many participants felt that normalizing FI and standardizing how the topic is approached and discussed would encourage people to be more forthcoming about their FI status. "Cameron", a parent of a toddler with CF, shared that it would be helpful to have a universal FI screener on registration that is "worded nicely... so that it would just be like, this is what we do. It's part of the process. We are not singling anyone out".

Almost every participant felt that the CF team should be screening for FI at every visit. "Dallas" stated that "there are a lot more people that are struggling [with FI] in the CF community than they [CF care team] know of". The majority also felt that nonverbal (i.e., questions on a handout or on a tablet) or a combination of nonverbal and verbal screening is preferred over verbal screening (i.e., staff asking on arrival to visit) alone. "Hayden", parent of a boy with CF, expressed that a nonverbal method of screening would be ideal because "some people are embarrassed about what they make ... and some people might not be open" if asked verbally.

Participants noted that FI screening can miss those who do not qualify for governmental assistance or whose income is irreflective of FI status. "Jesse", a parent of a girl with CF, expressed "[We were] never eligible to receive assistance because of [my] husband's salary. It's like even though he made a certain amount, it still didn't reflect how [we] were struggling".

The majority of participants felt that FI discussions should be with either the dietitian or social worker. "Remy", parent of a transgender male with CF, said "I feel comfortable with the social worker. They're there to help you and make sure everything's okay and they're less intimidating I think than maybe a doctor or nurse". "Sam", age 34, felt that "the dietitian should always be the one to initiate the conversation because they're the ones trained to understand the capacity of food, what food deserts are, what FI does from a nutritional standpoint".

Furthermore, about a third of participants felt that CF physicians have limited understanding of FI or should not be burdened with such non-medical aspects of care. "Riley" stated "[Physicians] should be more concerned about the problems that are going on, the things that can be done to help health-wise". "Emerson" also shared "it's hard to discuss FI ... and it's a difficult topic to get too personal with doctors, and frankly, I don't think they really understand either".

3.5. Theme 5: Comprehensive FI Resources Are Vital

The majority of participants felt that food banks, SNAP, and gift cards were helpful resources. "Hayden" stated, "The gift cards definitely helped because I can get gas, and that saves money for other things. We receive some food stamps ... that definitely helped during the pandemic". Most preferred that resources be provided via an electronic method (i.e., email) as opposed to paper. "Riley" explained how "electronically is easier because papers get lost. Emails you can always find".

Most participants felt strongly that family served as a great support network. "Morgan" expressed "I have a great support system. I think that's part of the reason why I'm alive. I have a family that has taught me to fight. I have a family that loves me and my significant other loves me very, very much".

4. Discussion

This is the first study to explore the perspectives and experiences of people with CF and families with children with CF related to FI. Participants expressed that FI is influenced by a variety of factors, ranging from the increased CF nutritional demands to competing financial barriers, and can lead to poor CF health outcomes. FI also elicited feelings of embarrassment and guilt that led to hesitancy in disclosing FI status. Given this, most participants felt that this was an important topic that should routinely be addressed with the CF care team to destigmatize and encourage individuals to be more forthcoming about their FI status so that they can be connected to the appropriate resources.

Our results highlight important barriers to optimal adherence to CF dietary recommendations that can be exacerbated by FI, including cost of food, increased caloric requirements, single income household, transportation, and cost of housing, utilities, and healthcare. A recent study by our team showed that people with CF living in and near food deserts, a risk factor for FI, had increased odds of with a non-ideal lower BMI or weight-for-length [12]. It has also been noted that FI contributes to greater healthcare costs and financial strain, particularly when looking at patients with chronic disease [17]. Food-insecure individuals with inflammatory bowel disease had higher "financial toxicity", defined as "financial hardship due to medical bills, personal and health-related financial distress, cost-related medication nonadherence, and healthcare affordability [18]".

FI is intricately associated with socioeconomic status (SES), and studies have shown an increased risk of mortality from CF before the age of 18 years in people with lower SES [19–21]. People with CF, particularly those with FI, report high rates of cost-related medication underuse, delayed health care visits, and eating less [22]. Additionally, a recent study linked respiratory health in people with CF to state- and area-level characteristics, particularly an association with area resource deprivation and overall state child health [23]. With these prior studies in mind, our results help us understand further why individuals with CF might have difficulty in adhering to dietary recommendations, which can lead to poor health overall. While SES screening can be challenging, universal FI screening may uncover those struggling to adhere to CF dietary recommendations.

Another important factor that most participants felt contributed significantly to their FI was the COVID-19 pandemic, which has led to unprecedented levels of FI due to increased food prices, disrupted community support networks, increased rates of unemployment and school/daycare closures [24,25]. The pandemic has also led to increased reports of worse mental health and self-care in food-insecure persons with CF [26]. The CFF recently conducted a national survey to assess trends in FI screening and interventions prior to and during the pandemic. FI screening increased during the pandemic and demonstrated an overall increase in FI rates, exceeding the national average in some centers. Some pediatric programs reported that 20% to 43% of their patients experienced FI during the pandemic [27].

We found that embarrassment about FI status often led to decreased disclosure to CF teams. The role of stigma in food inequities is understudied and is manifested at the individual and structural level [28]. A recent study examining the relationship between FI and adults' experiences of neighborhood safety and discrimination during the COVID-19 pandemic showed that food-insecure adults reported higher rates of interpersonal racism and disrespect compared to food-secure adults [29].

Patient-provider communication is key when discussing FI, and provider education is paramount. Physicians are trained to manage acute and chronic illness but are often less adept at screening and addressing social determinants of health (SDH), creating a significant gap in comprehensive care. As SDH, such as FI, contribute to health outcomes, medical training should incorporate SDH to improve provider comfort and competence in addressing FI thoroughly and sensitively. By building appreciation for the underlying social structures that contribute to poor health outcomes, such training can improve providers' cultural competency [30]. Recognizing the need for good patient-centered communication on this topic, the CFF recently published a discussion guide for CF providers related to FI.

This study did have limitations. We interviewed a small sample of participants (although thematic saturation was reached with our sample size of 20), which may limit generalizability. Additionally, one-quarter of our sample was recruited from CV, leading to self-selection bias with increased engagement, as these participants are active in CF advocacy and research. Additionally, there was unintentional lack of heterogeneity in participants' race/ethnicity, as most participants were White, resulting in a limited perspective from minority groups. Given that most of our participants were recruited from one center, there was little geographic diversity that might account for differences in how CF care centers approach FI screening and discussions; however, 25 percent of study participants were recruited through CV, which does include geographically diverse participants and can account for possible differences in how FI is approached by different CF care teams and regions. The positionality of the interviewer, particularly employment (i.e., physician), may have influenced the interview interactions, as participants may have been embarrassed sharing FI experiences; however, participants were aware that the interviewer was a physician prior to enrolling in the study, and many expressed gratitude that a physician was involved in a study exploring FI.

5. Conclusions

Results from this study will inform future larger investigations on the impact of FI on CF health and aide in the design and planning of targeted interventions and advocacy efforts. Findings may also be applicable to other chronic diseases that are impacted by nutrition. By gaining the perspectives of these individuals, we can develop patient-, provider- and systems-based interventions, including educational resources, provider training, and standardized screening practices.

Author Contributions: Conceptualization, M.A.C.-H. and T.M.K.; methodology, M.A.C.-H. and T.M.K.; validation, M.A.C.-H., I.M.Y. and T.M.K.; formal analysis, M.A.C.-H., S.E.A., O.M.S., D.J.W., I.M.Y. and T.M.K.; investigation, M.A.C.-H., S.E.A. and O.M.S.; resources, M.A.C.-H., O.M.S. and T.M.K.; data curation, M.A.C.-H.; software, M.A.C.-H. and O.M.S.; writing—original draft preparation, M.A.C.-H., S.E.A. and T.M.K.; writing—review and editing, M.A.C.-H., S.E.A., O.M.S., D.J.W., I.M.Y. and T.M.K.; visualization, M.A.C.-H. and T.M.K.; supervision, T.M.K.; project administration, M.A.C.-H. and T.M.K.; funding acquisition, M.A.C.-H., D.J.W. and T.M.K. All authors have read and agreed to the published version of the manuscript.

Funding: This research was funded by the Cystic Fibrosis Foundation (grant ID: CORBER20B0 and CORBER21D0).

Institutional Review Board Statement: This study was approved by the Institutional Review Board (IRB) of the University of Pittsburgh (protocol number: 20080145).

Informed Consent Statement: Patient written consent was waived and approved by the Institutional Review Board, but all participants verbally consented to the study. No written consent was obtained due to the study's anonymous low-risk nature.

Data Availability Statement: The data presented in this study are available on request from the corresponding author.

Acknowledgments: We would like to thank the Cystic Fibrosis Foundation for their support in funding this study. We would like to acknowledge the individuals from our CF Care Center and the CFF CV, who participated in our study. We would also like to thank the CF dietitians at our Care Center, who aided with recruitment. There was no conflict of interests regarding the publication of this data.

Conflicts of Interest: The authors declare no conflict of interest.

References

1. USDA Economic Research Service Definitions of Food InsSecurity. Available online: https://www.ers.usda.gov/topics/food-nutrition-assistance/food-security-in-the-us/definitions-of-food-security.aspx (accessed on 20 April 2020).
2. Food Security and Nutrition Assistance. Available online: https://www.ers.usda.gov/data-products/ag-and-food-statistics-charting-the-essentials/food-security-and-nutrition-assistance/#:~:text=In%202020%2C%2089.5%20percent%20of,from%2010.5%20percent%20in%202019 (accessed on 10 January 2021).
3. Parekh, N.; Ali, S.H.; O'Connor, J.; Tozan, Y.; Jones, A.M.; Capasso, A.; Foreman, J.; DiClemente, R.J. Food insecurity among households with children during the COVID-19 pandemic: Results from a study among social media users across the United States. *Nutr. J.* **2021**, *20*, 1–11. [CrossRef] [PubMed]
4. USDA Economic Research Service Food Insecurity, Chronic Disease, and Health among Working-Age Adults. Available online: https://www.ers.usda.gov/webdocs/publications/84467/err-235.pdf?v=7518.3 (accessed on 15 June 2021).
5. McDonald, C.M.; Christensen, N.K.; Lingard, C.; Peet, K.A.; Walker, S. Nutrition Knowledge and Confidence Levels of Parents of Children with Cystic Fibrosis. *ICAN Infant Child Adolesc. Nutr.* **2009**, *1*, 325–331. [CrossRef]
6. Yen, E.H.; Quinton, H.; Borowitz, D. Better Nutritional Status in Early Childhood is Associated with Improved Clinical Outcomes and Survival in Patients with Cystic Fibrosis. *J. Pediatrics* **2013**, *162*, 530–535. [CrossRef]
7. Konstan, M.W.; Butler, S.M.; Wohl, M.E.B.; Stoddard, M.; Matousek, R.; Wagener, J.S.; Johnson, C.A.; Morgan, W.J. Investigators and Co-ordinators of the Epidemiologic Study of Cystic Fibrosis. Growth and nutritional indexes in early life predict pulmonary function in cystic fibrosis. *J. Pediatr.* **2003**, *142*, 624–630. [CrossRef] [PubMed]
8. Stallings, V.A.; Stark, L.J.; Robinson, K.A.; Feranchak, A.P.; Quinton, H. Evidence-Based Practice Recommendations for Nutrition-Related Management of Children and Adults with Cystic Fibrosis and Pancreatic Insufficiency: Results of a Systematic Review. *J. Am. Diet. Assoc.* **2008**, *108*, 832–839. [CrossRef] [PubMed]
9. Borowitz, D.; Baker, R.D.; Stallings, V. Consensus Report on Nutrition for Pediatric Patients with Cystic Fibrosis. *J. Pediatr. Gastroenterol. Nutr.* **2002**, *35*, 246–259. [CrossRef] [PubMed]
10. Sullivan, J.S.; Mascarenhas, M.R. Nutrition: Prevention and management of nutritional failure in Cystic Fibrosis. *J. Cyst. Fibros.* **2017**, *16*, S87–S93. [CrossRef]
11. Brown, P.S.; Durham, D.; Tivis, R.D.; Stamper, S.; Waldren, C.; Toevs, S.E.; Gordon, B.; Robb, T.A. Evaluation of Food Insecurity in Adults and Children with Cystic Fibrosis: Community Case Study. *Front. Public Health* **2018**, *6*, 348. [CrossRef]
12. Corbera-Hincapie, M.A.; Kurland, K.S.; Hincapie, M.R.; Fabio, A.; Weiner, D.J.; Kim, S.C.; Kazmerski, T.M. Geospatial Analysis of Food Deserts and Their Impact on Health Outcomes in Children with Cystic Fibrosis. *Nutrients* **2021**, *13*, 3996. [CrossRef]
13. Community Voice. Available online: https://www.cff.org/get-involved/community-voice (accessed on 15 January 2022).
14. Braun, V.; Clarke, V. Using thematic analysis in psychology. *Qual. Res. Psychol.* **2006**, *3*, 77–101. [CrossRef]
15. Crabtree, B.F.; Miller, W.L. *Doing Qualitative Research*; Sage Publications: Thousand Oaks, CA, USA, 1999.
16. Sandelowski, M. Whatever happened to qualitative description? *Res. Nurs. Health* **2000**, *23*, 334–340. [CrossRef]
17. Food Insecurity and Chronic Disease. Available online: https://www.ifm.org/news-insights/food-insecurity-chronic-disease/ (accessed on 1 August 2021).
18. Nguyen, N.H.; Khera, R.; Ohno-Machado, L.; Sandborn, W.J.; Singh, S. Prevalence and Effects of Food Insecurity and Social Support on Financial Toxicity in and Healthcare Use by Patients with Inflammatory Bowel Diseases. *Clin. Gastroenterol. Hepatol.* **2020**, *19*, 1377–1386.e5. [CrossRef] [PubMed]
19. McColley, S.A.; Schechter, M.S.; Morgan, W.J.; Pasta, D.J.; Craib, M.L.; Konstan, M.W. Risk factors for mortality before age 18 years in cystic fibrosis. *Pediatr. Pulmonol.* **2017**, *52*, 909–915. [CrossRef] [PubMed]
20. Williams, W.A.; Jain, M.; Laguna, T.A.; McColley, S.A. Preferences for disclosing adverse childhood experiences for children and adults with cystic fibrosis. *Pediatr. Pulmonol.* **2020**, *56*, 921–927. [CrossRef]
21. Schechter, M.S.; Shelton, B.J.; Margolis, P.A.; Fitzsimmons, S.C. The Association of Socioeconomic Status with Outcomes in Cystic Fibrosis Patients in the United States. *Am. J. Respir. Crit. Care Med.* **2001**, *163*, 1331–1337. [CrossRef]
22. Hartline-Grafton, H. FRAC on the Move: 2019 North American Cystic Fibrosis Conference. *FRAC Chat*, 7 November 2019. Available online: https://frac.org/blog/frac-on-the-move-2019-north-american-cystic-fibrosis-conference (accessed on 10 July 2021).
23. Oates, G.; Rutland, S.; Juarez, L.; Ba, A.F.; Schechter, M.S. The association of area deprivation and state child health with respiratory outcomes of pediatric patients with cystic fibrosis in the United States. *Pediatr. Pulmonol.* **2020**, *56*, 883–890. [CrossRef]
24. Wolfson, J.A.; Leung, C.W. Food Insecurity During COVID-19: An Acute Crisis with Long-Term Health Implications. *Am. J. Public Health* **2020**, *110*, 1763–1765. [CrossRef]
25. Linkages Between Food Insecurity, Poverty, and Health During COVID-19. Available online: https://frac.org/wp-content/uploads/Linkages-_2021.pdf (accessed on 2 February 2022).
26. Lim, J.T.; Ly, N.P.; Willen, S.M.; Iwanaga, K.; Gibb, E.R.; Chan, M.; Church, G.D.; Neemuchwala, F.; McGarry, M.E. Food insecurity and mental health during the COVID-19 pandemic in cystic fibrosis households. *Pediatr. Pulmonol.* **2022**, *57*, 1238–1244. [CrossRef]
27. Bailey, J.; Brown, G.; Corbera-Hincapie, M.; Clemm, C.; Dasenbrook, E.; Durham, D.; Oates, G.; Reno, K.; Sapp, S.; Schechter, M.; et al. Food insecurity in the cystic fibrosis care center network during COVID19: Prevalence, screening, and interventions. *J. Cyst. Fibros.* **2021**, *20*, S161. [CrossRef]
28. Earnshaw, V.A.; Karpyn, A. Understanding stigma and food inequity: A conceptual framework to inform research, intervention, and policy. *Transl. Behav. Med.* **2020**, *10*, 1350–1357. [CrossRef]

29. Larson, N.; Slaughter-Acey, J.; Alexander, T.; Berge, J.; Harnack, L.; Neumark-Sztainer, D. Emerging adults' intersecting experiences of food insecurity, unsafe neighbourhoods and discrimination during the coronavirus disease 2019 (COVID-19) outbreak. *Public Health Nutr.* **2021**, *24*, 519–530. [CrossRef] [PubMed]
30. Klein, M.D.; Kahn, R.S.; Baker, R.C.; Fink, E.E.; Parrish, D.S.; White, D.C. Training in Social Determinants of Health in Primary Care: Does it Change Resident Behavior? *Acad. Pediatr.* **2011**, *11*, 387–393. [CrossRef] [PubMed]

Article

Usefulness of Muscle Ultrasonography in the Nutritional Assessment of Adult Patients with Cystic Fibrosis

Francisco José Sánchez-Torralvo [1,2,3], Nuria Porras [1], Ignacio Ruiz-García [1,2], Cristina Maldonado-Araque [1,2], María García-Olivares [1], María Victoria Girón [4], Montserrat Gonzalo-Marín [1,2], Casilda Olveira [4,*] and Gabriel Olveira [1,2,3,5]

Citation: Sánchez-Torralvo, F.J.; Porras, N.; Ruiz-García, I.; Maldonado-Araque, C.; García-Olivares, M.; Girón, M.V.; Gonzalo-Marín, M.; Olveira, C.; Olveira, G. Usefulness of Muscle Ultrasonography in the Nutritional Assessment of Adult Patients with Cystic Fibrosis. *Nutrients* 2022, 14, 3377. https://doi.org/10.3390/nu14163377

Academic Editor: Roberto Iacone

Received: 27 July 2022
Accepted: 15 August 2022
Published: 17 August 2022

Publisher's Note: MDPI stays neutral with regard to jurisdictional claims in published maps and institutional affiliations.

Copyright: © 2022 by the authors. Licensee MDPI, Basel, Switzerland. This article is an open access article distributed under the terms and conditions of the Creative Commons Attribution (CC BY) license (https:// creativecommons.org/licenses/by/ 4.0/).

[1] Unidad de Gestión Clínica de Endocrinología y Nutrición, Hospital Regional Universitario de Málaga, 29007 Malaga, Spain
[2] Instituto de Investigación Biomédica de Málaga (IBIMA), Plataforma Bionand, 29010 Malaga, Spain
[3] Departamento de Medicina y Dermatología, Facultad de Medicina, University of Malaga, 29010 Malaga, Spain
[4] Unidad de Gestión Clínica de Neumología, Hospital Regional Universitario de Málaga, 29010 Malaga, Spain
[5] Centro de Investigación Biomédica en Red de Diabetes y Enfermedades Metabólicas Asociadas (CIBERDEM), Instituto de Salud Carlos III, 28029 Madrid, Spain
* Correspondence: casi1547@separ.es

Abstract: Background: Muscle ultrasonography of the quadriceps rectus femoris (QRF) is a technique on the rise in the assessment of muscle mass in application of nutritional assessment. The aim of the present study is to assess the usefulness of muscle ultrasonography in patients with cystic fibrosis, comparing the results with other body composition techniques such as anthropometry, bioelectrical impedance analysis (BIA), dual-energy X-ray absorptiometry (DXA), and handgrip strength (HGS). At the same time, we intend to assess the possible association with the nutritional and respiratory status. Methods: This was a prospective observational study in adult patients with cystic fibrosis in a clinically stable situation. Muscle ultrasonography of the QRF was performed, and the results were compared with other measures of body composition: anthropometry, BIA, and DXA. HGS was used to assess muscle function. Respiratory parameters were collected, and nutritional status was assessed using Global Leadership Initiative on Malnutrition (GLIM) criteria. Results: A total of 48 patients were included, with a mean age of 34.1 ± 8.8 years. In total, 24 patients were men, and 24 patients were women. Mean BMI was 22.5 ± 3.8 kg/m^2. Mean muscular area rectus anterior (MARA) was 4.09 ± 1.5 cm^2, and mean muscular circumference rectus was 8.86 ± 1.61 cm. A positive correlation was observed between the MARA and fat-free mass index (FFMI) determined by anthropometry ($r = 0.747$; $p < 0.001$), BIA ($r = 0.780$; $p < 0.001$), and DXA ($r = 0.678$; $p < 0.001$), as well as muscle function (HGS: $r = 0.790$; $p < 0.001$) and respiratory parameters (FEV1; $r = 0.445$, $p = 0.005$; FVC: $r = 0.376$, $p = 0.02$; FEV1/FVC: $r = 0.344$, $p = 0.037$). A total of 25 patients (52.1%) were diagnosed with malnutrition according to GLIM criteria. Differences were observed when comparing the MARA based on the diagnosis of malnutrition (4.75 ± 1.65 cm^2 in normo-nourished vs. 3.37 ± 1.04 in malnourished; $p = 0.014$). Conclusions: In adults with cystic fibrosis, the measurements collected by muscle ultrasound of the QRF correlate adequately with body composition techniques such as anthropometry, BIA, DXA, and handgrip strength. Muscle ultrasound measurements, particularly the MARA, are related to the nutritional status and respiratory function of these patients.

Keywords: ultrasonography; muscle ultrasound; muscle mass; cystic fibrosis; malnutrition; GLIM criteria

1. Introduction

Cystic fibrosis (CF) is a disease caused by the alteration of a single gene located on the long arm of chromosome 7—the CFTR gene (regulator of the transmembrane conductance of cystic fibrosis). It consists of a multisystemic disease in which CFTR protein dysfunction causes alteration of ion transport in the apical membrane of epithelial cells in different organs and tissues. It is the most frequent life-threatening disease, with recessive Mendelian inheritance, in the Caucasian population [1].

Until a few years ago, CF was considered to be associated with malnutrition because it was practically always present at the time of diagnosis, and because the vast majority of patients suffered a deterioration in their nutritional status and died severely malnourished. Currently, the prevalence of malnutrition has decreased notably, although figures close to 25% continue to be reported in both children and adults [2], reaching up to 40% depending on the tool used [3].

Malnutrition affects the respiratory muscles and, thus, lung function, decreases exercise tolerance, and leads to immunological alteration. In both children and adults, it behaves as a predictive risk factor for morbidity and mortality and deterioration in quality of life [4–8]. In relation to lung function, there seems to be a clear interrelation between malnutrition and its deterioration, as well as chronic colonization (especially by *Pseudomonas*), which is accentuated with age. This is why nutritional intervention can, in addition to improving nutritional parameters, slow down the progressive decline in lung function.

Although, until recently, the recommendations focused mainly on achieving adequate weight, height, and body mass index (BMI) [9], the need to also assess fat-free mass and its functionality is currently highlighted, since its decrease is associated with a worse prognosis [4,5,7,8]. In fact, there is an increasing prevalence of overweight and obesity in CF, particularly in the era of modulator therapies [10]. Nevertheless, it is possible to be obese or overweight and have a depletion in fat-free mass. The decrease in fat-free mass has been associated with an increase in systemic inflammation that is observed in these patients [11].

Therefore, in patients with cystic fibrosis, it is recommended to collect body composition measurements [5,12]. To do this, different tools are available. The simplest and cheapest techniques would be the measurement of skinfolds and circumferences (at least the triceps skinfold and brachial circumference) and bioelectrical impedance analysis (BIA). Both techniques have been validated in CF [13,14]. Other useful methods to measure body compartments (less used due to their greater complexity and cost) are dual-energy X-ray absorptiometry (DXA), which could be considered the "gold standard" available to clinicians (also validated in CF), and other more complex ones such as computerized tomography (CT), magnetic resonance (MR), and the doubly labeled water technique [5,15–18].

In our series of adults with cystic fibrosis, we have been carrying out some of these body composition assessment techniques. Using BIA, 16.2% of our patients presented a low fat-free mass, compared to 17.7% using anthropometry and increasing to 37.6% using DXA [19].

In 2018, Global Leadership Initiative on Malnutrition (GLIM) criteria [20] were proposed for the diagnosis of malnutrition. These criteria include as a novelty the evaluation of muscle mass, proposing the use of the aforementioned techniques. However, there are other techniques for assessing muscle mass, both directly and indirectly, which are not presented as the first choice in the consensus. One of the techniques not originally included in the GLIM criteria is muscle ultrasonography of the quadriceps rectum (QRF), although its recommendation is recently beginning to be assessed [21]. This technique has been used since the 1980s to assess muscle status, but its use for diagnosing malnutrition is not as widespread. There are studies suggesting that this technique is reliable and less susceptible to errors due to swelling or inflammation than other techniques used in clinical practice such as BIA [22,23], in addition to having as an advantage over DXA or CT the absence of radiation [22]. In short, it is a simple, bloodless, and accessible technique to assess body composition in patients at risk of malnutrition or malnourished. And its use is on the rise [23,24], although there are presently no cutoff points described for the diagnosis of malnutrition, or its prognostic value in relation to complications has not been described [25]. Muscle ultrasonography has previously been used to assess muscle mass in children with cystic fibrosis [26], but there is currently no study that evaluated its use in the adult population.

Taking into consideration the above, the diagnosis and prevention of malnutrition are essential in patients with CF, paying special attention to muscular body composition, for which a simple, accessible, and reliable method is needed that can be performed as a routine analysis in consultation, such as muscle ultrasound of the QRF [27].

Our hypothesis is that muscle ultrasonography could be an appropriate technique for measuring muscle mass in the nutritional assessment of patients with cystic fibrosis, and that it would have good agreement with other diagnostic tools for estimating muscle mass and strength.

Thus, the aim of the study was to assess the usefulness of muscle ultrasonography in the diagnosis of malnutrition in patients with cystic fibrosis, comparing the results with other body composition techniques such as anthropometry, handgrip strength, BIA, and DXA. At the same time, we intend to assess the possible association with the nutritional and respiratory status during patient follow-up.

2. Materials and Methods

We designed a prospective observational study of routine clinical practice. Adult patients with a diagnosis of cystic fibrosis in a situation of clinical stability assessed at the Nutrition Unit of the UGC of Endocrinology and Nutrition of the Hospital Regional Universitario de Malaga were selected, coinciding with the annual study that is usually carried out.

2.1. Anthropometric and Body Composition Parameters

Weight was assessed through BIA (scale mode, weight function; TANITA MC980MA) and height was obtained using a stadiometer (Holtain limited, Crymych, UK). With these two values, BMI was calculated.

BIA was performed with TANITA MC980MA (TANITA Corporation, Tokyo, Japan), providing information about total body composition (phase angle and fat-free mass).

Dual-energy X-ray absorptiometry (DXA) was performed using a Lunar Prodigy Advance densitometer (General Electric Medical Systems). Fat-free mass was recorded. The software used was EnCore 12.3 (iDXA and Prodigy Advance).

The skinfolds assessed were the triceps, biceps, subscapularis, and supra-iliac using a Holtain constant pressure caliper (Holtain Limited, Crymych, UK). The same investigator (N.P.) performed the measurements in triplicate for each of the skinfolds assessed, and the mean was calculated. Fat mass and fat-free mass (FFM) were estimated according to the formulas of Siri and Durnin [28,29]. Age, sex, weight, and the sum of four skinfolds (triceps, biceps, supra-iliac, and subscapular) were taken into account in the formula.

The fat-free mass index (FFMI) was calculated for anthropometry, BIA, and DXA.

Muscle strength was assessed using a Jamar dynamometer (Asimow Engineering Co., Los Angeles, CA, USA) and was performed in the dominant hand, repeated three times. The mean was calculated.

2.2. Muscle Ultrasonography of the Quadriceps Rectus Femoris (QRF)

Muscle ultrasonography of the QRF was performed on the nondominant lower extremity with the patient in a supine position, using a 10–12 MHz probe and a multifrequency linear matrix color Esaote MyLab Gamma (Esaote, Genova, Italy). The probe was aligned perpendicular to the longitudinal and transverse axis of the nondominant QRF. The evaluation was performed without compression at the level of the lower third from the superior pole of the patella and the anterior superior iliac spine, measuring the X- and Y-axis (transverse muscle thickness), circumference (CMR), cross-sectional area (MARA), and transverse subcutaneous adipose tissue (cm) (SCAT) [30]. The index of the muscle to height (cm^2/m^2) (MARAI) was calculated. The ultrasonography was performed by the same person (G.O.) who was familiar with the technique. Three measurements were performed for each parameter, and the mean was calculated.

2.3. Assessment of Nutritional Status

A nutritional assessment was performed according to GLIM criteria. For achieving a diagnosis of malnutrition, the presence of at least one phenotypic criterion and one etiologic criterion was required [20]. The following phenotypic criteria were evaluated: more than 5% unintentional weight loss in 6 months, a low BMI (BMI below 20 kg/m^2), or a decrease in muscle mass determined by DXA (gold standard). Regarding etiological criteria, reduced intake and reduced assimilation were assessed, especially the existence of pancreatic insufficiency. Lastly, chronic disease-related inflammation was assessed using the Glasgow Prognostic Score [31].

2.4. Assessment of Respiratory Status

The exacerbations recorded during the annual examination were assessed, taking into consideration those happening in the year prior to the evaluation. Such exacerbations were classified into mild/moderate or severe (suggestive symptoms that worsen and require hospitalization and/or intravenous antibiotics on an outpatient basis). Moreover, patients underwent forced spirometry using a JAEGER pneumotachograph (Jaeger Oxycon Pro®, Erich Jaeger, Würzberg, Germany), following the Sociedad Española de Neumología y Cirugía Torácica (SEPAR) guidelines and determining the values of forced vital capacity (FVC), forced expiratory volume in 1 s (FEV1), and the ratio between both (FEV1/FVC). The values were expressed in absolute terms in mL and as percentages according to a reference population [32]. Initial colonization by microorganisms was analyzed, taking into account their appearance in sputum (at least three positive occurrences), regardless of their persistence at the time of the study. Measurement of the mean amount of sputum produced daily (in milliliters) was evaluated following the protocol of Martínez-García et al. [33].

2.5. Statistical Analysis

Quantitative variables were expressed as the mean ± standard deviation. The distribution of quantitative variables was assessed using the Kolmogorov–Smirnov test. Differences between quantitative variables were analyzed using Student's t-test and, for variables not following a normal distribution, using nonparametric tests (Mann–Whitney). The associations of the variables were evaluated by estimating the Pearson or Spearman correlation coefficient, according to normality. For calculations, significance was set at $p < 0.05$ for two tails.

Evaluations of the diagnostic performance of muscle ultrasound variables to detect malnutrition according to GLIM criteria were based on the receiver operating characteristic (ROC) curves and the area under the curve (AUC). We estimated the accuracy of these measurements using AUC by plotting sensitivity versus 1 − specificity. ROC curves were used to determine the optimal cutoff values by finding the point maximizing the product of sensitivity and specificity. Data analysis was performed using the JAMOVI software (version 2.2.2, Jamovi project, 2020).

2.6. Ethics

All subjects gave their informed consent for inclusion before they participated in the study. The study was conducted in accordance with the Declaration of Helsinki, and the protocol was approved by the Research Ethics Committee of Malaga on 30 March 2021 (reference number #30032021).

3. Results

A total of 48 patients were included. They had a mean age of 34.1 ± 8.8 years. Of the sample, a total of 24 were men (50%) and 24 were women (50%). Half of the individuals were heterozygous for ΔF508, and 77% had pancreatic insufficiency. Furthermore, 18 patients (37.5%) presented disease-related inflammation (Glasgow prognostic score > 0).

According to the BMI, only 8.3% of women and 12.5% of men had malnutrition (BMI less than 18.5 kg/m^2). In the application of the GLIM criteria for the diagnosis of malnutrition, we found that seven patients (14.5%) had lost more than 5% of their weight in the previous 6 months, and 13 patients (27%) had a BMI below 20 kg/m^2.

Using DXA as a determinant of muscle mass, 20 patients (41.7%) had an FFMI below the cutoff points (61.1% of women and 23.5% of men). With these data, 25 patients (52.1%) were diagnosed with malnutrition according to GLIM criteria (64% of women and 36% of men).

Table 1 shows the general characteristics and respiratory status of the sample, as well as its adjustment for nutritional status.

Table 1. General characteristics and respiratory status, adjusted by nutritional status.

		Overall	Normo-Nourished	Malnourished	p-Value
		n = 48	n = 23	n = 25	
Age (years)	m ± SD	34.1 ± 8.8	35 ± 8.8	32.9 ± 8.9	0.43
Gender	n (%)				
Men		24 (50)	15 (62.2)	9 (36)	
Women		24 (50)	8 (34.8)	16 (64)	0.02
Mutation	n (%)				
Homozygous for ΔF508		11 (22.9)	6 (26.1)	5 (20)	
Heterozygous for ΔF508		24 (50)	9 (39.1)	15 (60)	
Negative for ΔF508		13 (27.1)	8 (34.8)	5 (20)	0.26
Cystic fibrosis-related diabetes [34]	n (%)	25 (52.1)	11 (47.8)	14 (56)	0.68
Pancreatic insufficiency	n (%)	37 (77.1)	17 (73.9)	20 (80)	0.76
Bronchorrhea (mL)	m ± SD	22 ± 22.5	20.5 ± 24.1	22.3 ± 20.3	0.81
Total exacerbations	m ± SD	0.63 ± 0.93	0.73 ± 1.03	0.54 ± 0.83	0.51
Severe exacerbations	m ± SD	0.13 ± 0.4	0.18 ± 0.5	0.08 ± 0.28	0.41
FEV 1 (%)	m ± SD	59.4 ± 24.1	65.7 ± 22.6	52.5 ± 24.5	0.09
FVC (%)	m ± SD	67.6 ± 19.8	70.3 ± 18.4	64.5 ± 21.3	0.38
FEV1/FVC (%)	m ± SD	0.69 ± 0.11	0.73 ± 0.09	0.64 ± 0.11	0.01
Colonizations	n (%)	41 (87.2)	21 (91.3)	20 (80)	0.13
Pseudomonas aeruginosa		37 (77.1)	19 (82.6)	18 (72)	0.14
Staphylococcus aureus		38 (79.2)	20 (86.9)	18 (72)	0.10
Haemophilus influenzae		23 (47.9)	12 (52.2)	11 (44)	0.47

m: mean; SD: standard deviation; FEV1: forced expiratory volume in 1 s; FVC: forced vital capacity.

3.1. Body Composition and Other Anthropometric Measurements

Table 2 shows the values of the morphofunctional study performed on the subjects, including anthropometry, BIA, DXA, handgrip strength, and the results of the muscle ultrasound parameters.

3.2. Correlation between Ultrasound and Other Anthropometric and Body Composition Measurements

Table 3 shows the correlation between the muscle ultrasound parameters and other morphofunctional parameters.

Table 2. Body composition parameters.

		Men (n = 24)	Women (n = 24)	p-Value
BMI (kg/m^2)	Mean ± SD	23.2 ± 4.2	21.8 ± 3.4	0.21
Triceps skinfold (mm)	Mean ± SD	9.26 ± 4.4	17.3 ± 5.6	$p < 0.001$
Arm muscle circumference (cm)	Mean ± SD	24.8 ± 2.4	20.5 ± 2.4	$p < 0.001$
Fat-free mass (anthropometry) (kg)	Mean ± SD	54.1 ± 7.1	39.1 ± 5.8	$p < 0.001$
FFMI (anthropometry) (kg/m^2)	Mean ± SD	18.9 ± 1.9	15.6 ± 1.4	$p < 0.001$
Phase angle (°)	Mean ± SD	6.07 ± 0.6	5.02 ± 0.54	$p < 0.001$
Fat-free mass (BIA) (kg)	Mean ± SD	54.4 ± 8.1	38.6 ± 5.7	$p < 0.001$
FFMI (BIA) (kg/m^2)	Mean ± SD	18.9 ± 2.2	15.4 ± 1.5	$p < 0.001$
Fat-free mass (DXA) (kg)	Mean ± SD	51.8 ± 6	37.3 ± 6	$p < 0.001$
FFMI (DXA) (kg/m^2)	Mean ± SD	18.2 ± 1.4	14.6 ± 1.6	$p < 0.001$
Handgrip strength (kg)	Mean ± SD	38.3 ± 7.1	23.8 ± 5.5	$p < 0.001$
Muscular area rectus anterior (MARA) (cm^2)	Mean ± SD	4.97 ± 1.4	3.11 ± 0.89	$p < 0.001$
Muscular area index (MARAI) (cm^2/m^2)	Mean ± SD	1.73 ± 0.48	1.24± 0.34	$p < 0.001$
X-axis (cm)	Mean ± SD	3.96 ± 0.52	3.1 ± 0.41	$p < 0.001$
Y-axis (cm)	Mean ± SD	1.47 ± 0.33	1.16 ± 0.31	$p < 0.001$
Muscular circumference rectus (cm)	Mean ± SD	9.93 ± 1.26	7.73 ± 1.03	$p < 0.001$
Subcutaneous adipose tissue (SCAT) (cm)	Mean ± SD	0.55 ± 0.28	1.11 ± 0.38	$p < 0.001$

BMI: body mass index; SD: standard deviation; FFMI: fat-free mass index; BIA: Bioelectrical impedance analysis; DXA: dual-energy X-ray absorptiometry.

Table 3. Correlations between ultrasound and other morphofunctional parameters.

	MARA (cm^2)	MARAI (cm^2/m^2)	X-Axis (cm)	Y-Axis (cm)	Muscular Circumference (cm)	SCAT (cm)
BMI (kg/m^2)	r = 0.385 p = 0.008	r = 0.339 p = 0.02	r = 0.183 p = 0.217	r = 0.380 p = 0.008	r = 0.280 p = 0.057	r = 0.330 p = 0.023
Fat-free mass (anthropometry) (kg)	r = 0.747 $p < 0.001$	r = 0.574 $p < 0.001$	r = 0.688 $p < 0.001$	r = 0.554 $p < 0.001$	r = 0.736 $p < 0.001$	r = −0.381 p = 0.009
FFMI (anthropometry) (kg/m^2)	r = 0.712 $p < 0.001$	r = 0.642 $p < 0.001$	r = 0.605 $p < 0.001$	r = 0.568 $p < 0.001$	r = 0.659 $p < 0.001$	r = −0.286 p = 0.009
Fat-free mass (BIA) (kg)	r = 0.780 $p < 0.001$	r = 0.612 $p < 0.001$	r = 0.703 $p < 0.001$	r = 0.607 $p < 0.001$	r = 0.763 $p < 0.001$	r = −0.405 p = 0.006
FFMI (BIA) (kg/m^2)	r = 0.774 $p < 0.001$	r = 0.710 $p < 0.001$	r = 0.635 $p < 0.001$	r = 0.660 $p < 0.001$	r = 0.714 $p < 0.001$	r = −0.325 p = 0.029
Phase angle (°)	r = 0.695 $p < 0.001$	r = 0.675 $p < 0.001$	r = 0.578 $p < 0.001$	r = 0.623 $p < 0.001$	r = 0.632 $p < 0.001$	r = −0.589 $p < 0.001$
Fat-free mass (DXA) (kg)	r = 0.670 $p < 0.001$	r = 0.505 p = 0.002	r = 0.480 p = 0.004	r = 0.616 $p < 0.001$	r = 0.677 $p < 0.001$	r = −0.570 p = 0.005
FFMI (DXA) (kg/m^2)	r = 0.678 $p < 0.001$	r = 0.567 $p < 0.001$	r = 0.491 p = 0.003	r = 0.576 $p < 0.001$	r = 0.680 $p < 0.001$	r = −0.610 p = 0.002
Handgrip strength (kg)	r = 0.790 $p < 0.001$	r = 0.687 $p < 0.001$	r = 0.718 $p < 0.001$	r = 0.625 $p < 0.001$	r = 0.779 $p < 0.001$	r = −0.589 $p < 0.001$

MARA: muscular area rectus anterior; MARAI: muscular area index; SCAT: subcutaneous adipose tissue; BMI: body mass index; FFMI: fat-free mass index; BIA: bioelectrical impedance analysis; DXA: dual-energy X-ray absorptiometry.

The correlation between the muscle ultrasonography measurements and BMI, although statistically significant, was weak.

A statistically significant correlation was found between the fat-free mass determined by anthropometry and the muscle ultrasound variables, highlighting a good correlation with the muscular area rectus anterior (MARA) ($r = 0.747$; $p < 0.001$, Figure 1a) and rectus muscular circumference ($r = 0.736$; $p < 0.001$, Figure 1b). Likewise, a negative correlation was observed between subcutaneous adipose tissue and fat-free mass, a finding that also occurred in the remaining morphofunctional parameters (Table 3).

Regarding the BIA measurements, statistically significant correlation was found between the fat-free mass determined by anthropometry and the muscle ultrasound variables, highlighting a good correlation with the muscular area rectus anterior (MARA) ($r = 0.780$; $p < 0.001$, Figure 1c) and rectus muscular circumference ($r = 0.763$; $p < 0.001$, Figure 1d) (Table 3).

Fat-free mass determined by DXA also achieved a statistically significant positive correlation with ultrasound measurements, as did MARA ($r = 0.670$; $p < 0.001$, Figure 1e) and muscular circumference ($r = 0.677$; $p < 0.001$, Figure 1f) (Table 3).

3.3. Correlation between Ultrasound and Muscle Strength

Handgrip strength showed a good correlation with the ultrasound measurements of MARA ($r = 0.790$; $p < 0.001$; Figure 2a) and muscle circumference ($r = 0.779$; $p < 0.001$; Figure 2b).

3.4. Respiratory Variables

Table 4 shows the correlations between the ultrasound measurements and the different respiratory variables.

Table 4. Correlations between the ultrasound measurements and the different respiratory variables.

	Total Exacerbations	Severe Exacerbations	FEV1 (%)	FVC (%)	FEV1/FVC (%)
Muscular area rectus anterior (MARA) (cm^2)	$r = 0.019$ $p = 0.89$	$r = 0.161$ $p = 0.286$	$r = 0.445$ $p = 0.005$	$r = 0.376$ $p = 0.02$	$r = 0.344$ $p = 0.037$
Muscular area index (MARAI) (cm^2/m^2)	$r = 0.052$ $p = 0.730$	$r = 0.148$ $p = 0.326$	$r = 0.328$ $p = 0.044$	$r = 0.299$ $p = 0.068$	$r = 0.234$ $p = 0.164$
X-axis (cm)	$r = 0.019$ $p = 0.899$	$r = 0.092$ $p = 0.545$	$r = 0.279$ $p = 0.090$	$r = 0.270$ $p = 0.101$	$r = 0.107$ $p = 0.530$
Y-axis (cm)	$r = -0.057$ $p = 0.708$	$r = 0.038$ $p = 0.801$	$r = 0.444$ $p = 0.005$	$r = 0.398$ $p = 0.013$	$r = 0.350$ $p = 0.034$
Muscular circumference rectus (cm)	$r = -0.016$ $p = 0.917$	$r = 0.069$ $p = 0.646$	$r = 0.348$ $p = 0.032$	$r = 0.304$ $p = 0.064$	$r = 0.214$ $p = 0.203$
Subcutaneous adipose tissue (SCAT) (cm)	$r = 0.122$ $p = 0.420$	$r = 0.072$ $p = 0.634$	$r = -0.065$ $p = 0.698$	$r = -0.015$ $p = 0.931$	$r = -0.224$ $p = 0.183$

FEV1: forced expiratory volume in 1 s; FVC: forced vital capacity.

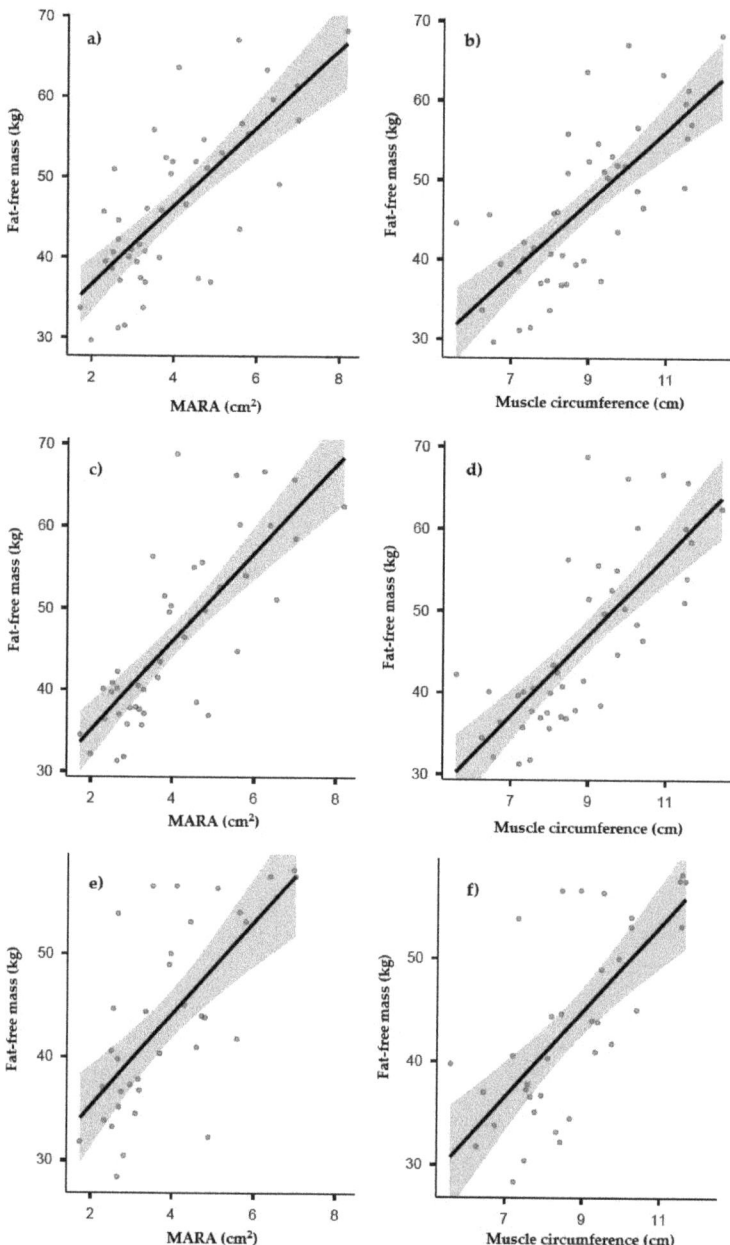

Figure 1. Correlation between MARA and muscle circumference and fat-free mass determined by BIA (**a**,**b**), anthropometry (**c**,**d**), and DXA (**e**,**f**). MARA: muscular area rectus anterior.

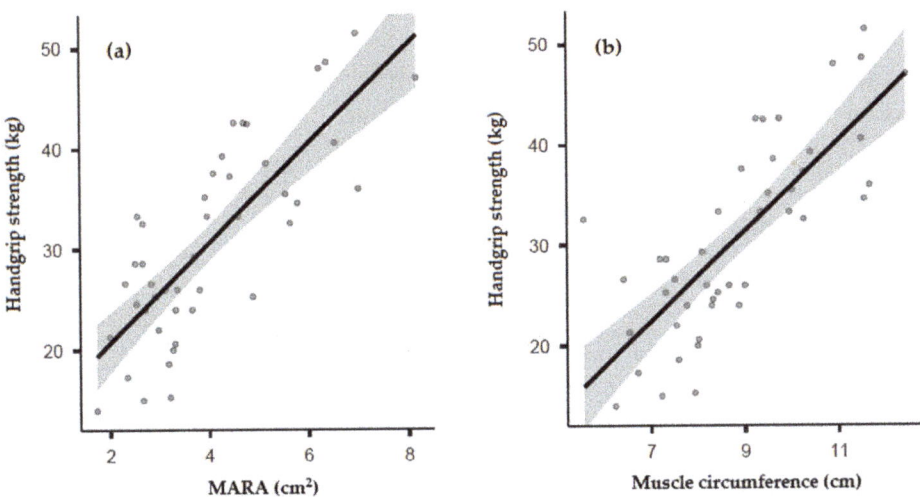

Figure 2. Correlation of handgrip strength with MARA (**a**) and muscle circumference (**b**). MARA: muscular area rectus anterior.

3.5. Nutritional Status

A comparison between muscle ultrasound parameters according to the nutritional status of the patient is shown in Table 5. Significant differences were found between normo-nourished and malnourished patients in the measures of MARA (4.75 ± 1.65 cm^2 vs. 3.37 ± 1.04 cm^2, p = 0.014) and Y-axis (1.45 ± 0.36 cm vs. 1.17 ± 0.28, p = 0.010).

Table 5. Differences between muscle ultrasound parameters according to nutritional status.

		Normo-Nourished (n = 21)	Malnourished (n = 27)	p-Value
Muscular area rectus anterior (MARA) (cm^2)	m ± SD	4.75 ± 1.65	3.37 ± 1.04	p = 0.014
Muscular area index (MARAI) (cm^2/m^2)	m ± SD	1.71 ± 0.51	1.28 ± 0.36	p = 0.016
X-axis (cm)	m ± SD	3.74 ± 0.65	3.37 ± 0.61	p = 0.279
Y-axis (cm)	m ± SD	1.45 ± 0.36	1.17 ± 0.28	p = 0.010
Muscular circumference rectus (cm)	m ± SD	9.48 ± 1.69	8.26 ± 1.34	p = 0.097
Subcutaneous adipose tissue (SCAT) (cm)	m ± SD	0.73 ± 0.37	0.95 ± 0.46	p = 0.872

m: mean; SD: standard deviation.

Using the ROC curve, we determined the MARA cutoff points for predicting malnutrition (Figure 3). ROC curve analysis showed that MARA had a significant discriminative ability to detect malnutrition. The MARA cutoff for malnutrition diagnosis was 2.97 cm^2, with AUC = 0.664 (sensitivity 72.7% and specificity 69.2%), in women and 4.71 cm^2, with AUC = 0.732 (sensitivity 68.8% and specificity 85.7%), in men.

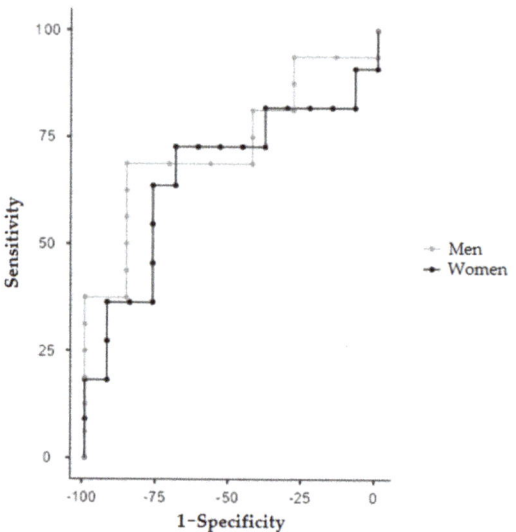

Figure 3. ROC curve analyses for muscular area rectus anterior (MARA) to detect malnutrition.

4. Discussion

To our knowledge, this is the first study to include the assessment of muscle ultrasonography of QRF in adult patients with cystic fibrosis. We found that muscle ultrasonography correlates with handgrip strength and body composition measurement techniques such as BIA, DXA, and anthropometry in patients with cystic fibrosis. In turn, we found an association between this tool and nutritional status and respiratory parameters.

Malnutrition and muscle mass depletion are related to an increase in complications and a worse prognosis in patients with cystic fibrosis. To assess the nutritional status of patients with cystic fibrosis, it is necessary to perform measurements of body composition beyond BMI. We used muscle ultrasonography to assess body composition as it is a simple, accessible technique, and free of radiation. Our study found a good correlation between muscle ultrasound measurements and the remaining morphofunctional assessment techniques, consolidating the results of previous studies [35]. As expected, there were differences in the ultrasound measurements of men and women in all the parameters measured.

Previous studies found that the prevalence of low FFM is high in adult CF patients, despite a normal BMI [3]. In our study, we observed a low correlation of the BMI with the ultrasound parameters. BMI was especially related to the amount of body fat mass. We also found an association of a similar magnitude with the subcutaneous adipose tissue of the leg, which suggests that this isolated measure does not adequately define the amount of total body fat.

However, muscle ultrasound measurements did show a good correlation with fat-free mass, using both BIA and anthropometry. The ultrasound parameters that best correlated with estimated FFM were MARA and muscle circumference. On the other hand, an association was also found between ultrasound and phase angle, a biomarker for malnutrition, hydration, and inflammatory status [36]. Muscle ultrasound could, therefore, be a reliable technique that helps to quantify the active cell mass, which could justify the evolutionary monitoring of the nutritional status in patients with cystic fibrosis together with the other morphofunctional assessment techniques.

Although DXA is considered the "gold standard" in the clinic for estimating fat-free mass, the ultrasound parameters showed slightly lower correlations than those observed with the other techniques. Other previous works have pointed out differences in the estimation of fat-free mass between DXA and BIA and anthropometry in patients with

CF [37], finding some overestimation in FFM using skinfold measurements and BIA [38]. Similar findings have been reported in patients with bronchiectasis [39].

Handgrip strength is a muscle assessment technique that is a good functional and global health status marker. It is a good marker of renutrition [40], and it has been shown that it can be used as a determinant of muscle mass in the application of the GLIM criteria [41]. In our study, we found a strong correlation with the muscle ultrasound measurements, even higher than with the techniques described above. These findings would support the claim that handgrip strength is a good estimator of muscle mass [42], despite not being recommended for muscle assessment according to the last guidelines [21]. In our experience, handgrip strength is associated with the nutritional status and respiratory parameters (lung function and exacerbations) in patients with CF [43], results similar to those found with muscle ultrasound in the present study. This fact postulates muscle ultrasound, as well as HGS, as a good predictor of the nutritional and respiratory status of patients with cystic fibrosis.

In our study, we found an association between muscle ultrasound parameters (especially the MARA and Y-axis) and respiratory function parameters. Even without being able to establish causal relationships, patients with lower muscle mass determined by ultrasound seemed to have worse FEV1, FVC, and FEV1/FVC values. This finding was previously described in patients with low muscle strength [44,45] or poor results in anthropometry, BIA, and DXA [8,15,40,46]. In this case, muscle ultrasonography is postulated as a good predictor of respiratory function, although further studies are needed.

To our knowledge, this is the first study to show the prevalence of malnutrition in adult patients with CF applying the GLIM criteria. Malnourished patients presented worse muscle ultrasound values, again best represented on the MARA and Y-axis. In previous studies, it was reported that malnourished patients with CF presented worse handgrip strength values [43]. Regarding the ability of the MARA to establish cutoff points for malnutrition according to GLIM criteria, it should be noted that the cutoff points offered by the ROC curve were close to the MARA means obtained in our measurements. This could be justified because almost half of the patients were malnourished according to GLIM criteria. This does not coincide with the prevalence previously reported [2], where muscle mass was not included. In our previous series, the prevalence of low fat-free mass was high (almost 50% especially in women) in adult patients with CF, despite having a normal BMI [3]. Therefore, the use of malnutrition diagnostic tools that include measurement of muscle mass is especially relevant in these patients.

This study had some limitations. It was a cross-sectional study, which prevented us from extracting causal conclusions; thus, we could only speculate on different associations. Moreover, it was a single-center study on a moderate number of participants. On the other hand, the muscle ultrasound measurement technique is not yet universally accepted, as nutritional ultrasound is considered to be a developing technique, lacking population reference values and widespread cutoff points [35].

Nonetheless, as strengths of our study, we highlight that all measurements were performed by a single, experienced observer using standardized methodology [30]. Furthermore, several body composition techniques were used, from the most accessible to the gold standard in clinical use, which enhances the reliability of the results and brings it closer to clinical practice.

In conclusion, in adults with cystic fibrosis, the measurements collected by muscle ultrasonography of the quadriceps rectus femoris correlate adequately with body composition techniques such as anthropometry, BIA, DXA, and handgrip strength. Muscle ultrasound measurements, particularly the MARA and Y-axis, are related to the nutritional status and respiratory function of these patients. It is a simple, accessible, reliable, low-cost, and noninvasive technique that could be used to assess the muscle mass of these patients, helping to diagnose malnutrition and monitor the evolution in patients with CF. However, further studies are required to provide information on normal values and widespread cutoff points.

Author Contributions: Conceptualization, F.J.S.-T., C.M.-A., I.R.-G. and G.O.; methodology, N.P. and G.O.; formal analysis, F.J.S.-T.; investigation, N.P., I.R.-G., C.M.-A., M.G.-O. and M.G.-M.; resources, M.V.G. and C.O.; data curation, F.J.S.-T., N.P. and M.G.-O.; writing—original draft preparation, F.J.S.-T.; writing—review and editing, G.O.; supervision, C.O. and G.O.; project administration, G.O.; funding acquisition, G.O. All authors have read and agreed to the published version of the manuscript.

Funding: This study was partially funded by an unrestricted grant from Vegenat Laboratories (Spain) and the Fundación SEEN-Nutricia 2021 Grant for the assessment of body composition by ultrasound.

Institutional Review Board Statement: The study was conducted in accordance with the Declaration of Helsinki and approved by Research Ethics Committee of Malaga on 30 March 2021 (reference number #30032021).

Informed Consent Statement: Informed consent was obtained from all subjects involved in the study.

Data Availability Statement: Not applicable.

Acknowledgments: The authors would like to thank all the individuals who participated in this study for their willingness to collaborate. The authors are grateful for the grant received from SEEN (Sociedad Española de Endocrinología y Nutrición), which allowed the realization of this study.

Conflicts of Interest: The funders had no role in the design of the study; in the collection, analyses, or interpretation of data; in the writing of the manuscript, or in the decision to publish the results.

References

1. Elborn, J.S. Cystic fibrosis. *Lancet* **2016**, *388*, 2519–2531. [CrossRef]
2. Culhane, S.; George, C.; Pearo, B.; Spoede, E. Malnutrition in Cystic Fibrosis: A review. *Nutr. Clin. Pract.* **2013**, *28*, 676–683. [CrossRef]
3. Contreras-Bolívar, V.; Olveira, C.; Porras, N.; García-Olivares, M.; Girón, M.V.; Sánchez-Torralvo, F.J.; Ruiz-García, I.; Alonso-Gallardo, S.P.; Olveira, G.F. Assessment of body composition in cystic fibrosis: Agreement between skinfold measurement and densitometry. *Nutr. Hosp.* **2022**, *39*, 376–382. [CrossRef]
4. Castellani, C.; Duff, A.J.; Bell, S.C.; Heijerman, H.G.; Munck, A.; Ratjen, F.; Sermet-Gaudelus, I.; Southern, K.W.; Barben, J.; Flume, P.A.; et al. ECFS best practice guidelines: The 2018 revision. *J. Cyst. Fibros.* **2018**, *17*, 153–178. [CrossRef] [PubMed]
5. Turck, D.; Braegger, C.P.; Colombo, C.; Declercq, D.; Morton, A.; Pancheva, R.; Robberecht, E.; Stern, M.; Strandvik, B.; Wolfe, S.; et al. ESPEN-ESPGHAN-ECFS guidelines on nutrition care for infants, children, and adults with cystic fibrosis. *Clin. Nutr.* **2016**, *35*, 557–577. [CrossRef]
6. Contreras-Bolívar, V.; Olveira, C.; Blasco Alonso, J.; Olveira, G. Actualización en nutrición en la fibrosis quística. *Nutr. Clínica Med.* **2019**, *13*, 19–44.
7. Gomes, A.; Hutcheon, D.; Ziegler, J. Association Between Fat-Free Mass and Pulmonary Function in Patients with Cystic Fibrosis: A Narrative Review. *Nutr. Clin. Pract.* **2019**, *34*, 715–727. [CrossRef] [PubMed]
8. Calella, P.; Valerio, G.; Thomas, M.; McCabe, H.; Taylor, J.; Brodlie, M.; Siervo, M. Association between body composition and pulmonary function in children and young people with cystic fibrosis. *Nutrition* **2018**, *48*, 73–76. [CrossRef] [PubMed]
9. Kerem, E.; Conway, S.; Elborn, S.; Heijerman, H. Standards of care for patients with cystic fibrosis: A European consensus. *J. Cyst. Fibros.* **2005**, *4*, 7–26. [CrossRef]
10. Thornton, C.S. Body Mass Index and Clinical Outcomes in Persons Living with Cystic Fibrosis—Is Bigger Always Better? *JAMA Netw. Open* **2022**, *5*, e220749. [CrossRef] [PubMed]
11. Olveira, G.; Olveira, C.; Gaspar, I.; Porras, N.; Martín-Núñez, G.; Rubio, E.; Colomo, N.; Rojo-Martínez, G.; Soriguer, F. Fat-Free Mass Depletion and Inflammation in Patients with Bronchiectasis. *J. Acad. Nutr. Diet.* **2012**, *112*, 1999–2006. [CrossRef] [PubMed]
12. Ionescu, A.A.; Nixon, L.S.; Luzio, S.; Lewis-Jenkins, V.; Evans, W.D.; Stone, M.D.; Owens, D.R.; Routledge, P.A.; Shale, D.J. Pulmonary Function, Body Composition, and Protein Catabolism in Adults with Cystic Fibrosis. *Am. J. Respir. Crit. Care Med.* **2002**, *165*, 495–500. [CrossRef] [PubMed]
13. Chomtho, S.; Fewtrell, M.S.; Jaffe, A.; Williams, J.E.; Wells, J.C.K. Evaluation of Arm Anthropometry for Assessing Pediatric Body Composition: Evidence from Healthy and Sick Children. *Pediatr. Res.* **2006**, *59*, 860–865. [CrossRef] [PubMed]
14. De Meer, K.; Gulmans, V.A.M.; Westerterp, K.R.; Houwen, R.H.J.; Berger, R. Skinfold measurements in children with cystic fibrosis: Monitoring fat-free mass and exercise effects. *Eur. J. Pediatr.* **1999**, *158*, 800–806. [CrossRef] [PubMed]
15. Calella, P.; Valerio, G.; Brodlie, M.; Donini, L.M.; Siervo, M. Cystic fibrosis, body composition, and health outcomes: A systematic review. *Nutrition* **2018**, *55–56*, 131–139. [CrossRef] [PubMed]
16. Schols, A.M.W.J.; Wouters, E.F.; Soeters, P.B.; Westerterp, K.R. Body composition by bioelectrical-impedance analysis compared with deuterium dilution and skinfold anthropometry in patients with chronic obstructive pulmonary disease. *Am. J. Clin. Nutr.* **1991**, *53*, 421–424. [CrossRef] [PubMed]
17. Andreoli, A.; Scalzo, G.; Masala, S.A.; Tarantino, U.; Guglielmi, G. Body composition assessment by dual-energy X-ray absorptiometry (DXA). *Radiol. Med.* **2009**, *114*, 286–300. [CrossRef]

18. Calella, P.; Valerio, G.; Brodlie, M.; Taylor, J.; Donini, L.M.; Siervo, M. Tools and Methods Used for the Assessment of Body Composition in Patients with Cystic Fibrosis: A Systematic Review. *Nutr. Clin. Pract.* **2019**, *34*, 701–714. [CrossRef]
19. Contreras-Bolívar, V.; Olveira, C.; Porras, N.; Abuín-Fernández, J.; García-Olivares, M.; Sánchez-Torralvo, F.; Girón, M.; Ruiz-García, I.; Olveira, G. Oral Nutritional Supplements in Adults with Cystic Fibrosis: Effects on Intake, Levels of Fat-Soluble Vitamins, and Bone Remodeling Biomarkers. *Nutrients* **2021**, *13*, 669. [CrossRef]
20. Cederholm, T.; Jensen, G.L.; Correia, M.I.T.D.; Gonzalez, M.C.; Fukushima, R.; Higashiguchi, T.; Baptista, G.; Barazzoni, R.; Blaauw, R.; Coats, A.J.; et al. GLIM criteria for the diagnosis of malnutrition—A consensus report from the global clinical nutrition community. *Clin Nutr.* **2019**, *38*, 1–9. [CrossRef]
21. Barazzoni, R.; Jensen, G.L.; Correia, M.I.T.; Gonzalez, M.C.; Higashiguchi, T.; Shi, H.P.; Bischoff, S.C.; Boirie, Y.; Carrasco, F.; Cruz-Jentoft, A.; et al. Guidance for assessment of the muscle mass phenotypic criterion for the Global Leadership Initiative on Malnutrition (GLIM) diagnosis of malnutrition. *Clin. Nutr.* **2022**, *41*, 1425–1433. [CrossRef] [PubMed]
22. Taskinen, M.; Saarinen-Pihkala, U. Evaluation of muscle protein mass in children with solid tumors by muscle thickness measurement with ultrasonography, as compared with anthropometric methods and visceral protein concentrations. *Eur. J. Clin. Nutr.* **1998**, *52*, 402–406. [CrossRef] [PubMed]
23. Özdemir, U.; Özdemir, M.; Aygencel, G.; Kaya, B.; Türkoğlu, M. The role of maximum compressed thickness of the quadriceps femoris muscle measured by ultrasonography in assessing nutritional risk in critically-ill patients with different volume statuses. *Rev. Assoc. Med. Bras.* **2019**, *65*, 952–958. [CrossRef] [PubMed]
24. Akazawa, N.; Okawa, N.; Hino, T.; Tsuji, R.; Tamura, K.; Moriyama, H. Higher malnutrition risk is related to increased intramuscular adipose tissue of the quadriceps in older inpatients: A cross-sectional study. *Clin. Nutr.* **2019**, *39*, 2586–2592. [CrossRef]
25. Mueller, N.; Murthy, S.; Tainter, C.; Lee, J.; Riddell, K.; Fintelmann, F.J.; Grabitz, S.D.; Timm, F.P.; Levi, B.; Kurth, T.; et al. Can Sarcopenia Quantified by Ultrasound of the Rectus Femoris Muscle Predict Adverse Outcome of Surgical Intensive Care Unit Patients as well as Frailty? A Prospective, Observational Cohort Study. *Ann. Surg.* **2016**, *264*, 1116–1124. [CrossRef]
26. de Souza, R.P.; Donadio, M.V.F.; Heinzmann-Filho, J.P.; Baptista, R.R.; Pinto, L.A.; Epifanio, M.; Marostica, P.J.C. The use of ultrasonography to evaluate muscle thickness and subcutaneous fat in children and adolescents with cystic fibrosis. *Rev. Paul Pediatr.* **2018**, *36*, 457–465. [CrossRef]
27. Hernández-Socorro, C.R.; Saavedra, P.; López-Fernández, J.C.; Ruiz-Santana, S. Assessment of Muscle Wasting in Long-Stay ICU Patients Using a New Ultrasound Protocol. *Nutrients* **2018**, *10*, 1849. [CrossRef]
28. Siri, W.E. Body composition from fluid spaces and density: Analysis of methods. 1961. *Nutrition* **1993**, *9*, 480–491; discussion 480, 492.
29. Durnin, J.V.; Womersley, J. Body fat assessed from total body density and its estimation from skinfold thickness: Measurements on 481 men and women aged from 16 to 72 years. *Br. J. Nutr.* **1974**, *32*, 77–97. [CrossRef] [PubMed]
30. García-Almeida, J.M.; García-García, C.; Vegas-Aguilar, I.M.; Pomar, M.D.B.; Cornejo-Pareja, I.M.; Medina, B.F.; Román, D.A.D.L.; Guerrero, D.B.; Lesmes, I.B.; Madueño, F.J.T. Nutritional ultrasound®: Conceptualisation, technical considerations and standardisation. *Endocrinol. Diabetes Nutr.* **2022**. [CrossRef]
31. Douglas, E.; McMillan, D.C. Towards a simple objective framework for the investigation and treatment of cancer cachexia: The Glasgow Prognostic Score. *Cancer Treat. Rev.* **2014**, *40*, 685–691. [CrossRef] [PubMed]
32. Máiz, L.; Baranda, F.; Coll, R.; Prados, C.; Vendrell, M.; Escribano, A.; Gartner, S.; de Gracia, S.; Martínez, M.; Salcedo, A.; et al. Normativa del diagnóstico y el tratamiento de la afección respiratoria en la fibrosis quística. *Arch. Bronconeumol.* **2001**, *37*, 316–324. [CrossRef]
33. Martínez-García, M.A.; Perpiñá-Tordera, M.; Román-Sánchez, P.; Soler-Cataluña, J.J. Quality-of-Life Determinants in Patients with Clinically Stable Bronchiectasis. *Chest* **2005**, *128*, 739–745. [CrossRef] [PubMed]
34. Moran, A.; Brunzell, C.; Cohen, R.C.; Katz, M.; Marshall, B.C.; Onady, G.; Robinson, K.A.; Sabadosa, K.A.; Stecenko, A.; Slovis, B.; et al. Clinical Care Guidelines for Cystic Fibrosis–Related Diabetes: A position statement of the American Diabetes Association and a clinical practice guideline of the Cystic Fibrosis Foundation, endorsed by the Pediatric Endocrine Society. *Diabetes Care* **2010**, *33*, 2697–2708. [CrossRef]
35. López-Gómez, J.J.; Plaar, K.B.-S.; Izaola-Jauregui, O.; Primo-Martín, D.; Gómez-Hoyos, E.; Torres-Torres, B.; De Luis-Román, D.A. Muscular Ultrasonography in Morphofunctional Assessment of Patients with Oncological Pathology at Risk of Malnutrition. *Nutrients* **2022**, *14*, 1573. [CrossRef]
36. Lukaski, H.C.; Kyle, U.G.; Kondrup, J. Assessment of adult malnutrition and prognosis with bioelectrical impedance analysis: Phase Angle and Impedance Ratio. *Curr. Opin. Clin. Nutr. Metab. Care* **2017**, *20*, 330–339. [CrossRef]
37. Alicandro, G.; Battezzati, A.; Bianchi, M.L.; Loi, S.; Speziali, C.; Bisogno, A.; Colombo, C. Estimating body composition from skinfold thicknesses and bioelectrical impedance analysis in cystic fibrosis patients. *J. Cyst. Fibros.* **2015**, *14*, 784–791. [CrossRef]
38. King, S.; Wilson, J.; Kotsimbos, T.; Bailey, M.; Nyulasi, I. Body composition assessment in adults with cystic fibrosis: Comparison of dual-energy X-ray absorptiometry with skinfolds and bioelectrical impedance analysis. *Nutrition* **2005**, *21*, 1087–1094. [CrossRef] [PubMed]

39. Doña, E.; Olveira, C.; Palenque, F.J.; Porras, N.; Dorado, A.; Martín-Valero, R.; Godoy, A.M.; Espíldora, F.; Contreras, V.; Olveira, G. Body Composition Measurement in Bronchiectasis: Comparison between Bioelectrical Impedance Analysis, Skinfold Thickness Measurement, and Dual-Energy X-ray Absorptiometry before and after Pulmonary Rehabilitation. *J. Acad. Nutr. Diet.* **2018**, *118*, 1464–1473. [CrossRef]
40. Papalexopoulou, N.; Dassios, T.G.; Lunt, A.; Bartlett, F.; Perrin, F.; Bossley, C.J.; Wyatt, H.A.; Greenough, A. Nutritional status and pulmonary outcome in children and young people with cystic fibrosis. *Respir. Med.* **2018**, *142*, 60–65. [CrossRef] [PubMed]
41. Contreras-Bolívar, V.; Sánchez-Torralvo, F.J.; Ruiz-Vico, M.; González-Almendros, I.; Barrios, M.; Padín, S.; Alba, E.; Olveira, G. GLIM Criteria Using Hand Grip Strength Adequately Predict Six-Month Mortality in Cancer Inpatients. *Nutrients* **2019**, *11*, 2043. [CrossRef]
42. Sánchez Torralvo, F.J.; Porras, N.; Abuín Fernández, J.; García Torres, F.; Tapia, M.J.; Lima, F.; Soriguer, F.; Gonzalo, M.; Martínez, G.R.; Olveira, G. Normative reference values for hand grip dynamometry in Spain. Association with lean mass. *Nutr. Hosp.* **2018**, *35*, 98–103. [CrossRef] [PubMed]
43. Contreras-Bolívar, V.; Olveira, C.; Ruiz-García, I.; Porras, N.; García-Olivares, M.; Sánchez-Torralvo, F.J.; Girón, M.V.; Alonso-Gallardo, S.P.; Olveira, G. Handgrip Strength: Associations with Clinical Variables, Body Composition, and Bone Mineral Density in Adults with Cystic Fibrosis. *Nutrients* **2021**, *13*, 4107. [CrossRef] [PubMed]
44. Bouma, S.F.; Iwanicki, C.; McCaffery, H.; Nasr, S.Z. The Association of Grip Strength, Body Mass Index, and Lung Function in Youth with Cystic Fibrosis. *Nutr. Clin. Pract.* **2020**, *35*, 1110–1118. [CrossRef] [PubMed]
45. Rovedder, P.M.E.; Borba, G.C.; Anderle, M.; Flores, J.; Ziegler, B.; Barreto, S.S.M.; Dalcin, P.D.T.R. Peripheral muscle strength is associated with lung function and functional capacity in patients with cystic fibrosis. *Physiother. Res. Int.* **2019**, *24*, e1771. [CrossRef]
46. Hauschild, D.B.; Barbosa, E.; Moreira, E.A.M.; Neto, N.L.; Platt, V.B.; Filho, E.P.; Wazlawik, E.; Moreno, Y.M.F. Nutrition Status Parameters and Hydration Status by Bioelectrical Impedance Vector Analysis Were Associated with Lung Function Impairment in Children and Adolescents with Cystic Fibrosis. *Nutr. Clin. Pract.* **2016**, *31*, 378–386. [CrossRef]

Article

α-Tocopherol Pharmacokinetics in Adults with Cystic Fibrosis: Benefits of Supplemental Vitamin C Administration

Maret G. Traber [1,*], Scott W. Leonard [1], Vihas T. Vasu [2,†], Brian M. Morrissey [3], Huangshu (John) Lei [4], Jeffrey Atkinson [4] and Carroll E. Cross [3,5]

1. Linus Pauling Institute, Oregon State University, Corvallis, OR 97331, USA
2. Department of Zoology, The Maharaja Sayajirao University of Baroda, Vadodara 390002, Gujarat, India
3. Adult Cystic Fibrosis Program, Department of Internal Medicine, University of California, Davis, CA 95817, USA
4. Department of Chemistry, Center for Biotechnology, Brock University, St. Catharines, ON L2S 3A1, Canada
5. Department of Physiology and Membrane Biology, University of California, Davis, CA 95616, USA
* Correspondence: maret.traber@oregonstate.edu
† Diseased.

Abstract: Background: Numerous abnormalities in cystic fibrosis (CF) could influence tocopherol absorption, transportation, storage, metabolism and excretion. We hypothesized that the oxidative distress due to inflammation in CF increases vitamin E utilization, which could be positively influenced by supplemental vitamin C administration. Methods: Immediately before and after receiving vitamin C (500 mg) twice daily for 3.5 weeks, adult CF patients ($n = 6$) with moderately advanced respiratory tract (RT) disease consumed a standardized breakfast with 30% fat and a capsule containing 50 mg each hexadeuterium (d_6)-α- and dideuterium (d_2)-γ-tocopheryl acetates. Blood samples were taken frequently up to 72 h; plasma tocopherol pharmacokinetics were determined. During both trials, d_6-α- and d_2-γ-tocopherols were similarly absorbed and reached similar maximal plasma concentrations ~18–20 h. As predicted, during vitamin C supplementation, the rates of plasma d_6-α-tocopherol decline were significantly slower. Conclusions: The vitamin C-induced decrease in the plasma disappearance rate of α-tocopherol suggests that vitamin C recycled α-tocopherol, thereby augmenting its concentrations. We conclude that some attention should be paid to plasma ascorbic acid concentrations in CF patients, particularly to those individuals with more advanced RT inflammatory disease and including those with severe exacerbations.

Keywords: vitamin E; carboxyethyl hydroxy chromanol (CEHC); stable isotope-labeled vitamin E

1. Introduction

Cystic Fibrosis (CF) is a genetic disorder, affecting approximately 40,000 people in the US and more than 160,000 world-wide [1,2]. Mutations in the Cystic Fibrosis Conductance Regulator (CFTR) gene give rise to decreased functional CFTR in epithelial cells, leading to desiccated and hyperviscous respiratory tract (RT) and gastrointestinal tract secretions [3]. Hallmarks of the disease include compromised RT mucociliary clearance, microbial infection and an exaggerated inflammatory response mainly by a florid RT neutrophilia and activations of neutrophil proteolytic and oxidative processes [4–6]. These latter processes are believed to play a central role in the progressive lung tissue destruction seen in CF, as indicated by strong evidence of active proteolytic [6–8] and oxidative [9–19] processes seen in RT of CF patients.

Nutritional deficiencies, in large part secondary to abnormalities of bile acid homeostasis and exocrine pancreatic insufficiency [20,21], represent important contributors to the CF clinical manifestations. Although current CF managements (replacement pancreatic lipase and lipophilic vitamin supplements) address the resulting severe lipid malabsorption and fat-soluble vitamin deficiencies associated with pancreatic insufficiency, these are only partially effective. There remain uncertainties as to the magnitudes of the absorption and plasma kinetics of lipophilic nutrients. Vitamin E is particularly relevant in CF because α- and γ-tocopherols represent major dietary chain breaking, lipophilic antioxidants. α-Tocopherol deficiency, in particular, is known to have deleterious clinical consequences in CF [22–24].

Many in vitro experimental models have convincingly documented interactions between the important lipophilic antioxidant, vitamin E, and the major hydrophilic antioxidant, vitamin C (ascorbic acid) [25–29]. The primary mechanism for this interaction is that the lower reduction potential of vitamin C is capable of reducing the accessible oxidative free radical of vitamin E (tocopheroxyl radical) in membranes [30], thus regenerating vitamin E [27]. This "recycling" phenomenon becomes of clinical relevance under circumstances of severe inflammatory processes known to generate reactive oxidants. This phenomenon could compromise ascorbic acid concentrations [31,32] and impact the capability to regenerate the tocopheroxyl radical. As ascorbic acid concentrations relate inversely to biomarkers of inflammation in CF [33], interplays of vitamin C and vitamin E could be of potential importance in this disease, particularly during periods of acute, infective exacerbations.

Our group has previously studied tocopherol pharmacokinetics in smokers, another group of subjects under considerable RT oxidative stress from inhalation of free radicals contained in cigarette smoke (CS) and subsequent RT inflammation-related oxidative injury [34–37]. Using deuterium-labeled α-tocopherol biokinetics similar to those used in the present studies, we showed that smokers compared to non-smokers exhibit a significant 25% decrease in the plasma α-tocopherol half-life ($t\frac{1}{2}$) [37]. We further showed in another group of smokers and non-smokers that the smokers $t\frac{1}{2}$ clearance rates for α-tocopherol were 40% faster than non-smokers, that the accelerated vitamin E plasma disappearance rates in smokers were inversely correlated with their ascorbic acid concentrations, and that the administration of a vitamin C supplement could decrease smokers' accelerated plasma disappearance rates so as to be not significantly different that those of non-smokers while also decreasing their markers of lipid peroxidation [35–37].

Considering this background, it seemed appropriate to further quantitate the pharmacokinetics and metabolism of α- and γ-tocopherols in a small, representative group of adults with CF and considerable inflammatory RT disease, using a repeated measures design where participants serve as their own controls. Our focus is on the interplay of vitamin C supplementation on α- and γ-tocopherols plasma pharmacokinetics.

2. Materials and Methods

2.1. Participant Characteristics

Participants with CF ($n = 6$, ages 23–31 years y, 3 females and 3 males) were recruited from the UC Davis Adult CF Clinic. All subjects met diagnostic criteria for CF. No other specific inclusion or exclusion criteria were specified. All participants were deemed to be compliant with their standard outpatient CF regimens and all appeared to be clinically stable. The study protocol was approved by the institutional review boards for the protection of human subjects at UC Davis (#200715896-3) and at Oregon State University (OSU IRB protocol #3988). All participants gave written, informed consent. Studies were conducted prior to patients receiving either CFTR potentiator or corrector therapy, as described below. The clinical trial was carried out in 2008–2009 prior to requirements for listing in Clinical Trials.gov.

2.2. Deuterium-Labeled Vitamin E

α-5,7-$(CD_3)_2$ tocopheryl acetate (d_6-α-tocopheryl acetate) and unlabeled α-tocopheryl acetate (d_0-α-tocopheryl acetate) were kind gifts from Dr. James Clark of Cognis Nutrition and Health. γ-3,4-(D) tocopheryl acetate (d_2-γ-tocopheryl acetate) was prepared from γ-tocopherol labeled with two deuterium atoms, as described [38]. The d_6-α- to d_2-γ-tocopherol molar ratio was determined by liquid chromatography/mass spectrometry

(LC/MS) to be 0.98. The internal standard used for mass spectroscopy, α-5,7,8-(CD$_3$)$_3$ tocopheryl acetate (d$_9$-α-tocopheryl acetate), was provided by Dr. Carolyn Good of The Bell Institute of Health and Nutrition, and was synthesized by Isotec, Inc. (Miamisburg, OH, USA).

2.3. Study Design and Blood Sampling

In brief, the protocol included a preliminary screening visit with an initial plasma inflammatory "profile" and clinical exam. The protocol was explained in detail to the six volunteers, who gave written informed consent at the UC Davis Translational Research Clinical Center at the Mather VA Hospital, Sacramento, CA, USA. A blood sample was obtained, and the participants were instructed to maintain their ordinary and usual lifestyle, CF management regimens, habitual diets, including their pancreatic-lipase, and A, D, E, and K vitamin preparations (no record of specific formulations is available). They were asked to not partake of any additional non-conventional complimentary nutritional supplements beyond their prescribed high-calorie nutritional supplements. General demographic data, pulmonary function, and respiratory tract cultures of the participants were obtained within 90 days of the 3 study visits and routine clinic hospital clinical laboratory blood measurements (Table 1). The clinical laboratory measured vitamin A as retinol and vitamin D as 25-hydroxyvitamin D by standard protocols.

Table 1. CF participant demographics.

#	Age (Years)	Sex	CF Genotype	BMI	Predicted FEV1 [1]	CRP [2] (mg/dL)	WBC	Hb	HbA1C	Vit. A [3] (mg/L)	Vit. D [4] (ng/mL)	Sputum [5]
1	28	M	508/508	20	38	—	5000	14.2	5.1	0.37	20	*Pseudomonas*
2	32	F	508/508	20	34	2.0	8700	9.8	5.3	0.40	25	*Pseudomonas*
3	31	M	508/1717 – G → A	22	66	0.2	8000	12.6	5.4	0.72	24	*Pseudomonas/S. aureus*
4	30	F	508/508	21	46	1.3	8000	12.7	6.5	0.34	29	*Pseudomonas/MRSA*
5	25	F	508/508	15	30	6.7	7700	12.8	6.6	0.20	34	*S. maltophilia*
6	23	M	508/711 + 1 G → T	20	47	0.6	15,600	12.4	6.1	0.30	49	*Pseudomonas/MRSA*

[1] forced expiratory volume in 1 s (FEV1), [2] C-reactive protein (CRP), [3] Vitamin A, [4] Vitamin D, [5] *Staphylococcus aureus (S. aureus)*, methicillin-resistant *S. aureus* (MRSA), *Stenotrophomonas maltophilia (S. maltophilia)*.

Two weeks after the preliminary screening visit, the subjects returned to the Research Clinical Center and were provided with a standardized breakfast (600 Kcal with 30% of the calories from fat), took their standard pancrealipase preparation followed by oral administration of a capsule containing an equal molar mixture of approximately 50 mg each of d$_6$-α- and d$_2$-γ-tocopherol acetates. Blood samples (*n* = 9 per trial) were collected prior to breakfast and at 3, 6, 9, 12, 24, 36, 48 and 72 h after taking the vitamin E capsule. At the 72 h visit to the Clinical Center the subjects were given vitamin C (500 mg tablets) to be taken twice daily for 3 ½ weeks, at which time they returned to the Clinical Center to repeat the 72 h vitamin E pharmacokinetic study.

Blood samples were collected from the antecubital vein into evacuated tubes containing either 0.05 mL 15% (wt:vol) EDTA or sodium heparin (Becton Dickinson, Franklin Lakes, NJ, USA). Plasma was promptly separated by centrifugation at 4 °C for 15 min at 500× *g* and stored at −80 °C until analyzed. After heparinized plasma was separated, an aliquot was acidified (1:1) with 10% PCA (perchloric acid) containing 1 mM diethylenetriaminepentaacetic acid (DTPA). This sample was then centrifuged (5 min, 15,000× *g*, 4 °C) the supernatant removed, frozen in liquid nitrogen, and stored at −80 °C until analysis.

2.4. Laboratory Analyses

Labeled and unlabeled plasma α- and γ-tocopherols were extracted [39] and measured by liquid chromatography/mass spectrometry (LC/MS) using negative atmospheric pressure chemical ionization (-APCI) as previously described [34,40]. Plasma ascorbic acid, following plasma acidification, was measured by HPLC with amperometric detection as previously described [26]. Plasma triglycerides and total cholesterol were determined by standard clinical assays (Sigma, St. Louis, MO, USA). Plasma vitamin E catabolites (CEHCs) were extracted using a modified method [41] and analyzed by LC/MS using negative electrospray ionization (-ESI) as previously described [34]. For use as standards 2,5,7,8-tetramethyl-2-(2′-carboxyethyl)-6-hydroxychroman (α-CEHC) and 2,7,8-trimethyl-2-(β-carboxyethyl)-6-hydroxychroman (γ-CEHC) were obtained (Sigma-Aldrich, St Louis, MO, USA).

Plasma malondialdehyde (MDA) concentrations were measured as described [42]. Quantitation was done using an external standard of 1,1,3,3-tetraethoxypropane (Sigma) prepared using the same method. The MDA-TBA adduct was extracted with butanol and measured by HPLC with fluorometric detection (532 nm excitation and 553 nm excitation). The TBA-MDA adduct was quantified against the MDA standards.

2.5. Mathematical and Statistical Analyeis

The maximum tocopherol and CEHC concentrations (Cmax) and the time of maximum concentration (Tmax) were determined by visual inspection of each participant's data. Plasma exponential disappearance rates of d_6-α-tocopherol, d_2-γ-tocopherol and d_2-γ-CEHC were estimated as previously described [34]. Half-lives of these compounds were calculated as $t\frac{1}{2} = \ln(2)/$exponential disappearance rate constant. One-sided, paired t-tests were used to compare values from the baseline trial to the vitamin C supplemented trial (Excel, Microsoft). Two-factor analysis of variance with repeated measures was used to assess changes over time within subjects (Prism 6 for Macintosh, GraphPad). Data are shown as means ± standard deviation (SD), $n = 6$, unless as otherwise noted.

3. Results

3.1. Baseline CF Subject Characteristics

Demographic profiles of the six CF participants are as depicted in Table 1. Note that four persons were homozygous for the most frequent 508/508 genotype [43]. All had moderate to severe lung function abnormalities based on forced expiratory volume and all had suboptimal nutritional status as measured by BMIs. Although one subject was a diabetic requiring insulin administration, all had normal HbA1c values. All subjects had pancreatic insufficiency, and all were taking conventional doses of pancreatic enzymes and a standard ADEK lipophilic micronutrient supplement. Vitamins A and D micronutrient concentrations, as determined in the clinical laboratories within several months of the study protocol, were generally in the low normal range, whereas inflammatory parameters, as reflected by the latest C-reactive protein (CRP) in juxtaposition to the study itself, showed a significant degree of variation. Table 2 depicts the average values of each of the subjects' total cholesterol, triglycerides and lipid (sum of cholesterol and triglycerides) concentrations during the trials, as well as plasma α-tocopherol concentrations prior to each of the two kinetic studies. Also shown are the calculated ratios of α-tocopherol per total cholesterol and α-tocopherol per total lipids.

Table 2. The participants' lipid concentrations during baseline and vitamin C interventions.

#	Triglycerides (mmol/L)		Cholesterol (mmol/L)		Total Lipids (mmol/L)		α-Tocopherol (μmol/L)		α-T/Cholesterol (mmol/mol)		α-T/Lipids (mmol/mol)	
	Baseline	Vitamin C	Baseline	Vitamin C	Baseline	Vitamin C	Baseline	Vitamin C	Baseline	Vitamin C	Baseline	Vitamin C
1	0.48	0.42	2.61	2.39	3.09	2.82	6.99	5.03	2.66	2.11	2.25	1.80
2	0.41	0.41	3.62	3.30	4.03	3.72	17.91	16.17	4.96	4.89	4.45	4.35
3	0.93	0.87	3.86	3.40	4.79	4.27	20.73	19.08	5.39	5.61	4.34	4.47
4	0.32	0.34	4.34	3.78	4.65	4.13	13.41	15.16	3.09	4.03	2.88	3.69
5	0.28	0.27	2.21	2.15	2.49	2.42	7.18	9.51	3.26	4.42	2.89	3.93
6	0.35	0.28	2.50	2.68	2.86	2.95	8.61	7.72	3.47	2.89	3.05	2.62

Cholesterol and triglyceride concentrations did not differ significantly between the two trials; however, their sum shown as total lipids (3.65 ± 0.97 vs. 3.38 ± 0.76) was slightly lower during the vitamin C (Vit C) intervention ($p = 0.0467$). Plasma unlabeled α-Tocopherol (α-T), α-T/cholesterol and α-T/lipids were not significantly different between the two interventions. Lipid levels were not used to modify vitamin E pharmacokinetics.

3.2. Efficacy of Vitamin C Supplementation

The baseline pharmacokinetic trial (control) was carried out, followed by 3.5 weeks of vitamin C supplementation, then the pharmacokinetic trial was repeated. Vitamin C supplements were effective in increasing plasma ascorbic acid concentrations ($p = 0.0023$, Figure 1a). MDA was measured to assess oxidative stress; these values were not significantly changed between the two trials (Figure 1b), although they were somewhat lover after vitamin C supplementation.

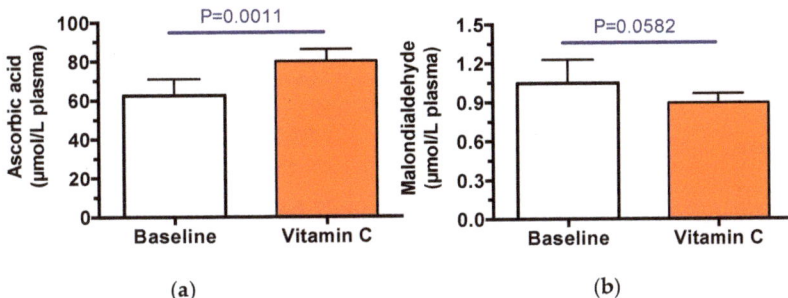

Figure 1. The efficacy of vitamin C supplementation in CF participants. Plasma concentrations of ascorbic acid (**a**) and malondialdehyde (**b**) during pharmacokinetic trials. Shown are the means (±SD, $n = 6$) of the average concentrations measured from plasma samples collected at 0, 3, 6, 9, 12, 24, 48, 72 h from each participant.

3.3. Vitamin E Pharmacokinetics

Plasma α- and γ-tocopherol concentrations measured during the trials were low (Table 3) and on average are in the deficient range [44]. Only one participant (#3) had values of ~20 µmol/L; the others ranged from 5 to 16 µmol/L, suggesting that the vitamin E supplementation was inadequate.

Participants consumed the deuterated tocopherols with a standard breakfast, then blood samples were taken up to 72 h. A representative subject's data illustrates that both deuterated α- and γ-tocopherols were similarly absorbed and appeared in the plasma with similar maximum concentrations (Figure 2). Vitamin C supplementation had no impact on the time of maximum concentrations (Tmax and Cmax) for either of the tocopherols (Table 3). Importantly, however, the exponential disappearance rates of α-tocopherol disappearance from the plasma were significantly slower ($p < 0.05$) during the vitamin C supplementation (Table 3).

The half-lives, calculated from the exponential disappearance rates, show that α-tocopherol has a half-life that is nearly double that of γ-tocopherol (Figure 3). Vitamin C supplementation prolonged the retention of α-tocopherol in the plasma, but not that of γ-tocopherol. Vitamin C supplementation had no effect on the rates of disappearance of either γ-tocopherol or its catabolite, γ-CEHC. It should be noted that not only were γ-tocopherol rates of disappearance rapid, they were also similar to those of γ-CEHC, suggesting the importance of vitamin E catabolism of γ-tocopherol, even in vitamin E deficient persons.

The entire data set were also calculated based on percentage labeled of either the total plasma α- or of γ-tocopherol concentrations, respectively. The purpose was to verify that the low plasma α-tocopherol concentrations observed in some participants did not affect the exponential rates of disappearance. The individual responses shown in Figure 3b,c illustrate the changes in response to vitamin C supplementation. Specifically, the exponential disappearance rates for d_6-α-tocopherol were slower, but those for d_2-γ-tocopherol were unchanged by vitamin C supplementation. These outcomes are similar to the responses to vitamin C supplementation that were observed when the rates were calculated based on plasma d_6-α- and d_2-γ-tocopherol concentrations.

Table 3. The vitamin E concentrations and pharmacokinetic parameters.

Plasma		Intervention Baseline	Vitamin C	Paired t-Test
α-Tocopherol	Average concentration (μmol/L plasma)	12.5 ± 5.9	12.1 ± 5.5	NS
γ-Tocopherol		0.52 ± 0.38	0.60 ± 0.51	NS
α-CEHC *	Average concentration (μmol/L plasma)	0.12 ± 0.09	0.19 ± 0.08	NS
γ-CEHC		0.47 ± 0.53	0.59 ± 0.80	NS
d_6-α-Tocopherol [1]	Cmax (μmol/L)	0.27 ± 0.15	0.32 ± 0.13	NS
	Tmax (h)	19.5 ± 7.0	17.5 ± 7.5	NS
	Disappearance rate (pools per day)	0.65 ± 0.14	0.50 ± 0.10	$p = 0.0263$
	AUC	9.78 ± 5.57	12.04 ± 4.38	NS
d_2-γ-Tocopherol	Cmax (μmol/L)	0.20 ± 0.10	0.30 ± 0.22	NS
	Tmax (h)	15.5 ± 6.7	11.5 ± 1.2	NS
	Disappearance rate (pools per day)	1.54 ± 0.50	1.42 ± 0.55	NS
	AUC	4.59 ± 3.14	5.66 ± 2.77	NS
d_2-γ-CEHC $n = 5$	Cmax (μmol/L)	0.20 ± 0.14	0.22 ± 0.10	NS
	Tmax (h)	16.2 ± 7.2	12.0 ± 0	NS
	Disappearance rate (pools per day)	1.25 ± 0.72	1.61 ± 0.79	NS

The data shown is calculated from $n = 6$ participants. Plasma labeled and unlabeled α- and γ-tocopherols were measured simultaneously in the same sample. Plasma labeled and unlabeled α- and γ-CEHCs were measured simultaneously in a separate analysis from the tocopherols. * Unlabeled γ-CEHC was detectable in all subjects' samples. No d_6-α-CEHC was detected in any plasma sample and several samples had no detectable unlabeled α-CEHC. During the baseline trial, α-CEHC was undetectable at multiple times in subjects 1 (once), 5 (at all times) and 6 (twice); while during the vitamin C trial α-CEHC was undetectable in all plasma samples in subjects 1, 5, and 6. During the baseline trial, subject 1 had no detectable d_2-γ-CEHC, so this subject's data is not included for d_2-γ-CEHC averages shown in the table ($n = 5$). [1] The d_2-γ- and d_6-α-tocopherol disappearance rates were different ($p < 0.001$) from each other in each trial, while the Cmax were not different, indicating similar absorption, but faster disposition of d_2-γ-tocopherol. The fast disposition was also supported by the d_2-γ-CEHC disappearance rates, which were similar to those of d_2-γ-tocopherol.

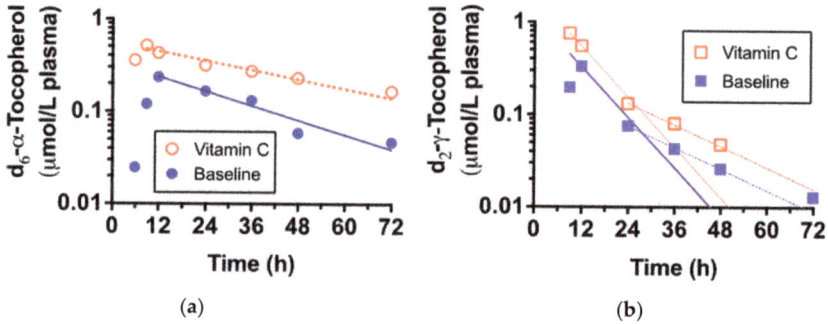

Figure 2. The plasma d_6-α- (**a**) and d_2-γ-tocopherol (**b**) concentrations at baseline and after vitamin C supplementation (representative participant, #3). Plasma-labeled and unlabeled tocopherols were measured by LC/MS from blood samples periodically collected up to 72 h. Filled symbols denote baseline, open symbols denote vitamin C pharmacokinetics trial. Lines indicate post-peak exponential decay curves. The d_2-γ-tocopherol rates of disappearance were so fast that the slopes were no longer linear after 36 h; thus, a second curve was fit to the data. Neither slope was altered by vitamin C status; only the slope from Tmax is reported.

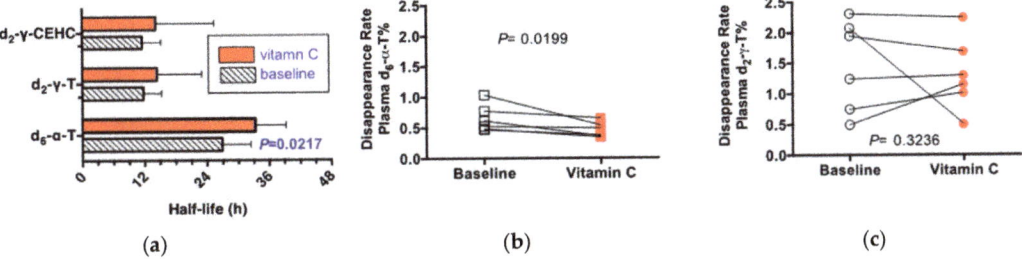

Figure 3. The plasma half-lives in hours for d_6-α- and d_2-γ tocopherols and d_2-γ CEHC. Half-lives indicate the length of time for half of the indicated compound to leave the plasma compartment (**a**). Individual data shows disappearance rates (pools per day) by person (**b**) d_6-α tocopherol; (**c**) d_2-γ tocopherols). The rates shown are based on the percentage labeled to demonstrate that the baseline tocopherol concentrations did not impact the outcomes.

4. Discussion

Vitamin C and E, representing the major dietary hydrophilic and lipophilic antioxidant micronutrients, respectively, have long been known to interact as elegantly demonstrated in chemical in vitro studies [27] and in selected clinical studies [35,37]. The current study demonstrates that, similar to our previous findings in cigarette smokers [35,37], this interrelationship can be demonstrated in moderately severe stable chronic suppurative CF RT disease patients. Importantly, the plasma concentrations observed in CF in response to a 50 mg dose of each d_6-α- and d_2-γ-tocopheryl acetates were much lower than observed in healthy subjects in other studies. For example, Bruno et al. reported that healthy adults with plasma α-tocopherol of approximately 20 µmol/L, who consumed 22 mg d_6-α-tocopheryl acetate with increasing levels of fat, showed a peak plasma d_6-α-tocopherol concentration of 5.6 ± 1.2 µmol/L (Table 2, ref. [45]). The participants with CF reported herein (Table 3) received a 50 mg dose of d_6-α-tocopheryl acetate, but their maximum d_6-α-tocopherol concentration was 10-times lower—approximately 0.5 µmol/L. Leonard et al. also tested a 50 mg dose of each d_6-α-tocopheryl acetate and d_2-γ-tocopherol in healthy adults. They reported that plasma d_2-γ-tocopherol peaked at lower concentrations ($p < 0.001$; 2.2 ± 1.2 µmol/L) than did d_6-α-tocopherol (6.0 ± 2.6 µmol/L) [34]. Again, these concentrations are 10-times higher than those reported herein, suggesting that vitamin E absorption in participants with CF was impaired. Notably, the average plasma α-tocopherol concentrations in the present study (Table 2) were at or near the deficient level, even when corrected for circulating lipid levels [44].

Vitamin E deficiency in humans can be devastating due to its neurologic consequences. Thus, supplemental vitamin E doses are larger than dietary recommendations for normal healthy persons [44]. Vitamin E deficiency symptoms were first described in children with fat malabsorption syndromes, principally abetalipoproteinemia, cystic fibrosis and cholestatic liver disease [46]. Subsequently, humans with severe vitamin E deficiency with no known defect in lipid or lipoprotein metabolism were found to have a defect in the gene for the α-tocopherol transfer protein (α-TTP). This syndrome is called "Ataxia with Vitamin E Deficiency" or AVED. The neurologic abnormalities due to vitamin E deficiency in AVED are described as a progressive sensory neuropathy that can be halted and in some cases reversed by vitamin E supplements. Enormous daily supplemental α-tocopherol amounts (>100 mg/kg body weight) given long-term can overcome the lack of apoB-lipoproteins in abetalipoproteinemia [47], to prevent neurologic disease progression [48] and to prevent oxidative damage [49]. Similarly, supplemental α-tocopherol (1000 mg/day) can prevent progression of neurologic defects in AVED [50]; one patient has been reported to be stable for over 30 years [51]. Persons with fat malabsorption due to impaired biliary secretion generally do not absorb orally administered vitamin E. They are treated with special forms of vitamin E, such as α-tocopheryl polyethylene glycol succinate, which spontaneously form micelles, obviating the need for bile acids [52]. Previous studies have suggested that vitamin E supplements needed to be 400 mg daily in order to raise CF plasma concentrations in adults to approximately 20 μmol/L [53].

Only one of the CF participants studied herein had plasma α-tocopherol concentrations at this normal level. Winklhofer-Roob et al. showed in children (average age 9.2 years) with CF that after an overnight fast, 100 mg α-tocopheryl acetate given with whole milk and an optimized dose of pancreatic enzymes (sufficient to correct fat maldigestion), showed a plasma α-tocopherol increase of nearly 10 μmol/L [54]. Despite the apparent low absorption in the present study, the fractional disappearance rates (0.5 pools/day) reported herein for CF post-vitamin C supplementation (Table 3) are similar to those reported for healthy participants (0.49–0.56 pools per day [45]). Previously, doses as low as 1 to 2 mg have been used successfully for measuring vitamin E kinetics [55,56]. Thus, the absorbed dose size, based on previous studies, does not impact disappearance rates.

As shown in Figure 2 and Table 3, we did indeed demonstrate significant synergism between α-tocopherol and vitamin C. It is particularly important to note that the stable CF patient group, prior to their vitamin C supplementation, had adequate plasma ascorbic acid concentrations (~60 μmol/L), which increased with supplementation to ~80 μmol/L. Moreover, none of the participants experienced clinical exacerbations during the study.

4.1. Nuances for the CF Community

The CF care community has long been aggressively focused on the needs for antioxidant micronutrient supplements [12–14,16,17,57–59], and particularly for the needs for lipophilic antioxidants such as vitamin E [17,18,23,24,60–63]. Under strong inputs from the CF Foundation scientific community [14,16], a revised version of a micronutrient antioxidant cocktail was designed and is now marketed with this increased need in mind [58]. Although studies have shown CF patients [33]; like smokers [64], have low or low normal levels of vitamin C, no augmentations of vitamin C were included in the new formulation. The recent report of significant improvements in the endothelial dysfunction of CF patients, as reflected by increases in measured brachial artery flow mediated dilatation after ingestion of an antioxidant cocktail containing 1000 mg of vitamin C, suggests that present supplemented antioxidants may not be optimized [65].

Although it is known that CF exacerbations are associated with increases in their baseline levels of oxidative stress [66,67] accompanied by further decreases in their levels of antioxidant micronutrients, including vitamin E [33,66,68], the existing CF exacerbation treatment guidelines do not address needs for increased antioxidant supplements during exacerbations [58,69–71].

Of note, the CF community has already recognized the value of standard of care monitored measurements of the lipophilic vitamins A, D, and E and probably would measure vitamin C, if easily determinable in a clinic or hospital setting. The emerging data from both smokers and critically ill patients [72–75] suggest the probable value of including vitamin C measurements and augmentations in CF patients undergoing hospital care for severe exacerbations due to their acute effects on chronic airway septic exacerbations. These severe CF exacerbations are frequently accompanied by nutritional deficiencies. To date, such measurements are usually not done because of the technological challenges of accurate ascorbic acid determinations in clinical laboratories which have not been widely used to determine ascorbic acid concentrations [75].

A prospective study to determine whether vitamin C supplementation in CF would be beneficial, particularly in those CF patients with advanced and/or exacerbated RT disease, should be considered, much as has been proposed for anti-inflammatory agents [76] and done for the case of doxycycline [77].

4.2. Limitations

Several limitations to the scope of the present study in a small number of CF participants should be mentioned. Most CF patients are now on revolutionary CFTR potentiator and/or corrector therapies and these are likely to exert impacts on vitamin E pharmacokinetic via potential effects on bile and pancreatic secretions, enterocyte functions and modulations of the intensity of RT inflammatory processes [78–85]. No attempts were made in the present study to relate plasma tocopherol kinetics to either parameters of RT or systemic inflammation/oxidative stress or to acute or chronic CF exacerbations. Finally, it should be recognized that interrelationships between vitamins E and C, as revealed by kinetics in the plasma compartment, are not likely to reflect the degrees of their interdependence in other extracellular sites. For example, considerable evidence supports the concept that oxidative processes are taking place in the RT of CF patients [10,16]. The goal of antioxidant micronutrient therapy in severe CF RT disease would seem more likely to restore proper levels in RT itself rather than plasma ascorbic acid concentrations. Caution needs to be exercised in extrapolations of their interdependence from blood to such complex redox milieu as exists in the inflamed CF RT.

5. Conclusions

Our major finding was that vitamin C potentiates the biological availability of vitamin E in the plasma compartment of CF patients with advanced but stable RT inflammatory disease. This outcome supports previous observations made in cigarette smokers who are known to have CS-related RT inflammatory changes. It would seem to be clinically prudent for CF clinicians and nutritionists to recognize the importance of dietary and supplemental vitamin C levels and their possible impact on plasma α-tocopherol kinetics in their patients who are undernourished and/or experiencing severe exacerbations of their CF RT inflammatory disease.

Author Contributions: M.G.T. and C.E.C. equally contributed to the conception and design of the research together with J.A. and B.M.M. Deuterated-γ-tocopherol was synthesized by J.A. and H.L. V.T.V., C.E.C. and B.M.M. were responsible for the acquisition of clinical data and blood samples. S.W.L. and M.G.T. analyzed the samples, collected the lab data and performed the statistical evaluation. C.E.C. and M.G.T. drafted the manuscript. All authors (except V.T.V., diseased) have read and agreed to the published version of the manuscript, agreed to be fully responsible for ensuring the accuracy of the work and approved the final manuscript.

Funding: This research was funded by grants from the UC Davis Center for Health and Nutrition Research Program, the NIH UCD Clinical and Translational Sciences Center (UL1 TR001860), and the Cystic Fibrosis Foundation to CEC. MGT was supported in part by the Linus Pauling Institute and Oregon State University Foundation fund. This research was also funded by Atkinson, NSERC Discovery Grant, 155187. α-5,7-$(CD_3)_2$ tocopheryl acetate (d_6-α-tocopheryl acetate) and unlabeled α-tocopheryl acetate (d_0-α-tocopheryl acetate) were gifts from Dr. James Clark of Cognis Nutrition and Health. The internal standard used for mass spectroscopy, α-5,7,8-$(CD_3)_3$ tocopheryl acetate (d_9-α-tocopheryl acetate), was provided by Dr. Carolyn Good of The Bell Institute of Health and Nutrition.

Institutional Review Board Statement: The study protocol was approved by the institutional review boards for the protection of human subjects at UC Davis (#200715896-3) and at Oregon State University (OSU IRB protocol #3988). The study was under continuous approval prior to subject recruitment, during the trial and sample analysis from 2007 to 2014.

Informed Consent Statement: Informed written consent was obtained from all subjects involved in the study.

Data Availability Statement: Not applicable.

Acknowledgments: We are grateful to the CF patients for their kind participation in these studies. We dedicate the study to the memory of one of the subjects and wife of another of the studied CF patients and to Vihas Vasu, who was intimately involved in collecting the patient plasma samples and interpreting data.

Conflicts of Interest: The authors declare no conflict of interest. The funders had no role in the design of the study; in the collection, analyses, or interpretation of data; in the writing of the manuscript; or in the decision to publish the results.

References

1. Cystic Fibrosis Foundation. 2021 Cystic Fibrosis Foundation Patient Registry Highlights. Available online: https://www.cff.org/medical-professionals/patient-registry (accessed on 16 August 2022).
2. Guo, J.; Garratt, A.; Hill, A. Worldwide rates of diagnosis and effective treatment for cystic fibrosis. *J. Cyst. Fibros.* **2022**, *21*, 456–462. [CrossRef]
3. Office of Science (OS); Office of Genomics and Precision Public Health; Centers for Diesease Control and Prevention. Cystic Fibrosis. Available online: https://www.cdc.gov/genomics/disease/cystic_fibrosis.htm (accessed on 16 August 2022).
4. Doring, G.; Gulbins, E. Cystic fibrosis and innate immunity: How chloride channel mutations provoke lung disease. *Cell. Microbiol.* **2009**, *11*, 208–216. [CrossRef] [PubMed]
5. Stoltz, D.A.; Meyerholz, D.K.; Welsh, M.J. Origins of cystic fibrosis lung disease. *N. Engl. J. Med.* **2015**, *372*, 351–362. [CrossRef] [PubMed]
6. Quinn, R.A.; Adem, S.; Mills, R.H.; Comstock, W.; DeRight Goldasich, L.; Humphrey, G.; Aksenov, A.A.; Melnik, A.V.; da Silva, R.; Ackermann, G.; et al. Neutrophilic proteolysis in the cystic fibrosis lung correlates with a pathogenic microbiome. *Microbiome* **2019**, *7*, 23. [CrossRef] [PubMed]
7. Bruce, M.C.; Poncz, L.; Klinger, J.D.; Stern, R.C.; Tomashefski, J.F., Jr.; Dearborn, D.G. Biochemical and pathologic evidence for proteolytic destruction of lung connective tissue in cystic fibrosis. *Am. Rev. Respir Dis.* **1985**, *132*, 529–535. [CrossRef]
8. Downey, D.G.; Bell, S.C.; Elborn, J.S. Neutrophils in cystic fibrosis. *Thorax* **2009**, *64*, 81–88. [CrossRef]
9. Brown, A.J. Acute effects of smoking cessation on antioxidant status. *Nutr. Biochem.* **1996**, *7*, 29–39. [CrossRef]
10. van der Vliet, A.; Eiserich, J.P.; Marelich, G.P.; Halliwell, B.; Cross, C.E. Oxidative stress in cystic fibrosis: Does it occur and does it matter? *Adv. Pharmacol.* **1997**, *38*, 491–513. [CrossRef]
11. Wood, L.G.; Fitzgerald, D.A.; Gibson, P.G.; Cooper, D.M.; Collins, C.E.; Garg, M.L. Oxidative stress in cystic fibrosis: Dietary and metabolic factors. *J. Am. Coll. Nutr.* **2001**, *20*, 157–165. [CrossRef]
12. Back, E.I.; Frindt, C.; Nohr, D.; Frank, J.; Ziebach, R.; Stern, M.; Ranke, M.; Biesalski, H.K. Antioxidant deficiency in cystic fibrosis: When is the right time to take action? *Am. J. Clin. Nutr.* **2004**, *80*, 374–384. [CrossRef]
13. Cantin, A.M. Potential for antioxidant therapy of cystic fibrosis. *Curr. Opin. Pulm. Med.* **2004**, *10*, 531–536. [CrossRef] [PubMed]
14. Cantin, A.M.; White, T.B.; Cross, C.E.; Forman, H.J.; Sokol, R.J.; Borowitz, D. Antioxidants in cystic fibrosis. Conclusions from the CF antioxidant workshop, Bethesda, Maryland, November 11–12, 2003. *Free Radic. Biol. Med.* **2007**, *42*, 15–31. [CrossRef] [PubMed]
15. Lucidi, V.; Ciabattoni, G.; Bella, S.; Barnes, P.J.; Montuschi, P. Exhaled 8-isoprostane and prostaglandin E(2) in patients with stable and unstable cystic fibrosis. *Free Radic. Biol. Med.* **2008**, *45*, 913–919. [CrossRef]
16. Galli, F.; Battistoni, A.; Gambari, R.; Pompella, A.; Bragonzi, A.; Pilolli, F.; Iuliano, L.; Piroddi, M.; Dechecchi, M.C.; Cabrini, G.; et al. Oxidative stress and antioxidant therapy in cystic fibrosis. *Biochim. Biophys. Acta* **2012**, *1822*, 690–713. [CrossRef] [PubMed]
17. Lezo, A.; Biasi, F.; Massarenti, P.; Calabrese, R.; Poli, G.; Santini, B.; Bignamini, E. Oxidative stress in stable cystic fibrosis patients: Do we need higher antioxidant plasma levels? *J. Cyst. Fibros.* **2013**, *12*, 35–41. [CrossRef]
18. Causer, A.J.; Shute, J.K.; Cummings, M.H.; Shepherd, A.I.; Gruet, M.; Costello, J.T.; Bailey, S.; Lindley, M.; Pearson, C.; Connett, G.; et al. Circulating biomarkers of antioxidant status and oxidative stress in people with cystic fibrosis: A systematic review and meta-analysis. *Redox Biol.* **2020**, *32*, 101436. [CrossRef]
19. Galiniak, S.; Molon, M.; Rachel, M. Links between Disease Severity, Bacterial Infections and Oxidative Stress in Cystic Fibrosis. *Antioxidants* **2022**, *11*, 887. [CrossRef]
20. Singh, V.K.; Schwarzenberg, S.J. Pancreatic insufficiency in Cystic Fibrosis. *J. Cyst. Fibros.* **2017**, *16* (Suppl. S2), S70–S78. [CrossRef]
21. van de Peppel, I.P.; Bodewes, F.; Verkade, H.J.; Jonker, J.W. Bile acid homeostasis in gastrointestinal and metabolic complications of cystic fibrosis. *J. Cyst. Fibros.* **2019**, *18*, 313–320. [CrossRef]
22. Koscik, R.L.; Farrell, P.M.; Kosorok, M.R.; Zaremba, K.M.; Laxova, A.; Lai, H.C.; Douglas, J.A.; Rock, M.J.; Splaingard, M.L. Cognitive function of children with cystic fibrosis: Deleterious effect of early malnutrition. *Pediatrics* **2004**, *113*, 1549–1558. [CrossRef]
23. Huang, S.H.; Schall, J.I.; Zemel, B.S.; Stallings, V.A. Vitamin E status in children with cystic fibrosis and pancreatic insufficiency. *J. Pediatr.* **2006**, *148*, 556–559. [CrossRef] [PubMed]
24. Okebukola, P.O.; Kansra, S.; Barrett, J. Vitamin E supplementation in people with cystic fibrosis. *Cochrane Database Syst. Rev.* **2017**, *3*, CD009422. [CrossRef] [PubMed]
25. Packer, J.E.; Slater, T.F.; Willson, R.L. Direct observation of a free radical interaction between vitamin E and vitamin C. *Nature* **1979**, *278*, 737–738. [CrossRef] [PubMed]
26. Frei, B.; England, L.; Ames, B.N. Ascorbate is an outstanding antioxidant in human blood plasma. *Proc. Natl. Acad. Sci. USA* **1989**, *86*, 6377–6381. [CrossRef]
27. Buettner, G.R. The pecking order of free radicals and antioxidants: Lipid peroxidation, alpha-tocopherol, and ascorbate. *Arch. Biochem. Biophys.* **1993**, *300*, 535–543. [CrossRef]
28. Buettner, C.; Phillips, R.S.; Davis, R.B.; Gardiner, P.; Mittleman, M.A. Use of dietary supplements among United States adults with coronary artery disease and atherosclerotic risks. *Am. J. Cardiol.* **2007**, *99*, 661–666. [CrossRef]

29. Sato, A.; Takino, Y.; Yano, T.; Fukui, K.; Ishigami, A. Determination of tissue-specific interaction between vitamin C and vitamin E in vivo using senescence marker protein-30 knockout mice as a vitamin C synthesis deficiency model. *Br. J. Nutr.* **2022**, *128*, 993–1003. [CrossRef] [PubMed]
30. Atkinson, J.; Marquardt, D.; DiPasquale, M.; Harroun, T. From fat to bilayers: Understanding where and how vitamin E works. *Free Radic. Biol. Med.* **2021**, *176*, 73–79. [CrossRef] [PubMed]
31. Ford, E.S.; Liu, S.; Mannino, D.M.; Giles, W.H.; Smith, S.J. C-reactive protein concentration and concentrations of blood vitamins, carotenoids, and selenium among United States adults. *Eur. J. Clin. Nutr.* **2003**, *57*, 1157–1163. [CrossRef] [PubMed]
32. Wannamethee, S.G.; Lowe, G.D.; Rumley, A.; Bruckdorfer, K.R.; Whincup, P.H. Associations of vitamin C status, fruit and vegetable intakes, and markers of inflammation and hemostasis. *Am. J. Clin. Nutr.* **2006**, *83*, 567–574. [CrossRef]
33. Winklhofer-Roob, B.M.; Ellemunter, H.; Fruhwirth, M.; Schlegel-Haueter, S.E.; Khoschsorur, G.; van't Hof, M.A.; Shmerling, D.H. Plasma vitamin C concentrations in patients with cystic fibrosis: Evidence of associations with lung inflammation. *Am. J. Clin. Nutr.* **1997**, *65*, 1858–1866. [CrossRef] [PubMed]
34. Leonard, S.W.; Paterson, E.; Atkinson, J.K.; Ramakrishnan, R.; Cross, C.E.; Traber, M.G. Studies in humans using deuterium-labeled alpha- and gamma-tocopherols demonstrate faster plasma gamma-tocopherol disappearance and greater gamma-metabolite production. *Free Radic. Biol. Med.* **2005**, *38*, 857–866. [CrossRef] [PubMed]
35. Bruno, R.S.; Leonard, S.W.; Atkinson, J.; Montine, T.J.; Ramakrishnan, R.; Bray, T.M.; Traber, M.G. Faster plasma vitamin E disappearance in smokers is normalized by vitamin C supplementation. *Free Radic. Biol. Med.* **2006**, *40*, 689–697. [CrossRef]
36. Bruno, R.S.; Leonard, S.W.; Li, J.; Bray, T.M.; Traber, M.G. Lower plasma alpha-carboxyethyl-hydroxychroman after deuterium-labeled alpha-tocopherol supplementation suggests decreased vitamin E metabolism in smokers. *Am. J. Clin. Nutr.* **2005**, *81*, 1052–1059. [CrossRef]
37. Bruno, R.S.; Ramakrishnan, R.; Montine, T.J.; Bray, T.M.; Traber, M.G. {alpha}-Tocopherol disappearance is faster in cigarette smokers and is inversely related to their ascorbic acid status. *Am. J. Clin. Nutr.* **2005**, *81*, 95–103. [CrossRef]
38. Lei, H.; Atkinson, J. Hydrogen-deuterium exchange during the reductive deuteration of alpha- and gamma-tocopherol chromenes. *J. Label. Compd. Radiopharm.* **2001**, *44*, 215–223. [CrossRef]
39. Podda, M.; Weber, C.; Traber, M.G.; Packer, L. Simultaneous determination of tissue tocopherols, tocotrienols, ubiquinols, and ubiquinones. *J. Lipid Res.* **1996**, *37*, 893–901. [CrossRef]
40. Vaule, H.; Leonard, S.W.; Traber, M.G. Vitamin E delivery to human skin: Studies using deuterated alpha-tocopherol measured by APCI LC-MS. *Free Radic. Biol. Med.* **2004**, *36*, 456–463. [CrossRef]
41. Lodge, J.K.; Traber, M.G.; Elsner, A.; Brigelius-Flohe, R. A rapid method for the extraction and determination of vitamin E metabolites in human urine. *J. Lipid Res.* **2000**, *41*, 148–154. [CrossRef]
42. Hong, Y.L.; Yeh, S.L.; Chang, C.Y.; Hu, M.L. Total plasma malondialdehyde levels in 16 Taiwanese college students determined by various thiobarbituric acid tests and an improved high-performance liquid chromatography-based method. *Clin. Biochem.* **2000**, *33*, 619–625. [CrossRef]
43. Ko, Y.H.; Pedersen, P.L. Cystic fibrosis: A brief look at some highlights of a decade of research focused on elucidating and correcting the molecular basis of the disease. *J. Bioenerg. Biomembr.* **2001**, *33*, 513–521. [CrossRef] [PubMed]
44. Food and Nutrition Board; Institute of Medicine. *Dietary Reference Intakes for Vitamin C, Vitamin E, Selenium, and Carotenoids*; National Academy Press: Washington, DC, USA, 2000; p. 529.
45. Bruno, R.S.; Leonard, S.W.; Park, S.; Zhao, Y.Y.; Traber, M.G. Human vitamin E requirements assessed with the use of apples fortified with deuterium-labeled alpha-tocopheryl acetate. *Am. J. Clin. Nutr.* **2006**, *83*, 299–304. [CrossRef] [PubMed]
46. Traber, M.G.; Head, B. Vitamin E: How much is enough, too much and why! *Free Radic. Biol. Med.* **2021**, *177*, 212–225. [CrossRef] [PubMed]
47. Zamel, R.; Khan, R.; Pollex, R.L.; Hegele, R.A. Abetalipoproteinemia: Two case reports and literature review. *Orphanet J. Rare Dis.* **2008**, *3*, 19. [CrossRef] [PubMed]
48. Burnett, J.R.; Hooper, A.J.; Hegele, R.A. Abetalipoproteinemia. BTI—GeneReviews(®). In *GeneReviews®[Internet]*; Adam, M.P., Everman, D.B., Mirzaa, G.M., Pagon, R.A., Wallace, S.E., Bean, L.J.H., Gripp, K.W., Amemiya, A., Eds.; University of Washington: Seattle, WA, USA, 2018; [Updated 2022 May 19]. Available online: https://www.ncbi.nlm.nih.gov/books/NBK532447/ (accessed on 16 August 2022).
49. Granot, E.; Kohen, R. Oxidative stress in abetalipoproteinemia patients receiving long-term vitamin E and vitamin A supplementation. *Am. J. Clin. Nutr.* **2004**, *79*, 226–230. [CrossRef]
50. Di Donato, I.; Bianchi, S.; Federico, A. Ataxia with vitamin E deficiency: Update of molecular diagnosis. *Neurol. Sci.* **2010**, *31*, 511–515. [CrossRef]
51. Kohlschutter, A.; Finckh, B.; Nickel, M.; Bley, A.; Hubner, C. First recognized patient with genetic vitamin E deficiency stable after 36 years of controlled supplement therapy. *Neurodegener. Dis.* **2020**, *20*, 35–38. [CrossRef]
52. Traber, M.G.; Thellman, C.A.; Rindler, M.J.; Kayden, H.J. Uptake of intact TPGS (d-a-tocopheryl polyethylene glycol 1000 succinate) a water miscible form of vitamin E by human cells in vitro. *Am. J. Clin. Nutr.* **1988**, *48*, 605–611. [CrossRef]
53. Winklhofer-Roob, B.M.; van't Hof, M.A.; Shmerling, D.H. Long-term oral vitamin E supplementation in cystic fibrosis patients: RRR-alpha-tocopherol compared with all-rac-alpha-tocopheryl acetate preparations. *Am. J. Clin. Nutr.* **1996**, *63*, 722–728. [CrossRef]

54. Winklhofer-Roob, B.M.; Tuchschmid, P.E.; Molinari, L.; Shmerling, D.H. Response to a single oral dose of all-rac-alpha-tocopheryl acetate in patients with cystic fibrosis and in healthy individuals. *Am. J. Clin. Nutr.* **1996**, *63*, 717–721. [CrossRef]
55. Traber, M.G.; Leonard, S.W.; Bobe, G.; Fu, X.; Saltzman, E.; Grusak, M.A.; Booth, S.L. alpha-Tocopherol disappearance rates from plasma depend on lipid concentrations: Studies using deuterium-labeled collard greens in younger and older adults. *Am. J. Clin. Nutr.* **2015**, *101*, 752–759. [CrossRef] [PubMed]
56. Violet, P.C.; Ebenuwa, I.C.; Wang, Y.; Niyyati, M.; Padayatty, S.J.; Head, B.; Wilkins, K.; Chung, S.; Thakur, V.; Ulatowski, L.; et al. Vitamin E sequestration by liver fat in humans. *JCI Insight* **2020**, *5*, e133309. [CrossRef] [PubMed]
57. Papas, K.; Kalbfleisch, J.; Mohon, R. Bioavailability of a novel, water-soluble vitamin E formulation in malabsorbing patients. *Dig. Dis. Sci.* **2007**, *52*, 347–352. [CrossRef] [PubMed]
58. Sagel, S.D.; Khan, U.; Jain, R.; Graff, G.; Daines, C.L.; Dunitz, J.M.; Borowitz, D.; Orenstein, D.M.; Abdulhamid, I.; Noe, J.; et al. Effects of an Antioxidant-enriched Multivitamin in Cystic Fibrosis. A Randomized, Controlled, Multicenter Clinical Trial. *Am. J. Respir. Crit. Care Med.* **2018**, *198*, 639–647. [CrossRef] [PubMed]
59. Ciofu, O.; Smith, S.; Lykkesfeldt, J. A systematic Cochrane Review of antioxidant supplementation lung disease for cystic fibrosis. *Paediatr. Respir. Rev.* **2020**, *33*, 28–29. [CrossRef]
60. Farrell, P.M.; Bieri, J.G.; Fratantoni, J.F.; Wood, R.E.; di Sant'Agnese, P.A. The occurrence and effects of human vitamin E deficiency. A study in patients with cystic fibrosis. *J. Clin. Investig.* **1977**, *60*, 233–241. [CrossRef]
61. Okebukola, P.O.; Kansra, S.; Barrett, J. Vitamin E supplementation in people with cystic fibrosis. *Cochrane Database Syst. Rev.* **2020**, *9*, CD009422. [CrossRef]
62. Nowak, J.K.; Sobkowiak, P.; Drzymała-Czyż, S.; Krzyżanowska-Jankowska, P.; Sapiejka, E.; Skorupa, W.; Pogorzelski, A.; Nowicka, A.; Wojsyk-Banaszak, I.; Kurek, S.; et al. Fat-Soluble Vitamin Supplementation Using Liposomes, Cyclodextrins, or Medium-Chain Triglycerides in Cystic Fibrosis: A Randomized Controlled Trial. *Nutrients* **2021**, *13*, 4554. [CrossRef]
63. Wagener, B.M.; Anjum, N.; Evans, C.; Brandon, A.; Honavar, J.; Creighton, J.; Traber, M.G.; Stuart, R.L.; Stevens, T.; Pittet, J.F. alpha-Tocopherol Attenuates the Severity of Pseudomonas aeruginosa-induced Pneumonia. *Am. J. Respir. Cell Mol. Biol.* **2020**, *63*, 234–243. [CrossRef]
64. Cross, C.E.; Traber, M.; Eiserich, J.; van der Vliet, A. Micronutrient antioxidants and smoking. *Br. Med. Bull.* **1999**, *55*, 691–704. [CrossRef]
65. Tucker, M.A.; Fox, B.M.; Seigler, N.; Rodriguez-Miguelez, P.; Looney, J.; Thomas, J.; McKie, K.T.; Forseen, C.; Davison, G.W.; Harris, R.A. Endothelial Dysfunction in Cystic Fibrosis: Role of Oxidative Stress. *Oxid. Med. Cell. Longev.* **2019**, *2019*, 1629638. [CrossRef] [PubMed]
66. Lagrange-Puget, M.; Durieu, I.; Ecochard, R.; Abbas-Chorfa, F.; Drai, J.; Steghens, J.P.; Pacheco, Y.; Vital-Durand, D.; Bellon, G. Longitudinal study of oxidative status in 312 cystic fibrosis patients in stable state and during bronchial exacerbation. *Pediatr. Pulmonol.* **2004**, *38*, 43–49. [CrossRef] [PubMed]
67. Horsley, A.R.; Davies, J.C.; Gray, R.D.; Macleod, K.A.; Donovan, J.; Aziz, Z.A.; Bell, N.J.; Rainer, M.; Mt-Isa, S.; Voase, N.; et al. Changes in physiological, functional and structural markers of cystic fibrosis lung disease with treatment of a pulmonary exacerbation. *Thorax* **2013**, *68*, 532–539. [CrossRef] [PubMed]
68. Hakim, F.; Kerem, E.; Rivlin, J.; Bentur, L.; Stankiewicz, H.; Bdolach-Abram, T.; Wilschanski, M. Vitamins A and E and pulmonary exacerbations in patients with cystic fibrosis. *J. Pediatr. Gastroenterol. Nutr.* **2007**, *45*, 347–353. [CrossRef]
69. Heltshe, S.L.; Goss, C.H. Optimising treatment of CF pulmonary exacerbation: A tough nut to crack. *Thorax* **2016**, *71*, 101–102. [CrossRef]
70. Rowbotham, N.J.; Palser, S.C.; Smith, S.J.; Smyth, A.R. Infection prevention and control in cystic fibrosis: A systematic review of interventions. *Expert Rev. Respir. Med.* **2019**, *13*, 425–434. [CrossRef]
71. Flume, P.A.; Mogayzel, P.J., Jr.; Robinson, K.A.; Goss, C.H.; Rosenblatt, R.L.; Kuhn, R.J.; Marshall, B.C.; Clinical Practice Guidelines for Pulmonary Therapies Committee. Cystic fibrosis pulmonary guidelines: Treatment of pulmonary exacerbations. *Am. J. Respir. Crit. Care Med.* **2009**, *180*, 802–808. [CrossRef]
72. Teng, J.; Pourmand, A.; Mazer-Amirshahi, M. Vitamin C: The next step in sepsis management? *J. Crit. Care* **2018**, *43*, 230–234. [CrossRef]
73. Nabzdyk, C.S.; Bittner, E.A. Vitamin C in the critically ill—Indications and controversies. *World J. Crit. Care Med.* **2018**, *7*, 52–61. [CrossRef] [PubMed]
74. Patel, J.J.; Ortiz-Reyes, A.; Dhaliwal, R.; Clarke, J.; Hill, A.; Stoppe, C.; Lee, Z.Y.; Heyland, D.K. IV Vitamin C in Critically Ill Patients: A Systematic Review and Meta-Analysis. *Crit. Care Med.* **2022**, *50*, e304–e312. [CrossRef]
75. Rozemeijer, S.; van der Horst, F.A.L.; de Man, A.M.E. Measuring vitamin C in critically ill patients: Clinical importance and practical difficulties-Is it time for a surrogate marker? *Crit. Care* **2021**, *25*, 310. [CrossRef] [PubMed]
76. Torphy, T.J.; Allen, J.; Cantin, A.M.; Konstan, M.W.; Accurso, F.J.; Joseloff, E.; Ratjen, F.A.; Chmiel, J.F.; Antiinflammatory Therapy Working, G. Considerations for the Conduct of Clinical Trials with Antiinflammatory Agents in Cystic Fibrosis. A Cystic Fibrosis Foundation Workshop Report. *Ann. Am. Thorac Soc.* **2015**, *12*, 1398–1406. [CrossRef]
77. Xu, X.; Abdalla, T.; Bratcher, P.E.; Jackson, P.L.; Sabbatini, G.; Wells, J.M.; Lou, X.Y.; Quinn, R.; Blalock, J.E.; Clancy, J.P.; et al. Doxycycline improves clinical outcomes during cystic fibrosis exacerbations. *Eur. Respir. J.* **2017**, *49*, 1601102. [CrossRef]

78. Hisert, K.B.; Heltshe, S.L.; Pope, C.; Jorth, P.; Wu, X.; Edwards, R.M.; Radey, M.; Accurso, F.J.; Wolter, D.J.; Cooke, G.; et al. Restoring Cystic Fibrosis Transmembrane Conductance Regulator Function Reduces Airway Bacteria and Inflammation in People with Cystic Fibrosis and Chronic Lung Infections. *Am. J. Respir Crit Care Med.* **2017**, *195*, 1617–1628. [CrossRef] [PubMed]
79. Barry, P.J.; Flume, P.A. Bronchodilators in cystic fibrosis: A critical analysis. *Expert Rev. Respir Med.* **2017**, *11*, 13–20. [CrossRef]
80. McCague, A.F.; Raraigh, K.S.; Pellicore, M.J.; Davis-Marcisak, E.F.; Evans, T.A.; Han, S.T.; Lu, Z.; Joynt, A.T.; Sharma, N.; Castellani, C.; et al. Correlating Cystic Fibrosis Transmembrane Conductance Regulator Function with Clinical Features to Inform Precision Treatment of Cystic Fibrosis. *Am. J. Respir. Crit. Care Med.* **2019**, *199*, 1116–1126. [CrossRef]
81. Sun, T.; Sun, Z.; Jiang, Y.; Ferguson, A.A.; Pilewski, J.M.; Kolls, J.K.; Chen, W.; Chen, K. Transcriptomic Responses to Ivacaftor and Prediction of Ivacaftor Clinical Responsiveness. *Am. J. Respir. Cell Mol. Biol.* **2019**, *61*, 643–652. [CrossRef]
82. Shanthikumar, S.; Ranganathan, S.; Neeland, M.R. Ivacaftor, not ivacaftor/lumacaftor, associated with lower pulmonary inflammation in preschool cystic fibrosis. *Pediatr. Pulmonol.* **2022**. [CrossRef]
83. Nichols, D.P.; Paynter, A.C.; Heltshe, S.L.; Donaldson, S.H.; Frederick, C.A.; Freedman, S.D.; Gelfond, D.; Hoffman, L.R.; Kelly, A.; Narkewicz, M.R.; et al. Clinical Effectiveness of Elexacaftor/Tezacaftor/Ivacaftor in People with Cystic Fibrosis: A Clinical Trial. *Am. J. Respir. Crit. Care Med.* **2022**, *205*, 529–539. [CrossRef]
84. Veltman, M.; De Sanctis, J.B.; Stolarczyk, M.; Klymiuk, N.; Bahr, A.; Brouwer, R.W.; Oole, E.; Shah, J.; Ozdian, T.; Liao, J.; et al. CFTR Correctors and Antioxidants Partially Normalize Lipid Imbalance but not Abnormal Basal Inflammatory Cytokine Profile in CF Bronchial Epithelial Cells. *Front. Physiol.* **2021**, *12*, 619442. [CrossRef]
85. Wang, Y.; Ma, B.; Li, W.; Li, P. Efficacy and Safety of Triple Combination Cystic Fibrosis Transmembrane Conductance Regulator Modulators in Patients With Cystic Fibrosis: A Meta-Analysis of Randomized Controlled Trials. *Front. Pharmacol.* **2022**, *13*, 863280. [CrossRef] [PubMed]

Article

Vitamin Status in Children with Cystic Fibrosis Transmembrane Conductance Regulator Gene Mutation

Paulina Wysocka-Wojakiewicz [1,*], Halina Woś [2], Tomasz Wielkoszyński [3], Aleksandra Pyziak-Skupień [4] and Urszula Grzybowska-Chlebowczyk [1]

1. Department of Pediatrics, Faculty of Medical Sciences, Medical University of Silesia, 40-752 Katowice, Poland
2. Faculty of Health Sciences, University of Bielsko-Biala, 43-309 Bielsko-Biała, Poland
3. Higher School of Strategic Planning and Laboratory Medicine Centre, 41-303 Dąbrowa Górnicza, Poland
4. Department of Children's Diabetology, Silesian Medical University in Katowice, 40-752 Katowice, Poland
* Correspondence: wysockap@gmail.com

Abstract: Background: The issue of vitamin metabolism in children with cystic fibrosis screen positive, inconclusive diagnosis (CFSPID) is not well known. The aim of this study was to determine the status of vitamins A, D, E, and C in the blood of a group of children with CFSPID. Material and Methods: A total of 89 children were enrolled in the study (Me: 3.6 years, 52.8% boys), as follows: 28 with CFSPID, 31 with CF (cystic fibrosis), and 30 HC (healthy children). Their blood concentrations of vitamins A, D, E, and C, and their dietary intake of these vitamins were analysed in the study groups on the basis of a three-day food diary. Results: The patients with CFSPID had significantly higher serum vitamin D ($p = 0.01$) and E ($p = 0.04$) concentrations, compared to the children with CF. None of the children with CFSPID revealed vitamin A or E deficiencies. Patients with CF had been consuming significantly higher vitamin D and E amounts ($p = 0.01$). The vitamin concentrations did not depend either on the pancreatic/liver function or on anthropometric parameters. In total, 32.14% of patients with CF did not cover the baseline recommended calorie intake, and 53.6% and 36% did not take the recommended vitamin E and vitamin A intake, respectively. Conclusion: Children with CF and CFSPID did not fully cover the dietary recommendations for vitamin supply, but vitamin deficiency was found only in CF.

Keywords: vitamin A, D, E, and C; CFSPID; cystic fibrosis; inconclusive diagnosis

1. Introduction

Cystic fibrosis (CF) is the most common autosomal, recessively inherited monogenic disease in the Caucasian population. It is estimated that the prevalence of the disease in Poland is 1:4394–5000 of live births [1]. So far, more than 2000 mutations of the CFTR gene have been detected, which can be divided into six classes, depending on the molecular abnormality, with some mutations belonging to more than one class [2,3]. In 2008, a classification was proposed, taking into account the clinical significance of the mutations [4,5]. The screening test model for newborn screening (CF NBS) has been changing over the years. Since 2011, the IRT (immunoreactive trypsinogen/trypsin)/DNA/EGA (extended gene analysis) model, which allows the detection of 95% of mutant alleles in the Polish population, has been implemented in Poland. Based on an extended DNA analysis, we can detect not only cystic fibrosis patients but also mutated gene carriers, patients with CFTR-dependent diseases, and children with CFSPID.

The definition of CFSPID (cystic fibrosis screen positive, inconclusive diagnosis) includes children with a positive neonatal screening, in whom cystic fibrosis cannot be unequivocally confirmed due to the following: the absence of symptoms of the disease; the presence of the CFTR gene mutation that is of, as yet, unknown clinical significance; and/or an abnormal sweat test result, but less than 60 mmol/L. Children with CFSPID constitute a challenge for CF specialists. The natural history of CFSPID is still unclear.

There are no standardized protocols or predictors of reclassification from CFSPID to CF or a CFTR-related disorder. The increase in sweat chloride concentration with age may be associated with a risk of reclassification [6]. There is clear evidence that infants with an intermediate sweat chloride value are more likely to convert to a CF diagnosis through a rise in subsequent sweat values or clinical features [7]. The reported rate of conversion or reclassification from a CRMS (cystic fibrosis transmembrane conductance regulator-related metabolic syndrome)/CFSPID designation to a CF diagnosis varies from 6% to 48% [8].

It is commonly known that there is a deficiency of fat-soluble vitamins in the group of children with cystic fibrosis. Fat-soluble vitamin deficiency is present even in 10–35% of children with CF and pancreatic insufficiency [9]. The issue of vitamin metabolism in children with cystic fibrosis screen positive, inconclusive diagnosis (CFSPID) is not well known. To our knowledge, none of the prior studies have comprehensively assessed the vitamin and nutritional status of children with CFSPID. So far, the diet and the implementation of nutritional recommendations in this group of children have not been analysed. Prognostic markers to determine the risk of reclassification from CFSPID to CF are still being researched. The early diagnosis of fat-soluble vitamin deficiency in this group of children would encourage the modification of dietary recommendations and the implementation of treatments earlier to prevent long-term complications. Currently, children with CFSPID are only followed up.

The aim of this study was to determine the status of vitamins A, D, E, and C in the blood of a group of children with CFSPID, in relation to children with CF and to healthy controls, and to analyse the relationship of those vitamins with selected markers of the pancreatic and liver functions of the studied patients.

2. Materials and Methods

A total of 89 children, aged 2 months to 17 years, were included in the analysis (Me (median) = 3.6 years.; the IQR (interquartile range) was 1.06–7.64; and 52.8% were boys). They were treated at the department of paediatrics and at the outpatient clinic for the treatment of cystic fibrosis, of the Upper Silesian Children's Medical Centre, in Katowice, Poland. The patients were divided into the following 3 groups: CFSPID (n = 28), CF (n = 31), and a control group—healthy children (n = 30). The group with CFSPID consisted of children with a positive CF NBS result, with two mutations of the CFTR gene, including at least one of which had unclear phenotypic consequences. Those children who did not meet the diagnostic criteria for CF, who were not chronically ill, and who had acute infections were excluded. The CF group included children with 2 CF-causing mutations in both alleles of the CFTR gene, confirmed by sweat test. Those children were in a good clinical condition, without exacerbations during the previous 3 months and without other concomitant diseases or *Peudomonas aeruginosa* colonisation.

In the patients with CFSPID and CF, several assays were carried out, including faecal elastase (by ELISA), serum pancreatic lipase, ALT, AST, GGT, bile acids, and total cholesterol level. The cholesterol concentration was used as an input to calculate a standardised serum vitamin E concentration, as α-tocopherol/cholesterol ratio, which should, ultimately, exceed 5.4 mg/g in paediatric patients with CF [10]. Biochemical tests were performed by standard methods. In addition, vitamin A, D, E, and C concentrations were determined in all the 3 study groups. The concentrations of vitamins A and E, vitamin C, and 25(OH)D were determined by HPLC, the kinetic-spectrophotometric method [11], and the chemiluminescent method, respectively (see the range of standards in Appendix A). A fasting blood sample was collected in the morning. Faecal elastase was measured at the time of admission to the study.

Children's dietary history was also analysed from the last 3 days before the samples for laboratory tests were collected and vitamin supplementation was assessed during the previous 3 months before inclusion into the study. Portion size and the weight of consumed meals and snacks were taken into account. Both the calorie content of the meals and the vitamin A, D, E, and C content in the diet were calculated. In the CFSPID group, energy requirements were assumed as for healthy peers with moderate physical activity and on the basis of dietary standards for the Polish population. The recommended daily allowance (RDA) for vitamins A, E, and C, and the adequate intake (AI) for vitamin D and infant vitamins were considered [12,13]. In the CF group, the percentage of use of the daily energy requirement, meeting the recommendations for CF patients, was considered, and was taken as 120% of the recommended energy supply vs. healthy peers [14]. No dietary history was obtained from 3 CF patients. A programme from the kcalmar.com platform was used to calculate the nutrient content of the diet.

Anthropometric measurements (body weight and body height) were taken in line with the current measurement techniques. Based on the following data: age, sex, weight, and height, the BMI z-score and z-scores for weight and height of the subjects were calculated. According to WHO, weight deficiency was diagnosed, when the BMI z-score was ≤ -2 SD, SD values ≥ 1 were taken as borderline for overweight, and SD ≥ 2 for obesity. A similar reference range of z-score values was adopted for weight and height [15].

The authors analysed the presence and type of mutations in the CFTR gene in children with CFSPID and CF, taking into account the clinical significance of the mutation and the co-occurrence of the F508del mutation (homo-, heterozygotes, and other/other).

Statistical analysis: Quantitative variables were presented as median and interquartile range (IQR) values, and qualitative variables were presented by means of absolute values and percentages. The normality of distribution was verified using the Shapiro–Wilk test. When comparing the differences in the assessed parameters between the study groups, in case of the normal distribution of numerical data, Student's t-test was used and, in cases when distribution deviated from normal, the analysis was performed using the non-parametric Mann–Whitney test. Comparing the differences in the assessed parameters between more than two study groups, in the case of normal distribution, one-way analysis of variance was used. For significant variables, the Bonferoni test was used, and with a distribution that was different from normal—the analysis was carried out using the non-parametric test. For statistical evaluation of differences in the frequency of the analysed characteristics, the chi2 test (with or without Yates' correction) or Fisher's exact test was used, depending on the size of the groups. For correlation analysis, depending on the distribution and type of variables, the Spearman correlation test or the Pearson correlation test was used. A p value < 0.05 was accepted as the threshold for statistical significance. The study was approved by the Biotics Committee of the Silesian Medical University in Katowice, Resolution No. KNW/0022/KB1/9/I/16, of 06.06.2016.

3. Results

3.1. Characteristics of the Groups

The study groups showed no statistically significant differences in terms of age and gender ($p > 0.05$). See Table 1 for the nutritional status data of the individual subjects. No significant correlation was observed among the concentrations of vitamins A, D, E, and C, or between the gender and the z-score of the weight, height, and BMI ($p > 0.05$) of the participants.

Table 1. Nutritional status of the study groups.

	Obesity n (%)	Overweight n (%)	Normal Body Weight n (%)	Body Underweight n (%)	p
CFSPID	0	6 (21.4)	17 (60.7)	5 (17.9)	
CF	0	4 (12.9)	18 (58.1)	9 (29.0)	0.574
HC	1 (3.3%)	5 (16.7)	20 (66.7)	4 (13.3)	

3.2. Results of Laboratory Tests

In all the children with CFSPID (100%), the faecal elastase (FE-1) concentrations were within their normal limits of >500 µg/g of faeces. In the CF group, 29 (93.54%) children had abnormal elastase levels—<200 µg/g of stools—which were indicative of pancreatic exocrine dysfunctions. In two children with CF, the faecal elastase concentrations were >200 µg/g of faeces. These were girls, aged three and seven years, with CFTR gene mutations, G542X/3272-26A→G and F508del/3272-26A→G, respectively. The other results were within the reference norms or slightly above the age limits. This was not relevant to the study. Neither in the CFSPID nor in the CF groups was there any significant correlation observed between pancreatic and liver function exponents and the blood levels of vitamins A, D, E, and C ($p > 0.05$).

3.3. Vitamins

In most of the children, vitamin A, D, E, and C concentrations remained in the normal range. None of the children with CFSPID showed any significant deficiency of fat-soluble vitamins, and only two children demonstrated suboptimal levels of vitamin D. Regarding vitamin D, the vast majority of the subjects with CFSPID and the healthy children had optimal vitamin D levels (30–50 ng/mL). Most deficient (<20 ng/mL) and suboptimal vitamin D concentrations (20–30 ng/mL) were observed among the children with CF, in 5 (16.1%) and 14 (45.2%), respectively. The differences were statistically significant ($p = 0.001$). No vitamin A excess was noted in any of the children. Vitamin E deficiency in the study group was found in only two (6.5%) children with CF (siblings); however, taking into account the α-tocopherol/cholesterol ratio, the percentage was higher and amounted to seven (22.6%). The vitamin A, D, E, and C status among the children in the study groups is shown in Table 2.

Table 2. Vitamin A, D, E, and C status among children in the study groups.

Vitamins	Vitamin Levels	CFSPID	CF	HC	p
A	Normal range n (%)	28 (100)	28 (90.3)	29 (96.7)	0.125
	Deficient n (%)	0	3 (9.7)	1 (3.3)	
	Excess n (%)	0	0	0	
D	30–50 ng/mL n (%)	22 (78.6)	12 (38.7)	22 (73.3)	0.001
	30–20 ng/mL n (%)	2 (7.1)	14 (45.2)	4 (13.3)	
	<20 ng/mL n (%)	0	5 (16.1)	1 (3.3)	
	>50 ng/mL n (%)	4 (14.3)	0	3 (10.0)	
E	Normal range n (%)	25 (89.3)	27 (87.1)	27 (90.0)	0.330
	Deficient n (%)	0	2 (6.5)	0	
	Excess n (%)	3 (10.7)	2 (6.5)	3 (10.0)	
C	Normal range n (%)	26 (92.9)	25 (80.6)	26 (86.7)	0.675
	Deficient n (%)	1 (3.6)	2 (6.5)	1 (3.3)	
	Excess n (%)	1 (3.6)	4 (12.9)	3 (10.0)	

Table 3 shows the distribution of the blood concentrations of particular vitamins.

Children with cystic fibrosis had significantly lower serum levels of vitamin D and E (although within the normal range), compared to those with CFSPID and the healthy patients, despite supplementation.

Considering the corrected α-tocopherol/cholesterol concentration, the values of <5.4 mg/g were observed in seven (22.6%) CF patients, whereas in all the CFSPID children the ratio was >5.4 mg/g. The vitamin D and vitamin E blood concentration among the CF, CFSPID and HC, the median and interquartile ranges are presented in Figures 1 and 2, respectively.

Table 3. Distribution of vitamin A, D, E, and C blood concentration values among different groups.

Vitamin	CFSPID		CF		Healthy Children		p	Post-Hoc
	Me	IQR	Me	IQR	Me	IQR		
A	436.7	370.0–496.3	409.4	320.1–514.1	408.8	351.4–509.1	0.669	
D	43.3	32.8–48.2	28.3	22.1–33.9	33.3	30.1–38.7	0.001	HC vs. CFSID $p > 0.05$ HC vs. CF $p = 0.004$ CFSPID vs. CF $p = 0.001$
E	12.2	10.4–13.3	7.6	6.3–13.0	12.2	10.4–14.7	0.004	HC vs. CFSID $p > 0.05$ HC vs. CF $p = 0.003$ CFSPID vs. CF $p = 0.006$
α-tocopherol/cholesterol ratio	8.2	7.0–10.0	6.9	5.5–9.12	-	-	0.038	
C	46.8	31.4–58.4	48.2	40.2–60.3	49.6	44.2–55.7	0.604	

Me—Median, IQR—Inter-Quartile Range.

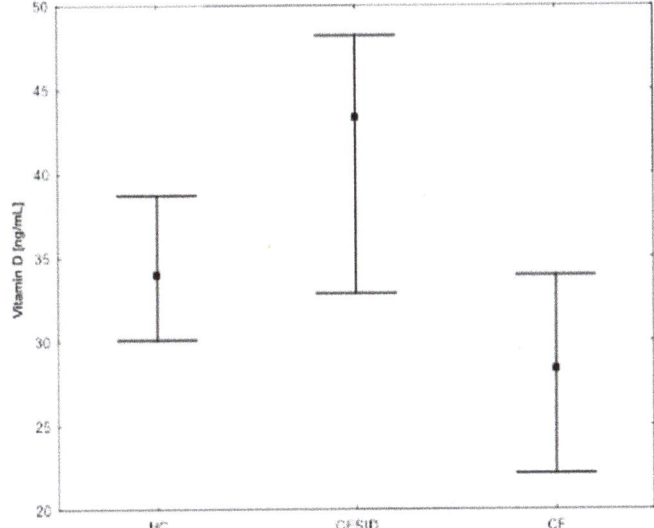

Figure 1. Vitamin D blood concentration among CF, CFSPID, and HC.

3.4. Dependence of Serum 25(OH)D Metabolite Concentrations on the Season of the Year

The majority of patients had their blood samples taken during the months of May to October (23 healthy subjects, 22 children with CFSPID, and 21 patients with CF). The differences in the frequency of the blood sample collection by season were not statistically significant ($p = 0.591$). The lowest 25(OH)D metabolite concentrations were observed in the children with blood samples taken during the winter months ($p = 0.024$).

3.5. Vitamins in Diet

An analysis of a three-day dietary diary of the CFSPID and CF patients showed the distribution of vitamins A, D, E, and C in their diets (see Table 4). The patients with CF consumed significantly more dietary vitamin D and E ($p = 0.001$). The dietary vitamin intake and supplementation had no significant effect on the assayed vitamin concentrations in their blood ($p > 0.05$). The data are presented in Table 5.

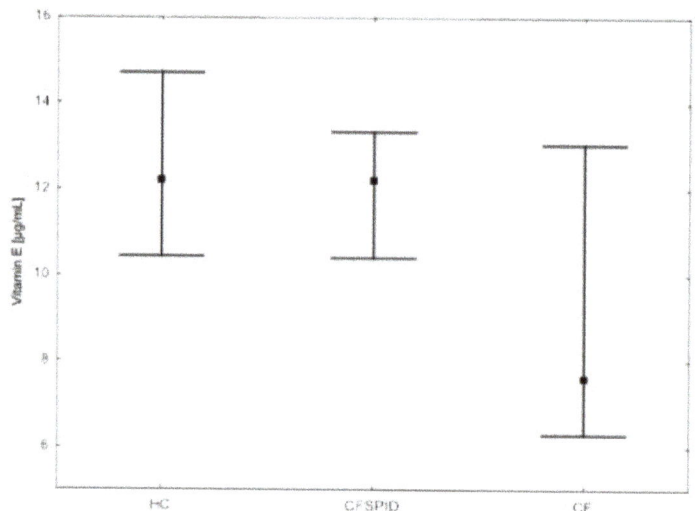

Figure 2. Vitamin E blood concentration among CF, CFSPID, and HC.

Table 4. Distribution of the content of vitamins A, D, E, and C in diets of particular groups.

	CFSPID		CF		p
	Me	IQR	Me	IQR	
μg of retinol equivalent/person/24 h	844.8	564.4–1141.4	1114.3	681.1–1502.0	0.171
mg of α-tocopherol equivalent/person/24 h	6.4	4.4–7.8	13.5	9.5–22.2	0.001
μg of cholecalciferol/person/24 h	1.7	0.8–3.8	5.6	2.7–8.4	0.001
mg of vitamin C/person/24 h	89.8	62.3–129.4	91.1	57.7–136.8	0.882

Me—median; IQR—inter-quartile range.

Table 5. Correlation between vitamin intake (diet and supplementation) and blood vitamin concentration in the groups.

Vitamin	CFSPID		CF	
	R	p	R	p
A	0.1	0.7	0.1	0.3
E	0.1	0.4	0.1	0.6
C	0.1	0.5	0.3	0.1
D	0.3	0.05	0.3	0.1

3.6. Adherence to Dietary Recommendations, Regarding Vitamin Intake

For the children with CFSPID, their dietary vitamin intake standards were established based on the estimated average requirement (EAR), the recommended dietary allowance (RDA), and the adequate intake (AI) for the healthy children in the Polish population [12]. These standards can be found in the Appendix A. For the children with CF, the 2002 guidelines were used for the statistical calculations regarding the children with CF, according to Borowitz et al. [16].

The adherence to the recommendations for the daily vitamin intake in the CF and CFSPID groups is shown in Table 6. No child with CFSPID was significantly deficient in vitamin A, D or E, although not all the children met their daily requirements for those vitamins. The differences in the total vitamin intake between the CFSPID and CF groups were significant only for vitamins D and E ($p = 0.001$).

Table 6. Adherence to recommendations for daily vitamin intake in CFSPID and CF groups.

	Vit. A	p	Vit. D	p	Vit. E	p	Vit. C	p
CFSPID n (%)	24 (85.7%)	0.122	3 (10.7%)	0.001	17 (60.7%)	0.283	25 (89.3%)	0.603
CF n (%)	18 (64.3%)		25 (89.3%)		13 (46.4%)		27 (96.4%)	

3.7. Adherence to Dietary Recommendations in CFSPID and CF Groups

European guidelines recommend that energy intake for people with CF range from 120% to 150% of the energy needs of the healthy population of a similar age, sex, and size [17]. Among the children with CFSPID, 46.5% did not meet their daily caloric requirements for healthy children, with moderate physical activity. The vast majority of children with CF did not take in the recommended number of calories per day during the study period. In total, 89.28% of patients did not meet 120% of the energy requirements of their healthy peers, with moderate physical activity. In total, 32.14% of the children with CF did not meet the baseline recommended caloric intake, yet they did not differ significantly in anthropometric parameters from the healthy children and those with CFSPID. Those patients periodically benefited from industrial diet support, but not during this study. A comparison of the caloric supply between the CFSPID and CF groups is presented in Table 7. The CF children consumed more calories, but the supply did not differ significantly from the caloric content in the diet of the children with CFSPID.

Table 7. The daily consumption of calories in the CFSPID and CF groups based on three-day food records.

	CFSPID		CF		p
	Me	IQR	Me	IQR	
Daily caloric intake	1051	780–1366	1487	825–2129	0.08
%RDA	105	83–112	101	92–107	0.96

3.8. Correlations in the Groups

When comparing the patients in the CFSPID group with the HC, no statistically significant differences were found in the concentration of vitamins and other laboratory parameters, apart from higher 25(OH)D levels in the children with CFSPID ($p = 0.043$). Children with CFSPID had a statistically significantly higher concentration of vitamin E, vitamin D, cholesterol, and lipase, and they had a higher α-tocopherol/cholesterol ratio compared to the CF patients ($p < 0.05$). All the collected clinical and laboratory parameters in the CFSPID group were analysed with the concentration of the determined vitamins. A statistically significant negative correlation was observed between vitamin A and GGTP (-0.641, $p = 0.018$), and between vitamin D and age ($R = -0.464$, $p = 0.015$). The other correlations were statistically insignificant ($p > 0.05$). A similar analysis was performed in the CF group. There was a positive correlation between vitamin A and cholesterol ($R = 0.409$, $p = 0.022$); a positive correlation between vitamin E, lipase ($R = 0.373$, $p = 0.039$), and cholesterol ($R = 0.495$, $p = 0.005$); a negative correlation between vitamin E and ALT ($R = -0.417$, $p = 0.022$); a positive correlation between vitamin D and body height ($R = 0.916$, $p = 0.020$); and the other correlations were not significant statistically ($p > 0.05$).

4. Discussion

In our study, neither the CF nor the CFSPID children were significantly different with respect to the healthy children in terms of their anthropometric measurements. In the group of patients with CF, this was probably due to the early enzyme replacement therapy, and to the adequate diet and vitamin supplementation. On the other hand, in the group of children with CFSPID, this resulted from higher parental awareness, a better care for a healthy lifestyle, and the absence of any diagnosed metabolic disorders. Despite the application of specific dietary recommendations, the deficiency of assayed fat-soluble vitamins mainly affected the children with CF.

4.1. Vitamin A

The determination of an optimal vitamin A supplementation dose for children with CF is still a problematic issue. Only since 2016, on the basis of ESPGHAN guidelines, it has been recommended that we supplement retinol in low doses, increasing them under serum concentration control until normal values are reached [17]. Previously, the supplemented dose depended on a child's age, so both vitamin A deficiency and excess were often found. A vitamin A deficiency among the CF patients (adults and children) in Poland was described in approximately 16% of them, and, in Australia, vitamin A deficiency was described in 11% of the examined children [9,18]. The results of our study are consistent with the reports of these authors: in our study, we found vitamin A deficiency in almost 10% of the examined children with CF. By contrast, in other publications prior to the recommendation change, almost no vitamin A deficiency was observed in children and adults with CF, while vitamin A excess was more often identified [19–21]. Due to the differences in the supplemented retinol doses, a direct comparison of those results is rather unfeasible.

In our study, we did not confirm any relationship between vitamin A concentrations and nutritional status, diet, or supplementation. Neither did vitamin A concentrations depend on pancreatic exocrine function or liver function. These results were consistent with the reports of other authors [18,19,21,22]. By contrast, in the study by Maqbool et al., retinol concentrations inversely correlated with standardised weight and height [21]; however, in the aforementioned study, the CF patients differed significantly in weight and height, while the age range of CF subjects also included adults up to 25 years of age.

In our study, most of the children followed the dietary recommendations for vitamin A intake, either in diet or in supplements, and a significant deficit of that vitamin was observed only in the group of children with CF. No one was found to have hypervitaminosis A. In the studies by Brei, Maqbool, and Graham-Maar, the total supply of vitamin A with food and vitamin preparations was much higher than recommended, resulting in elevated and even toxic serum vitamin A levels in some patients [20,21,23]. Nowadays, thanks to the change in the guidelines and the recommended annual monitoring of serum retinol levels, the toxic effects of vitamin A are no longer observed in children with CF.

4.2. Vitamin D

The guidelines of Polish scientific societies recommend a year-round vitamin D supplementation when skin synthesis is insufficient, especially between September and April. Depending on latitude, the severity of vitamin D deficiency among the studied children with CF varies, ranging from 24% in the French study by Munck, to 90% in the US study by Rovner, AJ [24,25]. Our results are close to the Polish results of Sands' study [26], where 25(OH)D concentrations <30 ng/mL in children with CF were found in 79% of the group and, in our study, in 61% of the children with CF. Munck's publication first investigated vitamin D concentrations in children with CFSPID, finding a deficit in 18% of the subjects [24]. In contrast, in our study, the children with CFSPID had significantly higher serum 25(OH)D concentrations than the other subjects, and suboptimal concentrations were found in only 7.4%. This was mainly due to parental care, adequate sun exposure, and proper supplementation.

Seasonal variability in vitamin D concentrations and an inverse relationship of vitamin D concentrations to age were observed both in our study and in previous publications [27–31]. In our study, we did not confirm any correlation between vitamin D levels and the nutritional status. This was because the tested groups did not differ significantly in that particular parameter. We did not confirm any association among vitamin D concentrations, and exocrine pancreatic and hepatic function. These results were consistent with the reports of other authors [29,32,33]. Data on the correlation of vitamin D with other fat-soluble vitamins are rather inconclusive. In our study, we demonstrated a positive correlation between vitamin D and vitamin E, as did Grey et al. [34], while other authors did not confirm it [29,33].

In our study, all the patients with CF consumed the recommended amount of vitamin D, yet most of them were found to be deficient in that vitamin. Compared to the synthesis in the skin, the diet covers a maximum of 20% of the daily vitamin D requirement [35], therefore, the analysis of the total dietary intake of vitamin D was not sufficient, either for CF patients or for the other groups. In Munck's study, a deficiency of vitamin D occurred in 24% of the patients in the cohort with CF—all receiving vitamin supplementation—and in 18% of the inconclusive CF cohort, 60% of whom were receiving a half dosage of fat-soluble vitamins [24]. There are few recommendations for children with CFSPID, so these children should consume/supplement vitamin D in doses appropriate for healthy children, taking into account the duration of their exposure to sun light.

Similarly, as in Rovner's study, we did not prove any relationship between the dietary vitamin D content and the serum levels in any of the studied groups [25]. Different results were obtained by Timmers, but he used the criterion of dividing the ingested dose of vitamin D kg/body weight [30]. Serum 25(OH)D concentrations also depend on the variation of the genes involved in vitamin D metabolism. In planning further studies on the pleiotropic effect of vitamin D, all the previously mentioned variables should be taken into account.

4.3. Vitamin E

Since α-tocopherol binds to lipids, in order to assess serum vitamin E levels, it seems more appropriate to calculate the ratio of α-tocopherol to total blood lipids or to cholesterol, which should, ultimately, exceed 5.4 mg/g in paediatric patients with CF [10,36]. This should be taken into account, especially among patients with liver disease and adiposity disorders [37]. In our study, vitamin E deficiency was found in 6.5% of preschool-aged children with CF (siblings) only, whereas, taking into account the α-tocopherol/cholesterol ratio, the percentage was higher—22.6%and mainly concerned the school-aged children. Most authors agree that the prevalence of vitamin E deficiency in CF patients depends on the age of the subjects and mainly affects older patients [9]. Sapiejka's results coincide with ours. α-tocopherol deficiency was found in 8.0% of subjects and in the group of children at 12–17 years, in 14.8% [38]. Recently, hypervitaminosis E has been more common than vitamin E deficiency. In our study, an excess of vitamin E was found in each of the study groups and in a similar proportion, as follows: CF (6.5%), CFSPID (10.7%), and healthy children (10.0%). These data are comparable to Sapiejka and Woestenenk's results in CF patients, being 11.4% and 12%, respectively [38,39]. This can be explained by Woestenenk's hypothesis that vitamin E deficiency is more related to chronic inflammation and exacerbations of lung disease than to dietary deficiencies [39]. Children with CFSPID and HC, despite their lack of supplementation for vitamin E, showed significantly higher serum α-tocopherol levels vs. the CF patients (although 89% were within the normal range).

In our study, we did not confirm the relationship between vitamin E concentrations and nutritional status, diet, supplementation, or liver function, thus, our results were consistent with the publications by other authors [10,38,40,41]. The results regarding the correlation of vitamin E concentrations with pancreatic exocrine function, demonstrate the highest differentiation. In our study, we did not confirm the relationship, either based on faecal elastase −1 or serum lipase levels, as did other authors [22,42]; however, some publications have documented such an association [9,40].

Both in our study and in Woestenenk's publication, the supply of vitamin E in supplements was higher than in the diet of the patients with CF. Although vitamin E intake among children with CF did not meet the dietary recommendations for CF, a α-tocopherol deficiency was rarely found [39]. In the presented material, the children with CF consumed significantly more dietary vitamin E than the children with CFSPID, and α-tocopherol deficiency was found only among the children with CF. By contrast, in Huang et al.'s study, most children with CF followed the CF Foundation guidelines regarding the volume of supplementation doses, demonstrating higher α-tocopherol concentrations in their blood serum and a higher α-tocopherol/cholesterol ratio than the children in a control group [10]. These differences resulted from the division of the obtained results into percentiles of the normal range, which was considered in other studies.

4.4. Vitamin C

There are few reports discussing the issue of water-soluble vitamins in children with CF and CFSPID. Most publications report no problems with vitamin C deficiency in CF patients [43]. According to the ESPEN-ESPGHAN-ECFS guidelines, vitamin C should be supplemented only in the case of its deficiency [17]. Among our patients, plasma vitamin C deficiency and excess were found in each of the study groups. In our study, we did not confirm any correlation between vitamin C levels and the nutritional status, or the pancreatic and liver function exponents, which was consistent with the reports by Brown RK and Winklhofer-Roob [44,45]. The results of previous analyses have shown that vitamin C intake covers the dietary standards for the groups studied in Poland [46,47]. In our study, three children in the CFSPID and one child in the CF group did not meet their vitamin C requirements. This was not reflected in their plasma vitamin C concentrations and was due to reduced fruit and vegetable intake vs. their peers. In our study, almost all the children with CF consumed the recommended vitamin C intake. Similar results were reported in Back's study, where the recommendations for vitamin C intake were met by 95% of the subjects [43]. Although the deficiency of water-soluble vitamins is a marginal problem among CF patients, it is still present and, thus, requires further research on the prevalence of this phenomenon in larger numbers of patients.

4.5. Diet Analysis of Children with CFSPID and CF

Our study also assessed adherence to dietary recommendations for energy intake and dietary vitamin supplementation among CF and CFSPID patients. An analysis of a three-day food diary showed that 46.5% of CFSPID and 32.14% of CF patients had not met the recommended daily energy requirements (RDA), respectively.

Those patients periodically benefited from industrial diet support, but not during the study. Nevertheless, the nutritional status of the studied children with cystic fibrosis was not significantly different from the other groups. This is, in general, possible by an early detection of the disease in neonatal screening and by the application of early nutritional interventions and enzyme replacement therapy. In the Calvo-Lerma study, up to 46% of children with CF did not take the standard daily dietary energy supply [48]. In our study, in the group of children with CF, the recommendations for the intake of vitamins E and A were met by only 46.4% and 64.3% of the patients, respectively, and the recommended intake of vitamins D and C were also not met by the vast majority of patients, amounting to 90% and more. In the group of children with CFSPID, the recommended amount of vitamin D in the diet was consumed by only 10.7%, while vitamin E was consumed by 60.7% of the subjects, and vitamins A and C were also consumed by almost 90% of the patients. The total vitamin intake in the group of patients with cystic fibrosis had no such significant effects on serum vitamin concentrations, as they did among the children with CFSPID. There are few publications that have reported on the adherence to dietary recommendations in a group of children with CFSPID.

It is important to draw the attention of children with CF and those with CFSPID to the need of adhering to a balanced, individually tailored diet, supported by appropriate vitamin supplementation.

5. Summary

The present study limitation includes the year-round food consumption—which shows varied meals that depend on the season of the year—and the lack of division into groups, according to vitamin dose and vitamin agent type used (β-carotene or retinol). In the future, a study should be attempted with a higher number of patients, with an evaluation of the effects of supplemented anti-oxidative vitamins on the functions of particular organs. For the CF patients, preparation should be standardized. In addition, in all groups of children, the vitamin D supplementation regularity and its preparation type needs to be assessed, and their time of exposure to sunlight needs to be considered.

Among the advantages of this study is the fact that this is one of the few studies that has described the phenotype of children with CFSPID [24,49,50]. The obtained data suggest that in asymptomatic children with CFSPID, despite the lack of routine vitamin supplementation, we did not observe significant deficits in this area. Currently, there are no indications for routine vitamin supplementation, other than vitamin D in children with CFSPID.

A further development of the knowledge about the effects of antioxidant vitamins A, E, and C, and of the pleiotropic effect of vitamin D on the CFSPID children's systems and on CF patients in general, may bring potential benefits and improve their quality of life.

6. Conclusions

Both the children with CF and those with CFSPID did not fully adhere to the dietary recommendations for vitamin supplies, but a significant vitamin deficiency (mainly of vitamins D and E) was only found in the group of children with CF. In addition to vitamin supplementation, cystic fibrosis patients' vitamin D and E body stores may be affected by pancreatic exocrine function and by mutations in the CFTR gene.

In children with CFSPID and CF, due to non-adherence to the recommended energy intake and total dietary vitamin intake, more attention should be paid to the necessity of adhering to the developed recommendations. Due to the presence of mutations with variable penetrance in the CFSPID group and the possibility that the cystic fibrosis phenotype may be revealed in the future, these children require further clinical evaluation, with an assessment of the pancreas and liver function, and an assessment of fat-soluble vitamins.

There were no significant differences between the children with CF and children with CFSPID in the anthropometric parameters.

Author Contributions: Conceptualization, P.W.-W. and H.W.; methodology, P.W.-W., H.W. and T.W.; software, P.W.-W. and A.P.-S.; validation, P.W.-W. and A.P.-S.; formal analysis, P.W.-W., H.W. and U.G.-C.; investigation, P.W.-W.; resources, P.W.-W.; data curation, P.W.-W.; writing—original draft preparation, P.W.-W.; writing—review and editing, H.W. and U.G.-C.; visualization, P.W.-W.; supervision: H.W. and U.G.-C.; project administration, P.W.-W.; funding acquisition, P.W.-W. and U.G.-C. All authors have read and agreed to the published version of the manuscript.

Funding: Research carried out with funds from the Medical University of Silesia in Katowice [grant number KNW-2-K28/D/7/N and KNW-1-149/K/6/K].

Institutional Review Board Statement: The study was conducted according to the guidelines of the Declaration of Helsinki and approved by the Bioethics Committee of Silesian Medical University, No. KNW/0022/KB1/9/I/16 dated 6 June 2016.

Informed Consent Statement: Informed consent was obtained from all subjects involved in the study.

Data Availability Statement: The statistical analysis and database used to support the findings of this study may be released upon application to the Medical University of Silesia, department of paediatrics, which can be contacted by the corresponding author.

Conflicts of Interest: The authors declare no conflict of interest.

Appendix A

The scope of reference values:

vit. A [ng/mL]: <1st year of life 200–800; 1–4 years of life: 250–800; >4th year of life: 300–800.

vit. E [μg/mL]: <4th year of life 3.8–16.0; 4–12 years of life: 4.0–16.0; >12th year of life: 5.0–20.0.

vit. C 28–85 µmol/L

The scope of reference values of vitamin D-(25)(OH)D):

Deficit: 0–20 ng/mL; suboptimal concentration: >20–30 ng/mL; optimal concentration: >30–50 ng/mL; high concentration: >50–100 ng/mL; potentially toxic concentration: >100 ng/mL.

Elastase-1 in faeces-normal result: 200 µg/g, lower values were considered abnormal and indicative of pancreatic exocrine insufficiency.

Table A1. Vitamin A—recommended dietary allowance (RDA); estimated average requirement (EAR); and adequate intake (AI) for polish healthy children.

Group Sex, Age (Years Old)	µg of Retinol Equivalent/Person/24 h (RAE)		
	AI	RDA	EAR
Infants			
0–0.5	350		
0.5–1	350		
Children			
1–3		400	280
4–6		450	300
7–9		500	350
Boys			
10–12		600	450
13–15		900	630
16–18		900	630
Girls			
10–12		600	430
13–15		700	490
16–18		700	490

Table A2. Recommended Vitamin E Intake for Children (RDA).

Age	mg of α-Tocopherol Equivalent/Person/24 h
0–6 months	4 mg
7–12 months	5 mg
1–3 years old	6 mg
4–8 years old	7 mg
9–13 years old	11 mg
14–18 years old	15 mg

Table A3. Vitamin D—adequate intake (AI) for Polish healthy children.

Group Age (Years Old)	µg of Cholecalciferol/Person/24 h
Infants	
0–0.5	10
0.5–1	10
Children	
1–3	15
4–6	15
7–9	15
10–12	15
13–15	15
16–18	15

Table A4. Vitamin C—recommended dietary allowance (RDA); estimated average requirement (EAR), and adequate intake (AI) for Polish healthy children.

Group Sex, Age (Years Old)	mg of Vitamin C/Person/24 h		
	AI	RDA	EAR
Infants			
0–0.5	20		
0.5–1	20		
Children			
1–3		40	30
4–6		50	40
7–9		50	40
Boys			
10–12		50	40
13–15		75	65
16–18		75	65
Girls			
10–12		50	40
13–15		65	55
16–18		65	55

References

1. Cystic Fibrosis Foundation; Borowitz, D.; Parad, R.B.; Sharp, J.K.; Sabadosa, K.A.; Robinson, K.A.; Rock, M.J.; Farrell, P.M.; Sontag, M.K.; Rosenfeld, M.; et al. Cystic Fibrosis Foundation practice guidelines for the management of infants with cystic fibrosis transmembrane conductance regulator-related metabolic syndrome during the first two years of life and beyond. *J. Pediatr.* **2009**, *155* (Suppl. 6), S106–S116. [CrossRef] [PubMed]
2. Ratjen, F.; Bell, S.C.; Rowe, S.M.; Goss, C.H.; Quittner, A.L.; Bush, A. Cystic fibrosis. *Nat. Rev. Dis. Primers* **2015**, *1*, 15010. [CrossRef] [PubMed]
3. Welsh, M.J.; Smith, A.E. Molecular mechanisms of CFTR chloride channel dysfunction in cystic fibrosis. *Cell* **1993**, *73*, 1251–1254. [CrossRef]
4. De Boeck, K.; Amaral, M.D. Progress in therapies for cystic fibrosis. *Lancet Respir. Med.* **2016**, *4*, 662–674. [CrossRef]
5. Castellani, C.; Cuppens, H.; Macek, M., Jr.; Cassiman, J.J.; Kerem, E.; Durie, P.; Tullis, E.; Assael, B.M.; Bombieri, C.; Brown, A.; et al. Consensus on the use and interpretation of cystic fibrosis mutation analysis in clinical practice. *J. Cyst. Fibros.* **2008**, *7*, 179–196. [CrossRef]
6. Ginsburg, D.; Wee, C.P.; Reyes, M.C.; Brewington, J.J.; Salinas, D.B. When CFSPID becomes CF. *J. Cyst. Fibros.* **2022**, *21*, e23–e27. [CrossRef]
7. Sinha, A.; Southern, K.W. Cystic fibrosis transmembrane conductance regulator-related metabolic syndrome/cystic fibrosis screen positive, inconclusive diagnosis (CRMS/CFSPID). *Breathe* **2021**, *17*, 210088. [CrossRef]
8. Barben, J.; Castellani, C.; Munck, A.; Davies, J.C.; Groot, K.M.D.W.; Gartner, S.; Kashirskaya, N.; Linnane, B.; Mayell, S.J.; McColley, S.; et al. Updated guidance on the management of children with cystic fibrosis transmembrane conductance regulator-related metabolic syndrome/cystic fibrosis screen positive, inconclusive diagnosis (CRMS/CFSPID). *J. Cyst. Fibros.* **2021**, *20*, 810–819. [CrossRef]
9. Rana, M.; Wong-See, D.; Katz, T.; Gaskin, K.; Whitehead, B.; Jaffe, A.; Coakley, J.; Lochhead, A. Fat-soluble vitamin deficiency in children and adolescents with cystic fibrosis. *J. Clin. Pathol.* **2014**, *67*, 605–608. [CrossRef]
10. Huang, S.H.; Schall, J.I.; Zemel, B.S.; Stallings, V.A. Vitamin E status in children with cystic fibrosis and pancreatic insufficiency. *J. Pediatr.* **2006**, *148*, 556–559. [CrossRef]
11. Ihara, H.; Matsumoto, N.; Shino, Y.; Aoki, Y.; Hashizume, N.; Nanba, S.; Urayama, T. An automated assay for measuring serum ascorbic acid with use of 4-hydroxy-2,2,6,6-tetramethylpiperidinyloxy, free radical and o-phenylenediamine. *Clin. Chim. Acta* **2000**, *301*, 193–204. [CrossRef]
12. Jarosz, M. *Normy Żywienia dla Populacji Polskiej—Red*; Instytut Żywności i Żywienia: Warsaw, Poland, 2017.
13. Suplementacja diety. *Wytyczne Postępowania u Dzieci, Kobiet Ciężarnych i Karmiących Piersią. Red*; Dobrzańska, A., Ed.; Medi Press Wydawnictwo: Warszawa, Poland, 2015.
14. Walkowiak, J.; Pogorzelski, A.; Sands, D.; Skorupa, W.; Milanowski, A.; Nowakowska, A.; Orlik, T.; Korzeniewska-Eksterowicz, A.; Lisowska, A.; Cofta, S.; et al. Zasady rozpoznawania i leczenia mukowiscydozy. Zalecenia Polskiego Towarzystwa Mukowiscydozy 2009. *Stand. Med.* **2009**, *6*, 352–378.

15. Kułaga, Z.; Rożdżyńska-Świątkowska, A. Grajda A i wsp. Siatki centylowe dla oceny wzrastania i stanu odżywienia polskich dzieci i młodzieży od urodzenia do 18 roku życia. *Stand. Med. Pediatr.* **2015**, *12*, 119–135.
16. Borowitz, D.; Baker, R.D.; Stallings, V. Consensus Report on Nutrition for Pediatric Patients with Cystic Fibrosis. *J. Pediatr. Gastroenterol. Nutr.* **2002**, *35*, 246–259. [CrossRef] [PubMed]
17. Turck, D.; Braegger, C.P.; Colombo, C.; Declercq, D.; Morton, A.; Pancheva, R.; Robberecht, E.; Stern, M.; Strandvik, B.; Wolfe, S.; et al. ESPEN-ESPGHAN-ECFS guidelines on nutrition care for infants, children, and adults with cystic fibrosis. *Clin. Nutr.* **2016**, *35*, 557–577. [CrossRef] [PubMed]
18. Sapiejka, E.; Krzyżanowska, P.; Walkowiak, D.; Wenska-Chyży, E.; Szczepanik, M.; Cofta, S.; Pogorzelski, A.; Skorupa, W.; Walkowiak, J. Vitamin A status and its determinants in patients with cystic fibrosis. *Acta Sci. Pol. Technol. Aliment.* **2017**, *16*, 345–354. [CrossRef]
19. Woestenenk, J.W.; Broos, N.; Stellato, R.K.; Arets, H.G.; van der Ent, C.K.; Houwen, R.H. Vitamin A intake and serum retinol levels in children and adolescents with cystic fibrosis. *Clin. Nutr.* **2016**, *35*, 654–659. [CrossRef]
20. Brei, C.; Simon, A.; Krawinkel, M.; Naehrlich, L. Individualized vitamin A supplementation for patients with cystic fibrosis. *Clin. Nutr.* **2013**, *32*, 805–810. [CrossRef]
21. Maqbool, A.; Graham-Maar, R.C.; Schall, J.I.; Zemel, B.S.; Stallings, V.A. Vitamin A intake and elevated serum retinol levels in children and young adults with cystic fibrosis. *J. Cyst. Fibros.* **2008**, *7*, 137–141. [CrossRef]
22. Lancellotti, L.; D'Orazio, C.; Mastella, G.; Mazzi, G.; Lippi, U. Deficiency of vitamins E and A in cystic fibrosis is independent of pancreatic function and current enzyme and vitamin supplementation. *Eur. J. Pediatr.* **1996**, *155*, 281–285. [CrossRef]
23. Graham-Maar, R.C.; Schall, J.I.; Stettler, N.; Zemel, B.S.; Stallings, V.A. Elevated vitamin A intake and serum retinol in preadolescent children with cystic fibrosis. *Am. J. Clin. Nutr.* **2006**, *84*, 174–182. [CrossRef] [PubMed]
24. Munck, A.; Bourmaud, A.; Bellon, G.; Picq, P.; Farrell, P.M.; DPAM Study Group. Phenotype of children with inconclusive cystic fibrosis diagnosis after newborn screening. *Pediatr. Pulmonol.* **2020**, *55*, 918–928. [CrossRef] [PubMed]
25. Rovner, A.J.; Stallings, V.A.; Schall, J.I.; Leonard, M.B.; Zemel, B.S. Vitamin D insufficiency in children, adolescents, and young adults with cystic fibrosis despite routine oral supplementation. *Am. J. Clin. Nutr.* **2007**, *86*, 1694–1699. [CrossRef]
26. Sands, D.; Mielus, M.; Umławska, W.; Lipowicz, A.; Oralewska, B.; Walkowiak, J. Evaluation of factors related to bone disease in Polish children and adolescents with cystic fibrosis. *Adv. Med. Sci.* **2015**, *60*, 315–320. [CrossRef] [PubMed]
27. Smyczyńska, J.; Smyczyńska, U.; Stawerska, R.; Domagalska-Nalewajek, H.; Lewiński, A.; Hilczer, M. Zmienność sezonowa stężeń witaminy D oraz częstość jej niedoboru u dzieci i młodzieży z regionu Polski Centralnej. *Pediatri. Endocrinol. Diabetes Metab.* **2019**, *25*, 54–59. [CrossRef] [PubMed]
28. Green, D.; Carson, K.; Leonard, A.; Davis, J.E.; Rosenstein, B.; Zeitlin, P.; Mogayzel, P. Current Treatment Recommendations for Correcting Vitamin D Deficiency in Pediatric Patients with Cystic Fibrosis Are Inadequate. *J. Pediatr.* **2008**, *153*, 554–559.e2. [CrossRef] [PubMed]
29. Chavasse, R.; Francis, J.; Balfour-Lynn, I.; Rosenthal, M.; Bush, A. Serum vitamin D levels in children with cystic fibrosis. *Pediatr. Pulmonol.* **2004**, *38*, 119–122. [CrossRef]
30. Timmers, N.K.L.M.; Stellato, R.K.; van der Ent, C.K.; Houwen, R.H.J.; Woestenenk, J.W. Vitamin D intake, serum 25-hydroxy vitamin D and pulmonary function in paediatric patients with cystic fibrosis: A longitudinal approach. *Br. J. Nutr.* **2019**, *121*, 195–201. [CrossRef]
31. Norton, L.; Page, S.; Sheehan, M.; Mazurak, V.; Brunet-Wood, K.; Larsen, B. Prevalence of Inadequate Vitamin D Status and Associated Factors in Children With Cystic Fibrosis. *Nutr. Clin. Pract.* **2015**, *30*, 111–116. [CrossRef]
32. Greer, R.M.; Buntain, H.M.; Lewindon, P.J.; Wainwright, C.; Potter, J.M.; Wong, J.C.; Francis, P.W.; Batch, J.A.; Bell, S.C. Vitamin A levels in patients with CF are influenced by the inflammatory response. *J. Cyst. Fibros.* **2004**, *3*, 143–149. [CrossRef]
33. Neville, L.A.; Ranganathan, S.C. Vitamin D in infants with cystic fibrosis diagnosed by newborn screening. *J. Paediatr. Child Health* **2009**, *45*, 36–41. [CrossRef] [PubMed]
34. Grey, V.; Lands, L.; Pall, H.; Drury, D. Monitoring of 25-OH Vitamin D Levels in Children with Cystic Fibrosis. *J. Pediatr. Gastroenterol. Nutr.* **2000**, *30*, 314–319. [CrossRef] [PubMed]
35. Carr, S.B.; McBratney, J. The role of vitamins in cystic fibrosis. *J. R. Soc. Med.* **2000**, *93*, 14–19.
36. Maqbool, A.; Stallings, V.A. Update on fat-soluble vitamins in cystic fibrosis. *Curr. Opin. Pulm. Med.* **2008**, *14*, 574–581. [CrossRef] [PubMed]
37. Farrell, P.M.; Levine, S.L.; Murphy, M.D.; Adams, A.J. Plasma tocopherol levels and tocopherol-lipid relationships in a normal population of children as compared to healthy adults. *Am. J. Clin. Nutr.* **1978**, *31*, 1720–1726. [CrossRef]
38. Sapiejka, E.; Krzyżanowska-Jankowska, P.; Wenska-Chyży, E.; Szczepanik, M.; Walkowiak, D.; Cofta, S.; Pogorzelski, A.; Skorupa, W.; Walkowiak, J. Vitamin E status and its determinants in patients with cystic fibrosis. *Adv. Med. Sci.* **2018**, *63*, 341–346. [CrossRef]
39. Woestenenk, J.W.; Broos, N.; Stellato, R.K.; Arets, H.G.; van der Ent, C.K.; Houwen, R.H. Vitamin E intake, α-tocopherol levels and pulmonary function in children and adolescents with cystic fibrosis. *Br. J. Nutr.* **2015**, *113*, 1096–1101. [CrossRef]
40. Hakim, F.; Kerem, E.; Rivlin, J.; Bentur, L.; Stankiewicz, H.; Bdolach-Abram, T.; Wilschanski, M. Vitamins A and E and Pulmonary Exacerbations in Patients with Cystic Fibrosis. *J. Pediatr. Gastroenterol. Nutr.* **2007**, *45*, 347–353. [CrossRef]
41. Wani, W.A.; Nazir, M.; Bhat, J.I.; Malik, E.-U.; Ahmad, Q.I.; Charoo, B.A.; Ali, S.W. Vitamin D status correlates with the markers of cystic fibrosis-related pulmonary disease. *Pediatr. Neonatol.* **2019**, *60*, 210–215. [CrossRef]

42. Dorlöchter, L.; Aksnes, L.; Fluge, G. Faecal elastase-1 and fat-soluble vitamin profiles in patients with cystic fibrosis in Western Norway. *Eur. J. Nutr.* **2002**, *41*, 148–152. [CrossRef]
43. Back, E.I.; Frindt, C.; Nohr, N.; Frank, J.; Ziebach, R.; Stern, M.; Ranke, M.; Biesalski, H.K. Antioxidant deficiency in cystic fibrosis: When is the right time to take action? *Am. J. Clin. Nutr.* **2004**, *80*, 374–384. [CrossRef] [PubMed]
44. Brown, R.K.; Wyatt, H.; Price, J.; Kelly, F. Pulmonary dysfunction in cystic fibrosis is associated with oxidative stress. *Eur. Respir. J.* **1996**, *9*, 334–339. [CrossRef] [PubMed]
45. Winklhofer-Roob, B.M.; Shmerling, D.H.; Schimek, M.G.; Tuchschmid, P.E. Short-term changes in erythrocyte α-tocopherol content of vitamin E-deficient patients with cystic fibrosis. *Am. J. Clin. Nutr.* **1992**, *55*, 100–103. [CrossRef] [PubMed]
46. Szczuko, M.; Seidler, T.; Stachowska, E.; Safranow, K.; Olszewska, M.; Jakubowska, K.; Gutowska, I.; Chlubek, D. Influence of daily diet on ascorbic acid supply to students. *Rocz. Panstw. Zakl. Hig.* **2014**, *65*, 213–220.
47. Dybkowska, E.; Waszkiewicz-Robak, B.; Piekot, E. Evaluation of Vitamins A, C and E Content in Diets of Adolescents Living in Warsaw, Poland. *Rocz. Panstw. Zakl. Hig.* **2014**, *65*, 21–25.
48. Calvo-Lerma, J.; Hulst, J.M.; Asseiceira, I.; Claes, I.; Garriga, M.; Colombo, C.; Fornés, V.; Woodcock, S.; Martins, T.; Boon, M.; et al. Nutritional status, nutrient intake and use of enzyme supplements in paediatric patients with Cystic Fibrosis; a European multicentre study with reference to current guidelines. *J. Cyst. Fibros.* **2017**, *16*, 510–518. [CrossRef]
49. Dolce, D.; Claut, L.; Colombo, C.; Tosco, A.; Castaldo, A.; Padoan, R.; Timpano, S.; Fabrizzi, B.; Bonomi, P.; Taccetti, G.; et al. Different management approaches and outcome for infants with an inconclusive diagnosis following newborn screening for cystic fibrosis (CRMS/CFSPID) and Pseudomonas aeruginosa isolation. *J. Cyst. Fibros.* **2022**. [CrossRef]
50. Castaldo, A.; Cimbalo, C.; Castaldo, R.J.; D'Antonio, M.; Scorza, M.; Salvadori, L.; Sepe, A.; Raia, V.; Tosco, A. Cystic Fibrosis-Screening Positive Inconclusive Diagnosis: Newborn Screening and Long-Term Follow-Up Permits to Early Identify Patients with CFTR-Related Disorders. *Diagnostics* **2020**, *10*, 570. [CrossRef]

MDPI
St. Alban-Anlage 66
4052 Basel
Switzerland
Tel. +41 61 683 77 34
Fax +41 61 302 89 18
www.mdpi.com

Nutrients Editorial Office
E-mail: nutrients@mdpi.com
www.mdpi.com/journal/nutrients